<u>Why this book?</u>
(and why a title in the form of a question that alleges nothing)

Text completely rewritten as of 01.06.2026 with the key idea that nothing is really acquired and that everything can be challenged once one accepts the idea that human beings are not capable of understanding everything.

This obvious fact is due to its limits which are numerous but not always accepted:

- biological (the human brain is a non-expandable organ, oriented towards efficiency and certain truths are beyond its reach. On the other hand, solicited by too much information and too many interactions, total understanding becomes unrealistic),
- cognitive (some structures of reality might be too complex to be represented mentally or even mathematically formalized in a manner intelligible to us),
- structural (certain behaviors could exist outside of time, space, causality, making them literally unthinkable for us),
- conceptual (our relativistic + quantum theories cease to be validated together at certain scales),
- of formatting into everyday language or mathematics (Kurt Gödel, showed that every formal system contains truths that it cannot demonstrate itself),
- observational (Werner Heisenberg's uncertainty principle, cosmological horizons, quantum entanglement...).

To say that the Universe is partially intelligible, but not totally explicable, is not a failure of science. It would rather be a fundamental property of our Universe that leads us to interpret at our convenience (decoherence), a reality fundamentally different from the one we perceive.

In line with this preamble, this book which ventures into unknown land, is therefore mainly conceptual. It will not be surprising if consistency with observations (galaxy formation, diffuse background, etc...) is not necessarily respected and if the ideas developed may prove difficult to formalize without exploitable mathematical support. They can prove to be incomplete, interpretative on many points and do not really provide a tested predictive framework.

1

This book is therefore rather atypical: it lies halfway between an essay on popularization, theoretical speculation, and personal reflection on cosmology.

It proposes a global and unified vision of the Universe, trying to overcome:
- the relativity
- quantum mechanics,
- and the limits of the standard model.

To do this, he develops several strong hypotheses, theoretically possible, although speculative:
- a cyclic or symmetrical vision of the Universe based on the existence of antimatter. In current physics, matter/antimatter symmetry is a real subject, studied within the framework of CP violation and fundamental laws.
- a questioning of the Big Bang as a 'hot' event.
- a central importance of thermodynamics and global energy evolution.
- a criticism of the current cosmological model deemed incomplete.
- The key idea of chirality affecting this quantum symmetry. Physicists have already proposed models of mirror universes that would exist on the margins of our space/time. It remains however to explain why no observable signature appears.

All that follows, is distinguished by an attempt at global reinterpretation, in a desire for synthesis trying to link gravitation, quantum mechanics, cosmology. He seeks to remain readable for a curious reader. By pushing to question the established models, he wants to be enriching in terms of ideas. This book contains insights that are implicit in current research, whether it is about thermodynamics, gravitation, vacuum, energy fields or what escapes observation and amounts to answers by default (dark matter, dark energy...). It makes us aware that we are ignoring the essentials of our Universe.

Let's resume briefly in this preamble, these few themes developed in the book, to emphasize the open and non-partisan nature of the book:

✓ **In this book, thermodynamics seems to play a fundamental role**, to try to explain the global evolution of the Universe and the permanent state transitions and more particularly:
- the cosmic microwave background,
- the formation of galaxies,
- the accelerated expansion,
- the observations of the space telescope (Hubble Space Telescope).

This book, while it proposes new ideas, conflicts with current cosmology on 3 essential points:

- It questions the hot Big Bang (~3000 K) that has become transparent. This model of the hot Big Bang, not directly observable but commonly accepted, would like to explain both the cosmic microwave background, the perfect thermal spectrum and the expansion of the Universe.
- He proposes ideas that are difficult to test (global symmetries, quantum chirality).
- There is a lack of verifiable mathematical formalization.

In the absence of measurable equations and predictions, the potentially symmetrical cold version proposed here cannot seem to reproduce all this in a similar way. He rubbed shoulders there:

- rigorous scientific concepts,
- personal intuitions,
- analogies (often useful...).

But above all, this book touches on real open questions:

- origin of the Universe,
- matter/antimatter asymmetry,
- unification of laws .....

and reminds one important thing: our cosmological model is not complete. Within the framework of current theories, we do not know how to unify Albert Einstein's general relativity and quantum mechanics. Criticism is therefore in no way illegitimate.

Serious models also offer a symmetrical Universe in time or load, for example:

- the symmetrical CPT models,
- some extensions of the Big Bang.

These ideas are based on fundamental principles such as CPT symmetry (load, parity, time).

✓ **<u>In this book, gravitation, an essential subject, is described as an interaction between energy fields,</u>** of a fundamentally electromagnetic nature where 'structures' or 'symmetries' interact with each other.

The purpose is to explain the 'curvature of space-time' to move towards a more dynamic, energetic, almost 'ondulatory' vision. In this book, gravitation would be an interaction between energy fields. General

Relativity precisely says that energy in all its forms curves space-time and not only mass.

The idea that gravitation involves energy fields is accepted. Also, the idea of linking different phenomena of an energetic nature seems relevant even if the absence of a mathematical framework is blocking and even if the proposed model does not clearly reproduce the observations. But we can ask ourselves if certain observations really have certainty value and if they are not subject to interpretation.

Wouldn't the gravitational effects of bodies be the result of a kind of energetic "depression" of vacuum, due to electromagnetic interactions between matter particles that make them gather in the framework of quantum mechanics?

This leads to consider that gravitation could emerge from more fundamental phenomena, such as quantum interactions or the 'vacuum' itself, considering that by interacting electromagnetically, particles gather locally. These groupings would create a kind of "vacuum energy depression" perceived through gravitational effects.

The idea that vacuum plays a role in gravitation therefore seems legitimate. Which amounts to saying that gravitation is not fundamental and opens up a real avenue for research, leading to questions about:

- the role of emptiness
- the idea of emergence
- the link energy / gravitation

Moreover, some current physicists are exploring similar ideas, despite an absence of formalism and incompatibility with observations, notably concerning:

- the emerging gravity,
- the entropic gravity,
- the links between information, thermodynamics and space-time.

Gravitation has a key property: it acts in the same way on everything that has energy. This is what is called the principle of equivalence, at the heart of Albert Einstein's theory.

If gravity comes from a vacuum depression related to electromagnetism and the presence of mass, then:

- why do neutral particles gravitate?
- why is the light deviated?
- why is gravity always attractive?
- and where are the equations that predict these effects?

The neutron is a composite particle composed of charged elementary particles and apart from the neutrino of negligible mass and the massless photon, all elementary particles of matter have an electrical charge.

Of course, a system can be electrically neutral while having mass, this is the case for atoms, macroscopic objects, planets.

But "basically, everything contains charge" or contributes to its intensity and the internal electromagnetic interactions that compensate each other on a large scale become negligible at a distance and the overall effect is apparently not electromagnetic.

This would explain that the neutrino, which has no electrical charge and has a mass (weak but not zero), is nevertheless subject to gravitation. It is the same for the photon of zero mass, and of zero charge which is nevertheless deflected by gravity (gravitational lenses) and follows the curvature of space/time by losing energy when it leaves a gravitational field.

✓ **In this book, another important topic is the nature of emptiness.**

Vacuum represents the propagation field of electromagnetic waves. Electromagnetic force structures matter made of charged particles by ensuring the cohesion of atoms and objects. Therefore, one can wonder if the gravitational effects would not be the emerging product of electromagnetism.

Atoms are almost entirely made up of 'vacuum'. Their nucleus is tiny and the electrons occupy very extensive orbitals.

But this 'vacuum' is not really empty. It corresponds to quantum fields, with an energy called 'vacuum'.

So, the idea that vacuum plays a fundamental role is totally legitimate, considering that talking about 'local vacuum depression' is an intuitive image, not a verified description.

In Albert Einstein's theory, it is the 'emptiness of atoms' as much as the total energy contained in a region that curves space-time,

This includes:
- the mass,
- kinetic energy,
- the fields,
- the pressure.

The 'void in atoms' is an electromagnetic and quantum structure.

Gravitation depends on the total energy of a space field that defines space/time and not just the 'internal void'.

Some works therefore explore:
- the link between quantum vacuum and space-time,
- the idea that space-time itself emerges from deeper structures.

It is therefore understandable that some approaches still speculative propose that gravity is not fundamental but emerges from a collective behavior (a bit like pressure in a gas).

To get closer to current physics, it is not really the 'void of atoms' that is widening, but more precisely the fundamental quantum structure of the void (still unknown). This void which undoubtedly plays a profound role in gravitation would give birth to the geometry of space-time.

Saying that emptiness is a set of evolving fields is correct. The fields interpenetrate. They coexist in the same space, interact, overlap in accordance with Albert Einstein's general relativity where all forms of energy contribute to gravitation.

With this formulation, we are much closer to modern physics which says that in quantum field theory:
- each particle is an excitation of a field,
- the 'vacuum' is the ground state of these fields,
- these fields exist everywhere, even without particles.

The quantum fields therefore participate well in the curvature of space-time. We obtain something very close to current research where space-time could emerge from more fundamental structures, considering that quantum fields fill the space and that their collective state would determine the structure of space-time.

Moreover, some approaches explore:
- The space-time emerging,
- The link between information and geometry,
- Gravity as a collective phenomenon.

Although equations are missing and some compatibility with observations (gravitational lenses, gravitational waves, cosmic expansion...), the ideas developed in these lines are intended to be progressive, partially aligned with modern physics and intellectually stimulating.

✓ **In this book, the idea that fields and energy are at the heart of the structure of reality is repeated.** But he takes a step further or too far by suggesting that they are enough to configure space-time.

Erik Verlinde's emerging gravity is probably the current framework that comes closest to what is developed here.

Verlinde proposes that:

6

- gravitation is not a fundamental force, but an emerging phenomenon, like temperature, pressure, or elasticity.
- Space, time and gravity would emerge from the information and entropy associated with microscopic states.

With common points joining the developments of this book:

- the 'void' contains information,
- the organization of this information produces what we perceive as gravity.
- Gravity would be an entropic force.
- In both cases, the vacuum is active and structured, gravity is a collective effect and energy plays a central role.

It is therefore highly plausible that:

- gravity is not fundamental,
- the void be structured,
- space-time in itself the emerging product.

Although mathematical tools, an information theory and experimental tests are missing, these ideas remain relatively close to some current research. Everything starts from analogical intuitions that let augur:

- that the vacuum is active
- that gravity could emerge from it
- that energy plays a fundamental role

<u>In this book, the fact that science works with local certainties and open questions is taken into account,</u> with two very different levels of perception of things:

- <u>What science formulates solidly</u>. Some things are known with remarkable precision:
  - the general relativity of Albert Einstein
  - quantum mechanics
  - the standard cosmological model

Not because they are 'true in themselves', but because they correctly predict observations, are experimentally tested, and withstand thousands of verifications.

- <u>What she doesn't understand yet</u>. For example, the ultimate origin of the Universe, the nature of 'nothing', the notions of infinity or finitude...
  Thus, we do not know exactly what dark energy is, dark matter ...
  But we think we know:
  - how gravity acts on light,
  - how the planets move,
  - how gravitational waves propagate.

A theory that contradicts this is considered a priori as fanciful, even if 95% of what represents energy in the Universe remains mysterious.

It will therefore not be surprising that the model proposed here enters into tension with presumed solid results such as the universality of gravitation, its independence from electrical charge and recognized cosmological observations.

We are far from understanding everything, but what we measure turns out to be particularly binding.

 "Dark matter, dark energy..." are major unknowns. But wouldn't they have been introduced solely for the purpose of being able to match theories with observations?

In this book, the entire issue lies in the embarrassment of validating part of its content. But this shows that we must learn to doubt, to ask questions and not to absolutize current models. On the other hand, it is not because everything is not understood that all theories are likely to be validated.

Astrophysics and cosmology are twin sciences that would like to be accurate. They are, however, too burdened with assumptions, assumptions, uncertainties, and imprecision for us to be convinced of. The following reflection, which is intended to be devoid of any intention of proselytism, has already been and will continue to be subject to frequent corrections and additions.

Free reading and downloading, it will not cease to be regularly updated on: lirenligne.net

These few pages aim to tell the story of our Universe as you have probably never imagined. One must "ask oneself questions like a child with an adult brain," said Albert Einstein. He might have added that the child's own— besides the fact that his mind is not yet encumbered by preconceived ideas — is to think without necessarily relying on words. The whole difficulty then consists in transposing these still-unformed intuitions to writing. On such a vast subject, the requirement for precision in formatting does not facilitate the rhetorical exercise.

This book will discuss what we think we know—barely 5% — and especially what we don't know — 95% — about our Universe. Claiming to bring together all observable and non-observable phenomena in a coherent and exhaustive synthesis, what physicists call a unified model, may seem, in the current state of our knowledge, unrealistic. Even with a more modest ambition, would it not be appropriate to speak rather of a contextual table or an informal assembly of data?

History is punctuated with truths that, from established certainties, have proven to be only erroneous interpretations. Contrary to widespread opinion, doubt and critical thinking have often opened new avenues of progress. Also, if you live with certainties, this book is undoubtedly not intended for you. Not without some awkwardness, this book, which claims to provide answers, in reality only asks questions, with no other ambition than to propose avenues for reflection. The affirmative tone is often implicitly joined to the conditional or interrogative tone. That is why a subtitle such as «Tales and legends of the Cosmos» has imposed itself: it confers on the subject an assumed freedom of interpretation, however likely to confuse.

9

This prerequisite posed and constantly present in the mind of the reader, how to represent our Universe, especially at the subatomic scale? To describe this invisible world, it seems that we cannot do otherwise than resort to specific concepts and vocabulary, invented to interpret what exceeds our perceptions. This scientific language responds to the need to build a physics in accordance with what we are given to observe. But in quantum mechanics — the physics of the infinitely small—it quickly becomes apparent that our vocabulary is becoming inadequate, and that it is often indispensable to reason by analogies, images, metaphors or allegories. The limits of our thought are then revealed at the same time as the insufficiency of words. The whole difficulty lies in the ability to imagine and conceive what happens at scales where phenomena no longer refer to our familiar reality. In quantum physics, talking about corpuscles, orbits, spatial positions or gravitational effects is no longer really relevant, but nevertheless remains a way to approach a Universe where the use of imagination becomes unavoidable.

What follows is the result of a reflection whose subtitle 'Tales and legends of the Cosmos' might suggest that it wants to be devoid of any scientific pretension. Yet, history has taught us that what reinforces our certainties is not always synonymous with truth. Everyone may find there their share of meaning, or, failing that, new avenues for reflection, in the extension of Einstein's relativity and beyond the current frameworks of quantum physics.

As a starting point, one way of imagining the Cosmos—described here as multiverse — would be to consider it as a timeless concept, a potential framework devoid of physical reality, in which energy cannot manifest. The very notion of container then becomes inappropriate. This concept of Cosmos, excluding the presence of any observer, would notably justify the existence of binary systems of entangled universes, in quantum symmetry (whose meaning will be specified later), duplicated to infinity and without links between them.

Both a source of cold and heat, luminous and of an unfathomable blackness, sometimes peaceful, sometimes of extreme violence, indecipherable in its causes as in its purposes: such is the feeling that inspires us a Universe which, despite all our advances, continues to leave us in ignorance of the essential. Since then, the profoundly counterintuitive idea of relativity has allowed us to glimpse a discrete symmetry with multiple facets, too inaccessible for observation to reach consensus.

10

Related to the idea of Cosmos Multivers, our Universe would only be one universe among an infinity of others, without any connection between them. In other words, an "event" of a disconcerting banality within a tangle of phenomena whose understanding, for lack of broader context, seems to remain at the state of thought exercise.

Faced with paradoxes and problems of scale, we are too often reduced to making assumptions. The major problem is that our scientific logic, based on a cause-and-effect principle that is difficult to refute although imbued with uncertainty, no longer seems sufficiently efficient, nor even fully appropriate. If it accounts for many observable phenomena, it would undoubtedly benefit from being rethought to free itself from a classical physics that has become almost emblematic.

The imaginary, never short of creativity, could not help us to revise a consensus based on a model still far from being unified and insufficient to satisfy our quest for the meaning of our existence? We can certainly welcome an increased understanding of phenomena at the atomic, molecular and structural nano scales. Nanotechnologies, exploring previously unimaginable dimensions — of the order of a hundred-thousandth of a millimeter —, are particularly promising. These advances, which explain the state and transformations of matter, constitute today's advanced chemistry. But concerning the origin, evolution and very finality of the Universe, our progress, for lack of sufficient means, remains indecisive and remains, to a large extent, in the state of hypotheses, to the point that we sometimes come to doubt the reality of what makes our Universe as we perceive it.

"Regarding matter, we were mistaken. What we call matter is only energy perceptible by our senses. There is no matter." — Albert Einstein.
"Everything we call real is made of things that cannot be considered real."
— Niels Bohr
This book, like a pavement thrown into the "frozen pond" of astrophysicists, tells an unfinished legend: that of our own history, since the dawn of time.

«I would write here, my thoughts without order but not in a confusion without purpose». According to Pascal

This reflection purposely proposes, without too much discursive order, a "global theory", by infiltrating the "hidden face", source of confusion, of our Universe.

To paraphrase Pascal

Trailer
(Image Shortcuts and Stops)

As living organisms, we perceive ourselves as occupying an infinitesimal portion of space, and our life is understood as a tiny fraction of time. Our reality comes down to what we are: a little space and a little time. It is through these two fundamental notions that we represent the Universe and everything it contains: a space-time continuum shaped by energy in its multiple forms.

We know how to design today beyond what our senses dictate to us, in order to give an intuitive and thoughtful meaning to what we are allowed to observe. The whole difficulty consists in not getting lost in an abstraction without guardrail, in a culture of the abstract. Without renouncing classical relativistic physics or quantum physics, which holds promise, how can we open new avenues of reflection and broaden the field of discourse?

Let's start with an inventory, based on two hypotheses considered as possible openings towards future advances.

- The first hypothesis, posed as a starting postulate, is that of a relative space — that is to say, varying jointly with time—circumscribed but devoid of accessible center and boundaries. This confusing space-time, which constitutes our Universe, would be made of an entity difficult to define energy, assumed here in quantum symmetry, or more precisely in symmetry break. Although indissociable, the two states that realize in symmetry matter and antimatter would remain non-fusional as long as certain required conditions were not met. Their possible conjunction would then mark the end of our Universe as it will be developed further.
- The second hypothesis, which follows from the first, posits a Universe likely to end as abruptly as it began. A discrete spatiotemporal shift, or quantum chirality, would initially have differentiated two symmetrical quantum states of matter, one of which representing—antimatter — would remain essentially concealed from our observation.

A postulate is, by definition, an assertion perceived as true but not demonstrated. It is therefore necessary to examine how these two postulates can be integrated into an advanced representation of the Universe, as coherent as possible. This would allow us to overcome the shortcomings and imperfections of the standard cosmological model. If the latter enjoys a broad consensus, let us recognize that it is largely a matter of membership by default. In reality, our draft cosmological model, as remarkable as it may be, remains incomplete, marked by a lack of unification, inconsistencies and links still poorly established.

We are the complex and evolved product of an organic chemistry capable of self-organization: living matter. As such, our observer status only gives us access to what we interpret as our reality. But what credibility should be given to this reality once we accept certain advanced properties of quantum mechanics, such as decoherence or wave-particle duality?

Our Universe could thus be reduced to two "dimensions" of space-time: one invested by matter, the other by antimatter, forming two complementary symmetries, correlated but imperfectly superposed in time and space. The Universe would not exist independently of this quantum symmetry of which it would be the expression.

That an elementary particle is considered as an indivisible entity of matter or as a quantum of energy implies that it cannot be assimilated to a portion of space nor endowed with its own temporality. It is energy transfers that make localization and relativization possible. In the absence of interactions between fermions (matter particles), space-time—this conceptual framework describing the dynamics of the Universe — would lose all meaning. The Universe would then be devoid of history, frozen, unchanging, deprived of mathematical representation for lack of observer.

A fraction of the primordial energy would have acquired mass properties ($E=mc^2$) during a phase known as radiative entanglement, modifying and structuring space-time. Elementary particles would possess a hidden structure, resulting from radiative anisotropies inherent to the Big Bang. This conformation, assimilated here to a packet of waves, would govern their behavior and give them properties interpreted as mass, charge or spin. Such quantum encoding suggests the existence of a symmetry that does not reveal itself at the scale of constructed matter. The particle could then be defined as a set of correlated discrete information, in a context devoid of temporal and spatial reference.

What is valid for matter should also be so for antimatter, which only manifests itself to us temporarily during nuclear reactions or processes of pair creation and annihilation. It could also be signaled indirectly by unexplained gravitational effects, often attributed to hypothetical dark matter. Space-time would thus consist of two fundamental components, inseparable and complementary, overlapping and interacting outside our field of observation.

At the present stage of cosmic evolution, it is probable that the conditions for the generalized annihilation of matter and antimatter are not met—by chance for us). However, the gradual disappearance of all radiative entanglement by destructive confrontation of symmetry could mark the ultimate end of the Universe. This final collapse would become inevitable when all particles, massive or not, have been accreted by a growing population of increasingly massive black holes, bringing matter back to its original state.

The Universe, deprived of interactions, frozen, without energy gradients, would no longer carry meaning as a space-time. He would then have completed a cycle whose reality existed only for an advanced form of life — contingent epiphenomenon—within the evolution of matter destined to reunify with its quantum symmetry.

The idea of multiverse is not new in cosmology. It is however taken up here in a different sense: not as a multitude of universes co-existing in a shared environment, but as universes without links between them. They ignore each other and can only be considered as unique entities. Otherwise formulated, our Universe does not physically fit into a set of universes to infinity.
The multiverse Cosmos underlies the idea of infinity—infinitely small as infinitely large — and the absence of temporality. These notions clash with our ways of thinking and escape our mathematical tools. We thus remain prisoners of a consensual but incomplete model, which struggles to open real alternatives.

In this perspective, the multiverse Cosmos is neither bearer of events nor of information exchanges and has nothing similar with the exponentially prolific one of "multiple worlds" proposed by H. Everett. It is not associated with any manifestation of energy. Even the notion of space loses its meaning. This multiverse Cosmos, devoid of physical properties and which has everything virtual, cannot be related to the idea of nothingness. This

15

latent energy out of our perceptive field is no more comparable to the potential energy that creates the space void, which could be shared between matter and a relegated antimatter? (see chapter XIV on dark matter).

Thus defined, the multiverse Cosmos would be both the potential receptacle of an infinity of binary systems of universes in quantum symmetry and a latent presence, hidden at the very heart of all energy.

The standard model leaves a feeling of incompletion. The binary system of universes in quantum symmetry considered here reconciles the sectoral space-time of restricted relativity, the uncertain space-time of quantum mechanics and the flexible space-time of general relativity. This discrete symmetry would represent the fundamental quantum property from which quantum physics itself derives, imposing on any form of massive energy a potential superposed, discrete state of opposite quantum numbers, out of reach of the observer.

The idea of an energy in breach of symmetry invites us to go beyond a certain form of intellection of physics that is familiar to us. Recently, we have learned to no longer refer to a space considered absolute nor to a time perceived as universal. One must then imagine energy as a 'state' in the broadest sense, or more precisely as an uncountable superposition of potential states. Any object, brought back to its most reductive level — that of the quantum «dimension» —seems fundamentally deprived of spatial reference and temporal development. In fact, time, correlated with space, becomes the business of each observer, related to the scale that is his/her own and which he/she has no choice but to take into consideration.

As soon as it does not reveal quantum symmetry, this energy evokes a multiverse Cosmos. In breach of symmetry, it would represent a two-component universe system, involving discrete interactions between matter and antimatter.

An event, an object, are described in terms of spatial coordinates and significant duration with respect to a circumscribed event context, in a logic of cause and effect. Big Bang and final collapse can hardly be qualified as events, insofar as they are necessarily devoid of spatial references, inscribed in a context impossible to conceive and devoid, for one, of anteriority, for the other, of future.

Wouldn't the true singularity then be our Universe itself, an event folded in on itself to the point of adopting a non-open «curvature»? We can therefore consider that, related to the concept of Cosmos multiverse, it is logical that this binary system of force fields in symmetry, brought back to quantum scale, is distinguished from a macroscopic reality which is familiar to us. This amounts to admitting that the world in which we perceive ourselves would only be an appearance, an interpretation of events whose history has neither beginning nor end capable of being linked to a broader clearly defined context.

How and why would our observable Universe fit into a binary system carrying quantum symmetry?

The simplest way is to represent the multiverse Cosmos as a continuum of ruptures and reconstructions of symmetry, carried by any binary system of universes. Our logic, built with reference to space and time, then loses its points of support. It's natural: we are evoking here a virtual multiverse Cosmos, in which space and time have no hold. Is it more difficult to conceive than a demiurge or a family of deities invented for the needs of the cause? For many of us, our knowledge has evolved with scientific advances and should lead to favor this type of extended cosmological model. We were blind; let's suppose that we are only one-eyed, perceiving only the most ostensible facet of the world around us.

How to imagine the story of a binary system of universes in quantum symmetry? It would ultimately be that of recognized matter and a particularly discreet antimatter. It is to one of these two forms of energy in breach of symmetry — matter—that we connect everything that constitutes our reality. Could it be that history ends when matter and antimatter are reunited, remedying a presumed chirality* that would explain our Universe in all its complexity? Such a process of reunification, however, remains beyond the scope of our current understanding. This mass symmetry can nevertheless be glimpsed through certain exchanges: thus, two massless photons can be substituted by a massive particle of negative charge (the electron) and its antiparticle of positive charge (the positron) ... and vice versa.

**Chiral: is said of an object and its mirror image when they constitute two different forms that cannot be superposed. The chirality can be illustrated by the simple example of gloves. Everyone has already faced a chirality*

17

problem when trying to put a right glove on the left hand, and vice versa. A glove is a chiral object, because it cannot be superimposed on its image in a mirror. This analogy, however, ignores time, which is not the case here, where time becomes a determining factor in this chirality (see developments in chapter III).

The chirality of symmetry is due to the fact that each symmetry has deviated from the other by endowing itself, in equal parts, with opposite electromagnetic charges, without doing so in a strictly shared time. It follows that matter and antimatter can only interact punctually—mainly during weak interactions and processes of pair creation or annihilation — and evolve in some sort of parallel, within their own dimensions, but called to join at the end of a constitutive process of the evolution of our Universe.

It should be noted that chirality is a very common property in organic chemistry, which studies carbon-based molecules, but it also concerns other complex molecules. Strikingly, a form of molecular asymmetry or chirality is omnipresent in biological phenomena. If chirality plays a predominant role, it does not seem to be limited to the chemistry of living things. Without chirality, point of remarkable symmetry. The idea of quantum chirality renders any potential, including this indefinable and deeply counterintuitive "thing", susceptible to take the form of a universe—from the point of view of the observer it harbors — and that we call energy.

Let's try to be more explicit about this idea of chirality. By coming together, matter calibrates space-time differently at every point of our Universe. By its presence, the assembled material gives the impression of lengthening time and shortening distances. Antimatter, which does not escape the principle of relativity, should produce similar effects. However, due to its differentiation during the radiative entanglement phase, nothing indicates that it must deform space-time in a strictly identical manner. From this initial pair creation would result an overlay of mutually influencing quantum frameworks, but whose effects largely escape our perception. It is this issue that we refer to here under the term chirality.

We do not have the ability to discern what occurs in such a context of universes in quantum symmetry. However, exchanges would take place in a discreet way, through interactions that are difficult to identify, at the border between the "most tenuous fraction" of these energies correlated by their symmetry and the multiverse Cosmos. The smallest part of energy contained in our Universe could thus exceed, in its meaning, what we consider to be its elementary constituents, as soon as we bring them back to the state of wave packets. The notion of particle refers to presumably invariant values, such as Planck units or the speed of light propagation in a vacuum. These constants make it possible to give energy a physical presence, perceived in the form of fields or quanta of energy.

How to define energy? It manifests itself in our reality as a protean dynamic of undetermined origin and with an unexplained foundation. But it can also be conceived as a superposition of potential states interacting with each other, of which we only perceive what our cognitive and conceptual abilities allow us to interpret. Energy, which we have so much trouble representing ourselves, is therefore not necessarily representative, under this definition, of what constitutes our observable reality. Neither the standard model nor some advanced theories in search of unification are today able to provide a fully satisfactory definition of what energy is fundamentally.

Our physics, with its problems and scale inconsistencies, can thus be considered as a conventional way of dressing a reality that belongs only to us, and whose interpretations are called to evolve over the advances.

Any particle of matter, any object, can then be understood as a node or a confluence point of more or less obvious interactions.

Quantum leaps, tunneling and quantum entanglement go against our need to locate everything and shake up the classical idea — from which we struggle to move away—of a three-dimensional space. Particles, these "nodes" or "points" of energy that are not truly locatable, can therefore be considered as necessary artifices to give an intelligible form to phenomena not directly observable. **One may thus wonder whether particles would not be primarily mathematical constructions, defined mainly in terms of fractionated electric charge, mass, spin and helicity, 'color' or 'flavor'.** The fact remains that without these indicators we would be unable to clearly describe the nature of interactions and the level of energy involved (see chapter XXIX). We explain them by the admitted presence of forces—electromagnetic, weak, strong — as well as by gravitational effects related to the massive nature of the bodies that roam our Universe.

Could these forces not be the result of discreet interactions, because indiscernible, between two quantum symmetries? In our reality, made of
19

space — or energy fields—occupied or not by matter, gravitation represents the phenomenon that gathers by deforming interstellar space. Wouldn't what we perceive as a constantly accelerating expansion actually reflect an energetic depression of space, induced by the gravitational clustering of bodies and the densification of matter?

Our standard model that satisfies many predictions is far from an exhaustive theory. In search of a certain coherence which it lacks, could we not hypothesize that the elementary particles of matter representing the primitive bricks of matter that are today, quarks and electrons, would be "packets" not separated by primordial waves? A short-term event, called here the phase of radiative entanglement and which could be assimilated to Planck's wall, would thus have marked the very beginning of our Universe. These "packets" of smoothed waves, without marked frequencies, latent in a way, by modifying their properties through radiative entanglement, would have become inseparable. Even though these qualifiers are somewhat inappropriate, they allow for the conceptualization of what originally had neither spatial nor temporal context and here refers to the notion of a multiverse Cosmos. **Such a configuration of entangled or intertwined waves can make one think of a Bose-Einstein condensate with zero viscosity but in which the temperature which is an indicator of entropy would no longer be a referent and where the energy would be in the latent state of a cold wave, frozen, as if wound on itself in a perpetual movement encountering no resistance. Related to this, the confusing string theory would therefore not be completely unfounded.** These waves packets have since made particles and antiparticles irreducible entities of matter and antimatter. Having lost in intensity, the non-entangled primordial waves have become the current electromagnetic waves to which we also give, like the elementary particle of matter (fermion), a corpuscular representation called photon. Property resulting from this primordial wave entanglement, the mass effect induces charge interactions. The latter ensure a certain durability to build matter (strong nuclear force for quarks within the atomic nucleus and photoelectric effect for electrons) which presages the existence of a link between electromagnetic force and gravitational effects.

The elementary particle would therefore be a cloud or field of inseparable photons, formed at the very beginning of the Universe from virtual pairs of potential particles/antiparticles. This preliminary phase marks the Planck era when photons had sufficient energies to intrigue each other locally in our eyes, in a constructive way. The elementary particles have no shape, no dimension, and are outside our spatiotemporal observation framework. Locating it punctually is therefore only an essential artifice of understanding to bring it into a mathematical study framework involving the wave/corpuscle duality and try to give it a trajectory and a determined position. The vacuum that makes space for the observer would therefore be filled with pairs of particles in power or "virtual", such as electrons and positrons, which are created to then annihilate. So, when an electron and a positron collide, they annihilate, giving mainly photons. This would also suggest that this void is a particularity that we attribute to the macroscopic Universe in a space/time context specific to classical relativistic physics. Fundamentally, quantum mechanics does not need vacuum. It therefore avoids the notions of distance, duration, position, and speed to form a whole in a whole where everything is linked with everything. But it is a point of view that conflicts with our understanding and marks our limits in conceptualizing outside the space/time frame.

The elementary particle could thus be defined as the configuration of corpuscular appearance of energy waves. This configuration corresponds to our feeling during any observational approach of matter in interaction. This fundamental concept in quantum mechanics is called wave packet reduction. It is an artifice that restrictively adapts the observed phenomenon to the specificities of our gaze. This definition of the elementary particle as a packet of waves in constant and differentiated interactions (depending on whether they are quarks, antiquarks, electrons or positrons) with an electromagnetic field that makes the so-called empty space, would lead to rethinking our standard model.

The elementary particle as an entangled wave packet, has no measurable dimension. It is supposed to be neither punctual nor really locatable.

The notion of an entangled wave packet joins, for lack of a better, the image of a closed-circuit flow of very high intensity gamma radiation related to the temperature of our Universe following the Big Bang. For the convenience of language, we will continue with wave packet expression.

Considering space/time as emerging from quantum interactions, would it not be more appropriate to define the elementary particle as a kind of standing wave, in self-sustained resonance and without significant dimension. On the fringe of space and not presenting any remarkable frequency, it is supposed to be inscribed out of time. Space and time

21

extrinsic to the elementary particle would, at the atomic scale, be the framework assigned to understanding quantum mechanics. It allows us to give a reality of cause and effect, to the observed interactions. But can we really, when it comes to elementary particles, speak of entangled waves in the classical sense of undulation fields? In any case, this amounts to giving meaning to the notion of energy so difficult to define, considering thus the elementary particle as the smallest quantifiable amount of energy. This quantum at quasi-smoothed frequencies and insignificant wavelengths, would hide an unsuspected amount of energy. This would explain why the curve establishing the ratio between body temperature, wavelength and intensity highlights for wavelengths below 400 nm an aberration of non-proportionality called ultraviolet catastrophe, compared to the experimental distribution.

Thus, considered on the fringes of space/time (which would also characterize black holes), the particle somewhat evokes a perfect black body. Even if the concept of a black body has all the characteristics of a theoretical model, these "virtual points" of energy, which are zero-sized but representative in particular of an electric charge and a spin, interact with each other. Their interactions translate locally into almost imperceptible variations in the curvature of space/time. This observation led to prescribe the need for vectors (bosons) representative of 2 so-called fundamental forces (electromagnetic force and nuclear forces) to model information movements. Indeed, for the observer that we are, any phenomenon observed must involve in one way or another, the occasional presence of corpuscles and massive objects, in a body dynamics managed above all by gravitational force. Wave-corpuscle duality, quantum decoherence and collapse of the wave function that shape our reality, are the illustration of how we build a macroscopic universe very real but which belongs only to us.

Matter would be nothing other than a change of state of part of the primitive energy resulting from the Big-bang after a phase called here radiative entanglement phase. The latter could be defined as the circumscribed imbrication of primordial waves with the appearance of charges (the state of spin determining the state of charge) which will be at the origin of electromagnetism when the electric field and the magnetic field are distinguished from each other. (see chap. IV). The phase of radiative entanglement then replaces that of decoupling radiation predicted in our standard model, which would like the native universe representative of the Big-bang to be, without precondition, filled with massive particles. The radiative entanglement of the beginnings of the Universe would have thus realized the particle of matter by giving it, in particular, a significant mass of intrinsic movements and corresponding to its inertial capacity. By realizing space/time and modifying its properties, this mass thus generated and carrying an electric charge and magnetic fields would, on a macroscopic scale of massive bodies, be the source of gravitational effects (see chapter XVIII). At the same time, to any nascent particle would be correlated an antiparticle. This imbrication of primordial waves or radiative entanglement in an elementary entity of mass would confer to the particle, as to its symmetry, the potentiality to manifest itself under several so-called superposed states. But in fact, a single state reveals itself to be consistent with the reality of the observer. The antimatter that excludes itself from this reality remains out of reach for it.

Antimatter does not reveal a remarkable physical presence in the palpable, tangible world that makes, within the limits of the observable, our environment. It does not manifest itself either in the so-called 'vacuum' energy. However, antimatter interacts with matter and is signaled to us especially during nuclear reactions. Although not detectable, it seems however likely to explain misunderstood gravitational effects, other than by the totally hypothetical presence of an unknown, invisible, undetectable matter called by default, dark matter. Particles and antiparticles when they meet annihilate and transform into pure energy, devoid of mass, in other words into electromagnetic radiation. In doing so, antimatter can be made to believe that it participates in vacuum energy. It would, however, seem more sensible and coherent to think that it has a hidden dimension, somewhat parallel or superimposed on the matter (see chapter XI). This leads us to admit that space/time, in a way filters what we observe and only reveals to us what we are able to understand or interpret. Which amounts to saying that if the square of the space-time interval, between observed referentials, can be considered as invariant, theoretically it would no longer be the case between referentials understood as fields of interactions of matter and unrecognized referentials relating to antimatter. It would also be necessary to accept the idea of a discrete reference frame for the latter and to be able to classically make a parallel that is not allowed by the chirality of symmetry between particles and antiparticles.

It is a phenomenon that, today, represents a recursive form of radiative entanglement. Under certain conditions, the photon, an energy quantum associated with EMW, is capable of transforming into a particle-antiparticle pair. There is therefore nothing to prevent thinking that the particles of matter, which we define mainly by their mass, could originally be the singular product, called here radiative entanglement, of potential energy in the latent state. This non-quantifiable, non-localizable energy, which has no reality for the observer that we are, would have given our Universe the tangible properties that we recognize. Past the Planck wall, the part of this potential energy not entangled and representative today of electromagnetic waves, will be made to interact with the mass particles thus created.

Virtual energy, representative of a multiverse Cosmos not definable in terms of space and time, would thus be at the origin of the first particles of matter that will become fermions and constitute organized matter, in interaction with the EMW.

Before this phase of radiative entanglement by creating particle/antiparticle pairs, and thus before matter exists, how to conceive space and time? Space-time is a notion that only makes sense because it allows, in terms of mass, charge and quantum numbers, to relativize what seems to participate in our environment. The notion of space-time constitutes our way of giving a framework to what we perceive as body movements, transformations of matter and the forces associated with them. On the other hand, no model allows today to describe the initial potential energy that would have generated our Universe.

The idea of Cosmos multiverse then allows to give a necessarily virtual framework to what our Universe would result from. Indeed, our perception does not go beyond a space-time that encloses us in a reality beginning with a singularity, the Big Bang not being considered an event in the classical sense of the term. This is supposed to mark the emergence of space and the starting point of time, reflecting a phase transition from a latent energy, devoid of remarkable properties, which we associate here with the idea of multiverse Cosmos.

This primordial energy, potential by nature, would have conferred its physical dimension to our Universe with the appearance of the first particles and the first charge interactions, generating electric and magnetic fields.

Past the initial phase of radiative entanglement, the free radiation involved in this phase but not entangled into particles and antiparticles of matter, due to lack of sufficient energy, would have quickly lost frequency due to
24

exchanges with the newly created matter. Since then, these residual waves, which have become the EMW that we know, have continued to interfere with these entangled wave packets, endowed with mass, which constitute matter (and antimatter). Having become vectors of kinetic energy, they now interact through diffraction, absorption and emission upon contact with matter. They operate by elastic diffusion, photoelectric effect, Compton diffusion or pair creation. In doing so, they contribute to placing the so-called "empty" space in a state of increasing energy depression.

On the other hand, the presence of matter influences the properties of these EMW — residual radiation from the Big Bang—by giving them a relativistic reference speed, called the speed of light, as well as a curved appearance trajectory, configured by the gravitational fields crossed. These gravitational wells will lead the EMW to be absorbed by "quantum" mega-singularities such as black holes.

Elementary entity emblematic of electromagnetic waves, the photon only has reality when it is perceived in a potential interaction with matter. It constitutes the corpuscular representation that we make, in quantum mechanics, of what seems to be constitutive of the EMW. This "grain" of light allows to apprehend certain electromagnetic phenomena, notably the photoelectric effect. The fact that the photon reflected by an atom is not identical to the one that has been absorbed, and that it is represented as redshifted, constitutes our most coherent way of approaching this mass-free energy vector.

Confronted with a macroscopic world made mainly of material and tangible objects, we dress thus, at our convenience, quantum phenomena that do not easily respond to our intuitive capacity for perception and understanding. The photon nevertheless remains a reality that cannot be rejected, in the sense that it is representative of a privileged state among others, which are not necessarily accessible to the observer. The superposition of states remains an advanced concept, disturbing, but now inseparable from quantum mechanics.

Kinetic energy and mass energy are potentially substitutable for each other ($E = mc^2$). This explains why a significant fraction of energy, essentially kinetic, is assimilated to the mass in the «weight» of a composite particle or an atomic nucleus. The same applies to any object. The energy that seems to be missing when summing up the constituents of an atomic nucleus is

25

actually found in force interactions ensuring the connection between the elementary constituents that are quarks.

Attraction of bodies and apparent inflation of space describe two phenomena likely to be part of a reductive vision of our Universe, related to our condition as observers being an integral part of any observation device. Gravitation and expansion could thus only be the duplicated image of what we will develop further under the term retrograde dispersion. Gravitation manifests itself mainly at the scale of stellar bodies and galaxies, while the depression of space, interpreted as expansion, becomes truly perceptible only on the macroscopic scale of galaxy sets. For these two phenomena, which can be considered as the two aspects of the same process, everything is a matter of scale of observation, in a Universe that seems to expand to better collapse.

Retrograde dispersion can be defined as an illusion of dispersion. The evolution she describes would lead the Universe to return to its initial state by gathering and unifying all forms of energy. It gives our Universe an inflationary appearance and implies a particularly remarkable homogeneity on a very large scale. In this perspective, the so-called problem of the horizon of the Universe turns out to be a false problem.

The retrograde dispersion also allows to solve the so-called flatness problem, insofar as the curvature of the space, related to the masses in presence, must be globally identical at any point, regardless of the observation location. The major difficulty lies in the fact that our gaze can only embrace the Universe by mixing distant past and present of proximity; the distant news escapes us.

This somewhat dissident evolution of our Universe then boils down to a process of deconstruction of what constitutes space-time, all forms of energy combined. Time, inseparable from space, will stop when all the energy of our Universe is about to return to its original state, at the end of a transitory change of state taking the form of a «black hole». A black hole would represent the final stage before confrontation with an antimatter awaiting reunification. Perhaps we could, in order to maintain a symmetrical terminology, evoke white holes, just as discrete in our eyes as antimatter. Energy would then no longer be in search of symmetry, as it is about to be restored. The final collapse of mega-massive black holes, after clustering and densification of matter in a concentrated and cooled Universe, constitutes an end scenario that cannot be excluded.

27

Thank you for your feedback on:

https://www.facebook.com/dominique.chardri

Foreword

Obviously, our Universe exceeds our understanding, both by its complexity and by its very nature. Seeking to explain its raison d'etre should, in all logic, be a strictly scientific approach. Let us recognize, however, that excluding any philosophical or metaphysical reflection from such a questioning would be a challenge.

With all the reservations of convenience, these few lines are not absent from 'marginal' considerations and critical annotations. The following statements would like to describe and put into coherence, in as simple as possible terms, what certain scientific theories and hypotheses that inspired this essay are trying to demonstrate. If they do it in a more elaborate way, they often remain less accessible and not free from paradoxes and shortcomings.

Popularizing ideas calling upon particularly abstract notions, or relying on hermetic mathematical formulations, can legitimately leave one perplexed. Available data are frequently insufficient and inevitably lead to interpretation and extrapolation. And, as always, going beyond what sets the precedent in the scientific world can be perceived, a priori, as an inclination towards speculation.

For the sake of clarity and convenience, references to mathematical formulations will remain deliberately limited. Developments related to nuclear physics and quantum interactions will be concise.

But can one really make it simple in such a complex field, still imperfectly understood and which remains, on many points, to be deepened? Black holes, dark matter, dark energy, unfinished unification theories, universes beyond the visible, superposition of states, virtual particles... Is everything truly so obscure and elusive?

One thing is certain: nothing can be taken for granted on such a vast and confusing subject. This reflection, which is intended to be global, may thus appear dissident. It is above all a compendium of objections, questions and suggestions. A number of advances are indeed based, partially if not fundamentally, on assumptions that would still need to be validated, while too many questions remain unanswered.

This essay, undoubtedly insufficiently developed and open to controversy, nevertheless proposes an original and relatively coherent approach of our Universe, in the continuity of current knowledge. It is also an invitation to relaunch a debate far removed from the concerns that structure our daily lives. Undeniably, our priorities evolve as our living conditions change, within a society more open, more critical and more curious, but always marked by deep inequalities and crossed by extreme behaviors and ideologies. However, it is necessary to recognize the obstacles to this evolution: the inability to go faster than allowed by established traditions, confronted with disturbing discoveries and new technologies. Fragile convictions, dogmatic beliefs and natural inertia in the face of change have always hindered the advancement of knowledge. Moreover, our daily concerns primarily meet basic needs — to feed oneself, protect oneself, and integrate socially. How to have leisure and share them, increase one's heritage, satisfy sometimes repressed desires, concretize projects often out of reach? As many concerns that tend to clutter the mind at the expense of issues deemed less urgent, such as self-awareness or the issue of the Universe.

Deepening a reflection on the origin and raison d'etre of the Cosmos requires an availability of mind frequently hindered by these priorities. Cosmology does not respond to any of these essential needs, the satisfaction of which is immediately rewarding. Recognize that questioning the Universe implies being able, for a time, to detach from a heavy reality and to free oneself from certain fears or apprehensions. How then to cultivate inventiveness and critical thinking, while becoming aware of our own limits, and find the time to take a truly interrogative look at the world?

These provisions, too rarely combined but necessary for understanding the physical phenomena that shape the Universe, are indeed selective—some would say elitist. It is therefore hardly surprising that the Universe remains today a marginal subject of reflection, when it is not relegated to the domain of religion, for the majority of the thinking heads of our planet.

It remains to be hoped that the growing pollution and unevenly distributed overexploitation of terrestrial resources will lead us to become aware of the fragility of the world, still largely unknown, which shelters us. Climate change is already an eloquent reminder of it. Uncontrolled overpopulation, which often leads to conflicts of interest, is hardly conducive to desirable developments.

29

The few commonplaces mentioned in titles I and II of this presentation, as well as some reminders of fundamental concepts, have no other objective than to facilitate the articulation of ideas that are sometimes delicate to formulate. The essence of this reflection aims to converge advances and tried but not always compatible theories, in particular by trying to reconcile quantum mechanics and gravitation. For this, it will frequently be question of forces, particles, dimensions and entities qualified as virtual.

The style, direct as it may be, does not claim to state any absolute truth, apart from presumed solidly established knowledge that has inspired these lines. The underlying logic is based on the principle that there cannot be several realities.

First of all, and we will return to this point (see in particular the parallel with the notions of nothingness and emptiness in chapter XXXI), it is important to specify the particular meaning given here to the term virtual. In its common use, this word refers to what pertains to the imaginary, as opposed to the real. For lack of a more appropriate term, the virtual as used in these pages deviates from this fictional meaning. It designates the indiscernible or inconceivable part of our Universe: its hidden interface, on which quantum mechanics would be based, within the framework of a multiverse Cosmos.

How, therefore, to establish a link between:

- on the one hand, the accessible information related to matter—what has a mass — and the interactions that govern it,
- and on the other hand, concealed information, qualified as virtual because freed from space and time? These unrecognized data, likely to represent the underlying and unrevealed cause of the physical Universe, evoke here a reality ontologically out of reach. This approach avoids the use of concepts without operational prolongation, such as nothingness, infinity, timelessness or singularity.

Breaking the line between our real world and this virtual substrate devoid of physical reality remains a challenge. Understanding the origin, intrinsic nature and raison d'etre of everything seems to remain, above all, an exercise in thought. What we perceive is very real—but we shape it through light, sounds, colors, thermal variations and all of our feelings. We gradually become aware that we are only embracing a tiny facet of a Universe observed, as it were, by the little end of the telescope. Given our condition as observers confined to an empirical and restrictive vision, could it be otherwise?

The virtual evoked in these lines does not directly appear in our materialized reality, to which we are linked by body and thought. The words to describe it remain, for some, to be invented. Also, to remain credible, we must rely on the tangible and the sensitive. This imposes :

- the use of parallels and metaphors that appear in purple characters (or italics for editions in N and B) to evoke what is virtual without sinking into pure abstraction. Thus, link/brane, energy node or bubble, funnel, barometric tide, rope, chirality..., are terms that are not precisely adequate but which nevertheless allow the development of an idea in a context relatively far removed from our reality.
- the use, when necessary, of terms enclosed in quotation marks to indicate their imperfect or approximate nature.

In cosmology, man is often considered a mere passive observer. It is forgotten that he can also be seen as the conscious resultant of everything that preceded him in the Universe he inhabits. He is not only in this Universe: he is, in his own way, the memory of it. Certainly, his limited field of vision and his own finitude justify a form of chronic amnesia. Yet, a recent awareness, more and more obvious, seems to revive certain memories... But is it really about memories, or projections and fantasies?

Note: Written in red or bold type (for editions in N and B): the key ideas... that may disturb!

I <u>A starting point witch something existential</u>
(Our advances sometimes only move questions)

Assigning a dimension to an object or phenomenon is equivalent to evaluating it in relation to another, generally of different size or extension. The same goes for assessing the duration of an event, which is based on comparative benchmarks. However, confronted with the notions of infinitesimally small or infinitely large, this relative approach quickly reveals its limits. The remark also extends to the chronology of events when considering a past devoid of identifiable beginning and a future without determinable deadline.

Moreover, the difficult to predict sequence of events, in a complex logic of causality, introduces an element of uncertainty both in spatial location and temporal evaluation. This uncertainty seems to exclude any rigorous and universal formalization of duration. The relativity highlighted by Albert Einstein, in a context of multiple and interdependent frameworks, illustrates the dependence of our observations on measurement conditions. It becomes all the more difficult to formalize mathematically as it implies the almost general impossibility of constant speed movement and simultaneity in a multi-referential universe. This results in a questioning of the idea of a stable, peaceful reality that is fully predictable on a human scale.

Nevertheless, as soon as the human mind abandons all reference to the supernatural, it manifests a remarkable ability to seek answers to a fundamental question, both philosophical and metaphysical, which one can formulate as follows: how to understand the reason for being that leads us to question our origin and condition? Or, in a more pragmatic approach: **what does the matter of which we are made represent and which shapes our perceptible universe?**

The logic that structures our thought can be defined as a mental process, partly innate, allowing to link sensory data and build an intelligible representation of our environment. This logic, often considered as the culmination of human cognitive evolution, however, tends to discard what seems counterintuitive or exceeds our immediate analytical capabilities. Essentially pragmatic, it quickly becomes restrictive, even subjective, by enclosing us in a representation of the real based on appearances.

Despite these limitations, this same logic today allows the exploration of a reality that is deeper and less directly accessible, based on the notions of superposition and plurality of states (see chapter XXIX). These concepts are central elements of quantum mechanics, which is based on principles of symmetry and probability. Classical relativistic physics can be seen as an extension of this scale-up approach. However, the unification of these two theoretical frameworks remains, on many essential points, unfinished.

From then on, how to apprehend a reality that one suspects is largely inaccessible to our usual ways of thinking? Should we free ourselves from intuitive logic, deeply rooted in human experience and feeling? As such a rupture appears difficult to conceive, it becomes necessary to admit that this field of study requires its own dialectic of abstraction, as well as tools yet to be invented. We will therefore do with the means on hand.

In this perspective, chapter XI, devoted to antimatter, departs from conventional approaches of astrophysics and proposes an approach that justifies its being addressed upstream. The same applies to chapter XII, which examines the consensual and subjective nature of a standard model that is highly dependent on conditional postulates.

Humanity has continued to question the reasons for being of an environment that it seeks, not without success, to transform to its advantage. For a long time, the Earth was considered the center of everything. To explain his origin and soothe the anguish related to his future, the human being imagined a supreme being, designed in his image without ever being able to be truly represented. This recourse, historically convenient, was often accompanied by mechanisms of domination and social control. Throughout the ages, many religious systems have been associated with forms of obscurantism, enslavement, and even dogmatic drifts. Yet, these symbolic constructions, largely based on beliefs and mystified narratives, continue to influence judgments and behaviors, precisely because they claim to explain or mask what remains misunderstood.

In recent decades, however, humanity has gradually shifted the center of its questions. Accumulated experience, the development of science, technological advances and the emergence of more rigorous methods of thought have enabled the exploration of a universe extending far beyond visible stars, while simultaneously probing the fundamental structures of matter. From then on, the Earth ceases to be the central point of a closed world, and our representation of reality is profoundly transformed.

Despite the resistances and taboos conveyed by certain mythological narratives traditionally opposed to the expansion of knowledge, human beings are now capable of imagining their solar system, then the galaxy that houses it. He realizes that this galaxy is itself only an infinitesimal element within immense galactic clusters, whose apparent expansion evokes a myriad of bubbles in interaction. This is, to date, one of the most commonly accepted scientific representations of the observable universe.

In a first approximation, this expansion seems to accelerate almost exponentially, suggesting the existence of an extremely dense and compact initial state. However, our capacity for imagination and modelling quickly reaches its limits when it comes to interpreting what is called the Big Bang. The question of the ultimate becoming of the universe also remains largely open.

Why, then, stop at this single hypothesis? The idea of plurality of universes and multiverse Cosmos invites to explore alternative theoretical frameworks, sometimes confusing, but likely to broaden our understanding. Such a perspective would make it possible to free oneself from a persistent anthropocentrism, which places humanity at the center of everything, and to consider more complex and nuanced cosmological models.

It is now necessary to examine how to deepen this broader vision of the universe, by freeing oneself from prejudices and overly restrictive descriptive schemes.

II <u>The Universe plays hide and seek</u>
(A mega game in search of partners)

Does the evolution of the intensity of astrophysical phenomena reflect an irreversible tendency towards disorder, or on the contrary does it manifest a search for order and balance, preferable to chance and confusion? The response proposed here admits these two perspectives, according to the scale considered. Entropy can thus be understood as a quantity synthesizing all the processes leading, in the long run, to the destructuring of the Universe. In this perspective, the Big Bang could be interpreted not as an isolated event, but as the response or replica to the collapse of a vanished universe.

Our understanding of the observable Universe is based on the idea of non-separability of space and time. These two indicators make it possible to describe the evolution of numerous cosmological phenomena. However, they seem to lose their relevance when it comes to accounting for extreme realities such as black holes or the very origin of the Universe, in the absence of a clearly established first cause. These events, unobservable by nature, are qualified as singularities: closed thresholds on the future for black holes, and without access to the past for the Big Bang. Beyond these thresholds, space and time would cease to be operative.

These singularities, difficult to integrate into the standard cosmological model, constitute a major obstacle to any attempt at unification of fundamental interactions. Could it nevertheless be that they come from the same physical reality interpreted differently? The question of the possible link between the opening of the Universe and its final collapse remains open. In the absence of a physical state prior to the emergence of time and a spatiotemporal context subsequent to the ultimate collapse, it even becomes problematic to speak of "events" marking a beginning or an end. The concept of timeless multiverse Cosmos, introduced here, aims precisely to overcome this restrictive framework.
In this context, a multiverse Cosmos model based on the terminal collapse of universes, reduced to the presence of black holes, obviates the need to choose between 2 competing opposite scenarios: that of a bouncing universe (Gasperini's model or De Sitter's space) and that of an eternal inflation resulting from an initial singularity (the most commonly accepted hypothesis today).

The ambition of this reflection is to propose a unifying context, reconciling widely validated theories but hardly compatible with each other.

For only a few decades, space-time has been conceived as an essentially event-based medium, structured by electromagnetic, nuclear and gravitational interactions. This representation, although fruitful, remains fragmentary and undoubtedly reductive. Although supported by the know-how of scientists touching on excellence and the most advanced technical means, the segmental representation we have, is it not ultimately rather reductive?

A central hypothesis of this work is that the observed phenomena could result from a quantum chirality between matter and antimatter. Numerous subatomic interactions suggest the existence of discrete interdependencies, interpretable as manifestations of a binary system of universes in quantum symmetry. In this perspective, entropy would not be the sign of a random disorder, but that of a programmed process starting from a spatiotemporal phase shift and eventually leading to the annihilation of matter and antimatter during the collapse of the system that connects them.

Relativity implies a diversity of spatiotemporal referentials, making any measure intrinsically dependent on a local context. This disparity could explain why the time associated with matter is not superposable to that of antimatter, due to non-agreeable coordinates. Thus, although directly unobservable, antimatter could manifest indirectly through broken symmetry effects. Several models, including the one called Janus, explore this idea of discrete symmetry, without however imposing itself in the scheme of the standard model.

A quantum symmetry involving matter and antimatter assumes an extended mathematical formalism, introducing variables inaccessible to direct observation. A four-dimensional space-time would be replaced by a seven-parameter space, including dimensions specific to antimatter and an imaginary time representing quantum interactions. This approach joins certain mathematical approaches (such as that proposed by René Thom) intended to describe predictable or probable but not observable processes.

Some scientists attribute the black hole to a vortex effect. In other words, a black hole would be like a kind of shortcut to connect our Universe to a parallel universe. The latter could be understood as an "anti-Union" with parallel quadrennial properties, making it totally discreet in our regard. This is the idea developed here, except that this portal between two universes constitutes in a way a metaphor inviting us to project ourselves into what would be the terminal phase of our Universe. The entirety of the energy that made up our space-time would then find itself confined within a population of mega massive black holes (MMBH), in superposition of close states with their symmetry. This prerequisite would lead to the final collapse as proposed in these lines.

The phenomena we observe would therefore result essentially from discrete processes aimed at correcting this chirality.

What clues do we have to advance such a hypothesis, which is based on the idea of two state symmetries without which our Universe could not exist?

• The idea of chirality of symmetry between our space-time of matter and a discrete space-time of antimatter allows us to avoid the hypothesis of an almost absence of antimatter, often inferred from the simple observation of its unobservability. How could the observer that we are, in an environment that seems to him essentially made of matter, recognize antimatter, knowing that it would instantly annihilate itself and lead to its own destruction? If antimatter is neither near nor far from us, it could simply be elsewhere, in a "dimension" somewhat parallel, discreetly superimposed on the spatiotemporal dimension of matter.

• Bringing matter and antimatter into a relationship leads to the decluttering of standing wave packets constituting mass particles (see chapter V). The confrontation of particles of matter with those of antimatter partly causes them to return to the state of EMW in open field. Potentially shared between symmetries, these EMW would eventually join one of the countless black holes populating our Universe.

• Einstein's relativity shows that what we call space-time is a patchwork of interwoven frames of reference. Everything is linked to the point that nothing can be defined in an absolute way. This disparity of measure makes everything dependent and shifted from the rest; a way to approach the idea of shifted symmetry, significant of hidden interactions. Relativity, based on the notion of a reference frame, establishes a direct link between the propagation speed of EMW (c) and the low-pressure level of the so-called

37

empty space. This depressive evolution would result from the concentration trend of matter under the effect of gravitation.

Relativity, indissociably linked to the notion of space-time, thus tends to demonstrate that the essentially cognitive perception we have, on a small as well as large scale, of our Universe is not completely reliable. We must therefore leave common sense and try to describe, by means of a more appropriate language and wanting to be free from all subjective consideration, the whole of the phenomena that make up our Universe. This language, excessively codified and transcribed in the form of mathematical formulations, has however its own limits. Even with the help of this tool, it seems that we do not have the ability to get to the bottom of things in understanding our Universe nor to properly interpret the mathematical formulations and complex models it inspires.

From the voices of mathematicians in particular, one often hears that reality would be fundamentally mathematical. Yet, certain symbols and signs used, certain so-called imaginary numbers, correspond to nothing directly representative of our physical reality. It is true that, up to a certain level of development, mathematics has proved its worth, even if the models derived from it are generally subordinated to circumscribed contexts which often deserve to be broadened. It must also be noted that with more and more complex manipulations, this tool sometimes tends to detach from reality to get lost in a confusing abstraction. Thus, the wave function seems to work, but no one really knows the reason for it, and the Schrödinger equation does not offer a precise description as soon as more than two particles are involved. Similarly, the calculation of the elliptic trajectories of several interacting bodies becomes more complex as their number increases, quickly becoming probabilistic or statistical over time.

Let us recall that to date, mathematics does not make it possible to make quantum mechanics and general relativity fully compatible, even though we use them simultaneously to describe our Universe.

The principle of causality implies a logical chronology of events. Interactions, correlations, exchanges of properties and, more generally, everything that contributes to the evolution of our Universe are supposed to be linked in a cause-effect relationship. Yet this principle, apparently unavoidable, relies on the use of benchmarks based on spatial location, movement, duration, occupation of space or even the force implemented.

38

Any mathematical reasoning, to remain coherent with our perception of reality, can only rely on units of measure such as the meter, the second, the light year, the angular degree, the local entropy, the density or the flux intensity. In other words, the use of mathematics amounts to relying on space-time. Therefore, how can this tool be applied to quantum phenomena involving out-of-time and dimensionless elementary particles?

It is only on a scale integrating atomic and molecular interactions that classical physics takes over. The mathematical formalization then allows to anchor these observed or anticipated interactions in the space of general relativity and in a non-reversible time. Are we really able to get out of this logic of thought and this methodology of information processing which, although effective, seem today to show their limits? The question is disturbing, because it leads us to doubt a reality that we now know is only a partial vision, dictated by our perceptions and by physical laws operative but contingent.

The image we make of the Universe is thus reductive and deeply subjective. It cannot integrate a deeper reality, not directly observable, made of superposed states, wave packets, chirality and quantum symmetries. It results from a phenomenon that we do not fully master, quantum decoherence, which inexorably brings us back to a macroscopic vision accessible to classical physics, but insufficient to explain the raison d'etre, evolution and ultimate foundations of our Universe.

Although the mathematical tool has allowed many advances leading to the expected practical applications, it is today insufficiently efficient or even inadequate to address some of the new questions raised by the most recent discoveries. Arising from this observation, and rather than taking refuge in denial, we cannot rule out the feeling that our theories on quantum and relativity need to be reinterpreted.

Despite everything, we continue to progress, even if it is done in stages. Future discoveries could result from the deployment of a new type of computing based on quantum manipulations. They will also benefit from the development of artificial intelligence: algorithms, oversized memory capacities and machine learning of innovative logical methods with automatic error correction will then take over. Our ego, even though it suffers from it, could see the machine become the essential extension of human intelligence and scientific applications realized.

39

We depend on an increasingly elaborate mathematical tool, developed from the observation of an environment whose foundations remain misunderstood. This valuable tool is perfectly suited for many practical and experimental applications. The problem is that it shows its limits at the subatomic scale and seems unsuitable for the analysis of the disproportionately large or the exploration of the infinitely small.

It happens that mathematics, applied to physical laws considered as proven, is unsuitable for transcribing phenomena that are difficult to observe but prescribed by other observations. We then speak of singularity, quantum uncertainty or indetermination, with all the ambiguity that underlies the notion of infinitely small or infinitely large. This highlights the incompleteness of this remarkable tool for helping to understand the phenomena that animate our Universe.

Thus, in a black hole (see chapter IV), although at this stage of evolution of matter it is difficult to speak of Planck units, the ratio between Planck's length and Planck's time ($0/0$) is assumed to be equal to 1 as much as 0. If $1 = 0$, the result cannot concern something physical and relates to something virtual. No doubt it was the same at the birth of the Universe, where distances and time were not quantifiable. In these two cases, the physical laws expressed in mathematical language, with reference to constants such as the speed of light measured today, are inapplicable.

Advanced theories, such as those of strings, superstrings or quanta of space, which claim to break the deadlock, unfortunately do not constitute mature mathematical proofs. They remain difficult to interpret and do not provide the expected insight. We are both too much a stakeholder and too much prisoners of our status to achieve this. But how to design an observation framework encompassing all the parameters of our Universe and offering a broader context?

To approach the true nature of our Universe, should we not strip ourselves of our feelings and agree to question knowledge considered today as sufficiently validated? This would possibly open the door to future advances. In recent decades, we have learned to think counter-intuitively, soliciting an imaginary sometimes confusing. Otherwise, how could we have talked about antiparticles, inverted time, space/time relativity, non-locality or quantum entanglement? It must be recognized that, in its foundations, our Universe is far from the image it inspires at first sight.

Today, we accept that the smallest elementary entity of matter, although not representative of an occupied space, can only be described as a point of space. It seems that it does not really move in space. But it creates, through its potential presence and diversity, what we perceive as space in a macroscopic reality. The measurement of time refers to our experience and can be conceived as a causality tracer for phenomena which, for the most part, do not seem to involve the existence of antimatter. Why would matter/antimatter interactions escape the linear time we know? If quantum mechanics cannot be thought in terms of space, it excludes itself from any event environment. We then encounter a scale threshold.

Space could be defined as an energy field where everything is potentially possible. It is not sanctionable. Time has no direction in quantum mechanics and can be forgotten. The problem is that we seek to explain the chirality of symmetry in relation to time, relying on spatial dimensions. Always this reference to a lived reality from which we cannot extract ourselves, but which nevertheless allows us to move forward.

Talking about quantum symmetry leads to imagining a kind of parallel or overlapping dimension, in a time different from ours. This imaginary time would consider the symmetry anomaly between particles and antiparticles, called here chirality. Maybe we should introduce the notion of quantum symmetry in a standard model to be expanded. **This symmetry has nothing geometric, and the chirality matter/antimatter prescribed here does not imply any plane of symmetry in space or time. It is essentially quantum and does not merge with the enantiomorphs superposition of objects, such as the image reflected by a reflective flat surface.**

To any particle is potentially associated an antiparticle. In most nuclear reactions, antimatter is suspected to interact with matter, but it remains difficult to observe due to its nature and the absence of a shared spatiotemporal dimension.

Quantum mechanics delivers its mysteries sparingly. Thus, the quark, an elementary particle at the heart of matter, only exists paired with others within an atomic nucleus. A quark cannot be observed in isolation; our observation capacity does not allow access to such a level of scale. On the contrary, the electron, by constantly interacting with OEMs and ensuring bonds between atoms, can be detected, although its presence depends on the composition of the nucleus and is only a probability of localization in the

41

electronic cloud. Certain electroweak interactions (beta-plus decays in particular) thus leave clues about the emergence of antielectrons observable as traces in so-called "fog" chambers.

The proton, a composite particle made up of quarks, represents a particularly stable "quantum brick". Its presence, although directly unobservable, can be felt within the atomic nucleus. A single proton forms the nucleus of hydrogen, the simplest atom representing more than 90% of the atoms in the Universe. This probably explains why the proton seems unable to spontaneously decay, an essential condition for the existence of matter.

As with every quark, an antiquark can exist, and every proton could have its antiproton, unless quantum symmetry does not organize in the same way for composite particles due to chirality. The antiproton appears "under cover" during certain interactions affecting the nucleus. This permanence of the proton, perceptible indirectly, allows its confinement in a kind of vacuum chamber, isolated by magnetic fields. Similarly, it would be possible to confine emerging antiprotons during nuclear reactions without violating the principle of quantum symmetry. The non-local correlation or quantum entanglement, which abstracts time and space, prevents a classical definition of elementary particles in terms of time and space, unlike composite particles or atomic nuclei.

Quantum entanglement, at the origin of the concept of non-locality, implies that certain elementary particles, inseparable from a global context forming space-time, can instantaneously share certain properties, regardless of their distance. Any change affecting one immediately modifies the other. Time then reduces to a change of state, and this correlation reveals a shared before and after without transmission delay. But how to reconcile these non-local exchanges with the concept of space-time, which involves travel and transport duration? That entangled particles exchange information without physical transfer seems so counterintuitive to us that one often concludes that quantum mechanics relies on randomness and probability.

For changes in quantum properties by contact or proximity, the exchanges mainly concern observable or prescribed charge interactions involving a certain duration. These local interactions can alter the degree of quantum correlation and, on a larger scale, limit non-local exchanges.

It can be assumed that originally all particles were closely correlated in a common state of symmetry. Many elementary particles remain intrinsically linked since their division or appearance, without loss of properties. The Universe has thus gradually become more local, in an evolving context of space/time where our reality as a non-quantum observer fits. Quantum entanglement does not guarantee that these links escape the grip of time after force interactions between particles of opposite charges.

One cannot directly observe an atom, given its size smaller than the wavelengths of visible light. It is however possible to reconstruct its image with a scanning tunneling microscope. We can even confine a light atom by cooling it and isolating it in a cavity devoid of gravitational effects. With heavier atoms, the difficulty increases. Isolating antiatoms would require considerable amounts of energy. Isolating an antimolecule is a challenge, especially since there is no guarantee that antimatter on this scale is configured like matter (see CPT symmetry, chapter XXVII). Antimatter cannot spontaneously join matter, explaining the rarity of annihilation phenomena.

The chirality in particle physics would result from an imperfectly shared quantum symmetry between matter and antimatter, depending on spacetime relativity. Particles and antiparticles, as packets of waves not definable in terms of space and time, should not fundamentally manifest chirality. It must be considered that only interactions, perceived as information exchanges, would be the cause. A slight spatio-temporal asymmetry, related to the shifted evolution of composite and antiparticle particles during the combination into atomic nuclei, fact that classical nuclear reactions convert only a tiny part of the particles and antiparticles into gamma radiation without mass or charge. Quantum chirality and relativity are closely related, like time and space. Quantum symmetry becomes an integral part of relativity. Molecular matter could only annihilate with its symmetry at the end of endless exchange processes and gravitational collapses shaping the evolution of the Universe.

Let's stay on the idea that matter and antimatter are not in perfect quantum symmetry, but that a certain chirality limits their interactions and prevents us from perceiving the latter. This amounts to conceiving that matter and antimatter belong to two distinct dimensions, representative of two space-times that are somehow parallel or superposed. We know that matter is only a tangible form of energy, called to complete its evolution to the first state, within and in the form of black holes. At this stage, the recognized properties of matter (mass, charges, spin...) disappear. One can reasonably imagine that it is the same for antimatter.

Elementary particles and antiparticles are totally deconstructed, and the energy they represented has nothing more to do with potential or even kinetics. To speak of singularity, as far as the black hole is concerned, is an elegant way of making a statement about a form of latent energy detached from space-time that is ours and which is probably close to the energy present in the first moments of our Universe. This undifferentiated form of energy, common to what was matter and antimatter, cannot be declined in terms of time or space occupation. This explains why mathematics is no longer appropriate to describe the black hole, this singularity on the margins of our space-time. A black hole 'fed' with matter is then no longer distinguishable from a black hole 'fed' with antimatter. Both escape, at this stage, the general relativity on which our reality is based.

Without however making the hypothesis of white holes, a black hole would ultimately represent the prerequisite for the coalescence of what was matter and antimatter. Matter and antimatter therefore come together within these singularities which, outside of a gravitational context and emerging from a spatiotemporal multi-referential framework, leave nothing to be imagined about the shared destiny of antimatter. Quantum symmetry, which is a matter of massive particles, just like the state of chirality that characterizes it, disappears. The black hole represents a transient state without symmetry. This state cannot be discovered but leads to the erasure of what makes our reality: an environment marked by gravitational effects and nuclear and electromagnetic interactions.

The final collapse will be realized when what was matter and antimatter, completely destructured and therefore deprived of any form of interaction within black holes, has left a Universe empty of any electromagnetic field. We could consider that the black hole is in a state of permanent present, without any significant past or future. Detached from our space-time, it can, for the observer that we are, only register in a distant future, marking the ultimate phase of the evolution of our Universe. As such, it escapes any kind of observation as well as our ability to understand. We cannot, body and mind, extract ourselves from a space-time that encloses us and only allows speculation about a before as well as an after. This makes that Big Bang and

44

final collapse can only be the domain of speculation. We are reduced to observing, in deep space, the degraded vestiges of a truly established ageless past. We can only predict possible, uncertain and limited futures.

This evolution would be inevitable, in a way self-programmed, without any possible and irreversible alternative. It could be summarized as follows:

Matter gathers under the effect of the four so-called fundamental forces to form the most massive stars. These will eventually collapse on themselves (supernova), forming in general a neutron star in which the electrons, by joining the atomic nuclei, transform the protons into neutrons. These neutron stars will finish their collapse, one way or another, until they become black holes. The latter then have no future but to absorb energy (matter and radiation) within the reach of their irresistible gravitational power. The elements not retained by the neutron star or the black hole created during such events will gather together to later form new stellar bodies, heralding future supernovae. It has probably not always been the same when our Universe was at its very beginning, with the formation without preconditions of primordial black holes. The latter would have had the effect of generating particularly extensive areas, depressionary in energy, almost devoid of matter and of any form of interaction.

The Universe strips of neutrons thus deconstructed at the heart of black holes. Now, neutrons are components necessary for the evolution of matter. However, some protons, by rallying electrons and neutrinos to them through nuclear reactions, will transform into new neutrons, necessary for the evolution of matter. The stability of the atom with these substituted neutrons is thus preserved. But in this game of musical chairs, the population of electrons, neutrinos, protons, neutrons and other composite particles continues to decrease to the benefit of a population of increasingly massive black holes. In an "empty" space, which will end up deprived of any other form of energy than MMBH (massive mega black holes), space and time will no longer have much meaning. One can then hardly imagine another final outcome than a global collapse of all these MMBH in convergence, by recessive depression of the so-called empty space. This upside-down Big Bang would thus mark the end of our Universe.

The hypothesis of an expanding Universe accelerated from a so-called singular point should not normally, even corrected for the relativistic aspect, present a perfect uniformity of energy density. For an imagined Universe in

45

open expansion, the Minkowski metric (measurement method supposed to consider the effects of relativity) can only be retained on the reduced scale of a space confined within the limits of the observable. The finality of the Universe then seems unpredictable. Talking about expansion for something without a delimited edge or center, "generated" by a virtual nature multiverse Cosmos, seems inappropriate.

The hypothesis of an expanding Universe accelerated from a so-called singular point should not normally, even corrected for the relativistic aspect, present a perfect uniformity of energy density. For an imagined Universe in open expansion, the Minkowski metric (measurement method supposed to consider the effects of relativity) can only be retained on the reduced scale of a space confined within the limits of the observable. The finality of the Universe then seems unpredictable. Talking about expansion for something without a delimited edge or center, "generated" by a virtual nature multiverse Cosmos, seems inappropriate. We equate a continuous expansion, discerned from an extended scale of observation, with an increase in volume likely to be occupied. But this observation seems contrary to the idea that time, like space, fades as we approach the quantum world. In the latter, everything becomes a matter of variable fields that intermingle without restraint, requiring abstraction of time and space. At this level of introspection, we realize our limits. In all likelihood, the illusion of expansion is due to the fact that our capacity for understanding does not allow us to model, through the mathematical tool, other than in spatial positioning data (particles) and duration data (interactions). In our reality, the one we are given to observe, the Universe thus appears well expanding.

We describe electromagnetic waves in the form of undulations traversing space and marked with peaks and valleys. This is how we imagine the topography of kinetic energy fields, which constitute the backdrop to our Universe. This vision inspired the idea, now abandoned, of an ether serving both as a support and medium for the movement of electromagnetic waves and stellar objects. It seems rather that this energy medium, falsely said to be empty, represents, as a framework, the space-time required by a reality which belongs only to the observer in the mathematical representation he makes of it.

A non-perceptible defect of «synchronization», called here chirality, creates this organized disorder that is our Universe, in a form of determinism difficult to identify. Of course, invoking chaos theory would be the easy answer, but it explains nothing truly logical.

Knowing that it is only an image and detaching from a familiar reality that guides all observation, how to define these discrete exchanges between two universes of quantum symmetry?

Consider a tight and permeable weft, woven in three dimensions to evoke space. Let's give a spongy, crumpled, and moving look to this fabric to represent the weather. Let's imagine, in the superposition of images, a clearly visible place aspect and a non-visible reverse aspect, an exact copy representing its symmetry. Interways now facing and facing with an array of osmotic interactions not recognized. This artifact can be seen as an exchange area or a mirror effect. As the shadow waits for the meridian hour to merge with its subject, this process of reunification through discrete interactions between particles and antiparticles will find its completion in the collapse by coalescence of what was matter (which makes our reality) and antimatter.

This amounts to agreeing that these universes in symmetry only have physical reality in the potential, but thwarted, confrontation of their opposite states. We can draw a parallel with $+X$ and $-X$ which cancel each other out in arithmetic, without completely excluding that the sum of $+X$ and $-X$ (to represent the final collapse) is different from zero. But we then change the «register»: the latent energy of the multiverse Cosmos can only be of a virtual nature.

This aphorism of energy in symmetry break being posed, it must be recognized that our Universe is perceived above all as a "bubble" of energy devoid of measurable dimensions and symmetry, teeming with waves and particles (at the convenience of the observer), in a context of non-reversible time. This notion of irreversibility leads one to think that what is done cannot be undone, except in very exceptional cases (cancellation of particle/antiparticle pairs), in the same inverse process. But why should it not be by some kind of more general loop mechanism, which would bring back, as described later, to the «initial box»?

Particles and antiparticles are supposed to, in order to 'coexist', not be in total direct interaction. When they meet, they lose their distinctiveness. Their decay generates gamma radiation, incidentally, accompanied by a few

47

short-lived mass particles. It seems that we are unable to observe directly such interactions between symmetries, given their evanescence.

This idea of quantum symmetry also solves the problem of infinite divergences, reducing them to loop phenomena, which may have inspired string theory and loop quantum gravity in an attempt at unification.

String theory postulates that elementary particles would not be dimensionless points, but one-dimensional strings. These ropes, by vibrating at different frequencies, would be the cause of gravitational effects not recognized on this scale. String theory thus confers a status to quantum gravity, space becoming a kind of quantum network. Exchanges of information would then manifest themselves in the form of links weaving an interactive network. From this dynamic would emerge space-time, which would induce the existence of a gravity of electromagnetic nature at the subatomic scale. The problem is that by trying to measure gravitational interactions based on quantum physics, the equations can produce results devoid of physical meaning, except by introducing additional dimensions.

To say that the space-time "emerges" means:
    -That space and time would be "interfaces" between our mode of perception and a more abstract reality.
    -That on a small (fundamental) scale, it does not really exist. Space is no longer a "place" and time is no longer a "universal flow"
    -That on a large scale, it appears as an approximation, a bit like temperature emerges from the movement of molecules.
In summary:

The end of the Planck era is marked by the disunification of forces allowing to model all the physical phenomena observed in the Universe. Composite particles (grouping of elementary particles) result from this decisive step where electromagnetic interactions predominate. The most stable (protons and neutrons), by associating in a perennial way, will become the nucleus of the atoms to come. In these early quantum exchanges of our Universe, the presence of these same composite particles called baryons and electrons seems to predict a causal context

based on state transitions. Such a perspective leads to consider that at the atomic scale; an adapted observation framework is required. Otherwise, how could we relativize spin, movements, and information sharing, which are indicative of these post-Big-bang quantum interactions? This conceptual framework of space/time, possibly emerging from the quantum world, would take on meaning as a representative of reality only with the emergence of living things, an exotic and precarious form of constructed matter. This is how, moved by a kind of awareness, the observer that we are, shapes a macroscopic reality in his own image. Yet this observable world is essentially based on this notion of space/time from which it cannot be extracted but which is fully justified and is not just a pure product of the observer's imagination. This is what makes all the difference and seems to make classical relativistic physics incompatible with a more fundamental underlying quantum mechanics.

Loop quantum gravity requires that the space-time be, in a way, pixelated in the form of loops forming spin networks. This theory leads to the thought that our Universe would be part of an endless cycle, marked by successive Big Bang and Big Crunch. Does this mean that our Universe would have known a previous state?

The theory of loop quantum gravity suggests that the Universe would undergo a phase of contraction before expanding again, thus bouncing indefinitely. This vision joins, in certain aspects, that of a multiverse Cosmos composed of a multitude of Universes, born and disappearing over the course of cycles without identifiable beginning or end.

A binary system of universes in quantum symmetry cannot be assimilated to a Universe with six spatial dimensions. In a space with more than three dimensions, the gravitational force would indeed tend to fold back on itself, compromising any stability of the architecture of matter. The disruption of the cosmological equilibrium would rather be a discrete and out-of-phase superposition of two opposite, inseparable states. Essential point, this hypothesis allows to clarify the enigmatic deficit of energy and matter observed (see chapter XIV).

Any observation necessarily implies a reference to time and space, through the relativization of measured distances and event durations. By his mere presence, any observer—whoever he is and wherever he is — embodies a fraction of time and a portion of space that serve as archetypes of measurement. What cannot be defined, in one way or another, in terms of space and time, cannot belong to our reality, except as part of the fiction stemming from our imagination, such as the idea of Cosmos Multivers.

Space-time is not a physical entity in itself, but it constitutes an essential framework for analysis. What would be the reason for it in the absence of the material that, precisely, constitutes us? Without the interaction of matter, a point in time; without the displacement of bodies, a point in space. If any form of matter is doomed to dematerialize into black holes, entities not representative of space, then the very concept of space-time would lose all meaning when our Universe reaches the end of its evolution, for lack of an observer to witness it.

But how, as privileged observers in the evolution of living things, have we come to establish this space-time context as an unavoidable framework for thought? A tacit consensus, often adopted as an initial postulate, assumes that the absence of space and time preceded the point of beginning of our Universe. Space, even described as a void, only has meaning in relation to its occupation by matter, in one form or another, that is to say, the presence of elementary particles endowed with mass.

The concept of absence of space/time is excessively counterintuitive. How could it immediately fit into the framework of any thought exercise? Yet it will be discussed about the beginnings of our Universe, the supposed uncharted limits of it as well as the deep nature of black holes.

Symmetry constitutes a remarkable property of matter and is expressed in various aspects: matter/antimatter symmetry, charge conjugation symmetry, parity symmetry, or even time inversion symmetry. Quantum symmetry does not mean here an exact correspondence, comparable to a mirror on either side of an axis. The Big-bang, as a breach of cosmological equilibrium, would be marred by an unrecognized chirality, revealing an imperfect symmetry without which particles and antiparticles could not have coexisted durably and perpetuated matter.

This original singularity would thus mark the emergence of a temporality inseparable from the notion of space. The energy intensity of the free radiation that filled space during the Big-bang has no equivalent in the current Universe. By interacting, these radiations would be at the origin of the first radiative entanglements, heralding a quantum symmetry. Some of these high-energy radiations, entangled into proto-particles of matter and antimatter, would confer on space-time its fundamental properties. The radiation of lesser energy would constitute the EMW of the current Universe.

This phenomenon, characteristic of the very first moments of our Universe, would attribute specific properties to the "packets" of waves thus transformed into elementary particles. These new entities, similar to state vectors, would be endowed with mass, spin and, for some, charge. They would constitute the irreducible components — the elementary particles— at the basis of contemporary physics. If we could trace back to this primitive phase, space would seem unlimited and the flow of time too fast to be measurable. At this initial stage, relativity cannot yet integrate gravitational effects that remain to come.

Like the wave/corpuscle duality used to describe a particle—and using, for lack of better, imperfect terms — the multiverse Cosmos can be apprehended in several ways, depending on the point of view adopted:

- A purely virtual "thought artifact", consisting of a latent energy of non-quantifiable intensity, devoid of physical representation, mass, revealed symmetry, location—lack of occupation of space — and interaction, in the absence of relation to time.
- A cosmological equilibrium, likely to be described as a continuum of ruptures and reconstitutions of unrecognized symmetry. These confrontations — collapses and Big-bangs—countless, involving "universe" pairs in quantum symmetry, do not possess a directly remarkable physical reality. This is precisely what gives the multiverse Cosmos its "virtual" legitimacy.

The multiverse Cosmos is neither emptiness nor nothingness and, not occupying space, it is not physically apprehensible. It cannot be confused with what we call quantum fields or force fields, which relate to the energy space specific to our Universe. The fields of forces represent the space where, with varying intensities and in interdependence, the three

51

fundamental forces of interaction are exerted in a gravitational context. Absolute emptiness or nothingness are pure abstraction, without place in any cosmological model. In summary, the multiverse Cosmos cannot be assimilated either to a support or to a substrate of the Universe.

This interpretation of the multiverse Cosmos conceals an unobservable, entirely virtual entity. This qualifier is also frequently used by scientists to designate quantum particles likely to change status. It reflects the difficulty of locating them in space or conceiving their volatility when they seem to move without inscribing themselves in time. What is virtual is not directly detectable; it remains presumed, while providing justification for measurable or potentially feasible phenomena.

The multiverse Cosmos does not fall under either quantum mechanics or classical relativistic physics. In other words, the nuclear, electromagnetic and gravitational interactions would essentially correspond to circumscribed exchanges, specific to a binary system of universes in quantum symmetry.

It is difficult to conceive a beginning of the Universe without being able to imagine its end. More than an infinite story, a cyclical scenario appears to be modellable: *like the mythological phoenix, condemned to rise from its ashes.* But such a cycle presupposes a detonator and an exceptional escape (see chapter VII), which could be the destiny of this stellar monster that is the black hole. Devoid of color, it would open, in the very distant future, a "door" to a new binary system of universes.

Several clues allow us to sketch what a black hole might be, although it refuses any direct introspection. Imagine a region of space not comparable to an electromagnetic force field: no photovoltaic effect, no magnetic excitation, no electromagnetic interaction would manifest there. However, EMW result in vibrations or distortions of the vacuum, without which we could not represent space. These undulations, by interacting, constantly modify the energetic properties of space and give it a meaning as a field of interactions between charged particles, through photon exchange.

Deprived of interactions with EMW, the quantum states of electrons would become unstable, leading to the collapse of molecules and, consequently, matter, in a space devoid of electric and magnetic fields. The bodies that would be there, emitting no spectral rays, would escape our observation. The strong, weak and gravitational interactions would cease to be remarkable there. Don't we find there the very characteristics of a black hole? This singular object, collapsed on itself, does not emit any electromagnetic signal and transmits no information, except those coming from its accretion disk and a powerful magnetic field. Space-time would, in a way, be absorbed there.

The quantum state of electrons deprived from interactions with EMW, would become unstable, causing the collapse of molecules and therefore of matter in an "empty" space of electric and magnetic fields. The bodies that would be there, not emitting spectral rays, can only escape our sight. Strong, weak nuclear and gravitational interactions are no longer remarkable. But wouldn't those be precisely the characteristics of a black hole? This singular body collapsed on itself, emits no electromagnetic or other signals and transmits no information (except those emitted by its accretion disk and a powerful magnetic field). Space-time is somehow absorbed by the black hole.

The magnetic dipole of an active black hole logically implies its rotation. In any system subject to gravitational interactions, the more massive the objects involved, the higher the rotation speed of the star resulting from their fusion. This is particularly the case for neutron stars when they fuse after forming a binary or even ternary system, as with some pulsars. Active black holes, like most stellar objects, thus acquire their rotational speed from the orbital velocity of the bodies they have orbited before their fusion.

The observed rotation of the accretion disk is probably not synchronous with that of the central singularity, whose entropy could be uniformly zero. This dissociation would explain the emission of particle jets that spring from both sides of the accretion disk—and not from the heart of the black hole — like waterspouts projected into space. The entropy of a black hole thus seems to be summarized by its event horizon, and possibly by surface phenomena.

A fraction of the energy that accumulates in the accretion disk is projected towards the poles. After an accelerated orbital path, this excess energy is expelled in the form of ionized plasma and radiation. These twisted jets approximately follow the geomagnetic axis of the black hole, determined by the configuration of its accretion disk. When the event horizon — this observable limit area—is not saturated, the black hole has nothing to reject.

The black hole then appears as an unfathomable well, inexorably absorbing the Universe that houses it. The tidal forces reach their peak there. All the «information» having crossed the event-based horizon merge into it and are irreversibly lost for the external observer. They could however reconfigure differently during a «second generation» Big-bang (see chapter VII), consecutive—although not directly linked — to an event of extreme violence: the final collapse.

A Big-bang could thus be considered as the result, out of space and out of time, of the ultimate collapse of a binary system of universes reduced to the state of black holes—and, by analogy, of white holes — in symmetry. But can we really speak of continuity between two events that open and close on the multiverse Cosmos?

It is necessary to distinguish the stellar black hole, resulting from the disintegration of a massive star, from the supermassive black hole, which constitutes its extension and generally resides at the center of galaxies, whose contents it gradually phagocytes. In the long term, this entire population of black holes could evolve into mega-massive black holes (MMBH) in a cooled Universe. These MMBH do not yet exist; they will probably populate our Universe at the end of its 'life cycle'.

However, the existence of isolated colossal black holes—seeming to reach several million or even billions of times the mass of those located at the center of galaxies — appears highly probable. The major difficulty then lies in their detection when they are devoid of accretion discs. Indeed, although the Universe appears homogeneous on a large scale, it also reveals vast regions belonging to the past, already apparently poor in matter, but likely to host such singularities. This hypothesis could partly explain the observed material deficit, as the effect of gravitational lensing does not always allow for their presence to be betrayed.

The temperature — or rather the absence of temperature — of a black hole could correspond to true absolute zero, synonymous with total absence of interaction. The black hole would contain a quantity of "deconstructed" energy so considerable that its content would appear to be densified to the maximum, in the absence of any interstitial space. Yet these two conditions — extreme density and absence of interaction—are precisely those required for superconductivity.

54

Moreover, as no differentiation is possible there, no charge interference could occur due to the disaggregation of electrons, protons, ions and atomic nuclei constituting baryonic and leptonic matter. It would then become inappropriate to talk about resistivity or magnetic field inside the black hole. It is different with the event horizon, this frontier zone where matter deconstructs before joining the singularity, expression of an energy in a ground state.

The absence of any motion, at whatever scale, would allow the superposition—or combination — of two a priori contrary states, nonconductivity and superconductivity, into a single state specific to black holes. This singular property seems to defy all classical logic. But the Universe continues to confront us with such apparent contradictions. The electrons would have forced the intimacy of the protons, in a way 'neutralized', before melting into a dark and cold homogeneity, evoking the state of our Universe just before Planck's wall. All the constituents of what was matter, would then have lost their particularities.

The black hole could thus be apprehended as a quantity of non-fractionable energy, on the fringe of space-time, in which vacuum no longer has room. Such a conception makes it, in a certain way, a quantum object. On many points, the parallel between elementary particle and black hole is necessary. Both are:

- carriers of an energy equivalent to a mass,
- of unknown composition,
- not divisible,
- on the fringe of space-time,
- and if one marks the beginnings of the Universe, the other seems to announce its end.

Their kinetics, and more particularly their possible rotation, raise questions: we talk about spin for the particle, and tidal effects for the black hole.

In a black hole, energy accumulates without limit. One might be tempted to imagine an infinitely hollow cavity, but this image becomes inoperative when one considers that energy is freed from space-time. Under these conditions, how to talk about dimension, and therefore about container?

55

This leads to the belief that time does not have more control over the intrinsic properties of the particle than it does over those of the black hole. Exiting a black hole would require such energy that it would amount to a return to the past, implying a modification of the history of our Universe. Such a paradox led to consider the possibility of a passage towards another universe, endowed with its own history. This scenario would amount to a form of «teleportation» towards another binary system of universes in opposite symmetry, thus joining the theories of parallel worlds and wormholes. In the perspective of a virtual multiverse Cosmos, such an idea can hardly surpass the status of an image belonging to science fiction.

Everything suggests that it is not within our reach to penetrate the intimacy of a black hole. Nor does it seem possible to pierce, other than in the conditional, the mystery of what is hidden within the smallest constituent of matter, behind a sometimes undulatory appearance, sometimes corpuscular — perfectly coherent duality if one admits the existence of entangled wave packets.

Everything seems to oppose a mega-massive black hole devoid of an accretion disc to the singularity of the Big-bang, or more precisely to what precedes the "wall" of Planck. This wall marks the emergence of a context of energy in symmetry break, non-localizable, not open, and of strong entropy. Our Universe can then manifest by generating heat and brightness, in response to the nascent constitution of a still embryonic matter.

What is further developed suggests that the set of these TNMM could finally converge towards an ultimate confrontation, marking the collapse of our Universe. In a David Copperfield-worthy staging, all black holes would disappear from the cosmic scene and reappear, on the reverse side of the curtain, as a quantum primo-singularity—the term Big-bang then offering a pictorial representation of this event.

The Big-bang would not only be the starting point of a Universe of matter, but also that of a mirror Universe of antimatter.

The question of a necessary chirality between Universe and "anti-Unions" will be taken up further on (see chapter X). Deprived of hindsight and perspective, trapped in our physiological condition, we remain too constrained to imagine reasonably beyond what our observation, analysis and deduction capabilities allow. These are based on a logic that we have built from our perceptions and our feelings.

Let us nevertheless try to contravene it, by developing point by point this reflection within a theory that would like to be global, dedicated to the foundations of a multiverse Cosmos.

III <u>The Universe guilty of speeding!</u>
(Except to admit a misinterpretation of our physical laws)

A body called «at rest» can be defined as a body whose inertia is not distinguished from that of the finite system to which it belongs. Such a situation constitutes an essentially theoretical limit case, in which the body would be its own inertial reference frame, invariant. In practice, the relativity of motion implies that any observation remains sensitive to interactions external to the system under consideration. Moreover, within any physical system, the gravitational forces experienced and generated continuously modify the velocities, trajectories and, consequently, the effective inertial mass of bodies.

However, let's imagine a particle at rest. It is then admitted that: $E=mc2$. This famous simplified formula of Einstein implies:
- That matter (m) and energy (E) are closely related in various forms and substitutable with each other.
- That the speed of light (c) would be a constant of Space-time, a limit that could not be broken by any massive particle, while accepting...
- That the passage of time which makes it possible to measure changes affecting all forms of energy (see chaps. XIX and XXX) is, like the occupation of space, without any absolute value (see chap. XXV).

The energy (E), which any observer drains, and the gravitational effects (m) that affect it, make everyone possess without being perceptible, their own notion of time and therefore their own value of (c). The speed of light (c) for invariant as a displacement/time ratio, remains nonetheless evolving, depending on gravitational contexts that keep changing, called referential. It constitutes a limit for the transmission of information in space-time, but this local invariance does not exclude a global contextual dependence related to the dynamic evolution of the Universe.

The speed of light is considered unsurpassable by any form of information transmission or exchange. Considered as invariant, it nevertheless remains relative to the extent that any observer, like any observed event, continually changes its reference frame under the influence of a "neighborhood" understood in the broadest sense. One could thus affirm that any reference system varies locally, due to the very fact of the process of deconstruction of our Universe, inscribed in a context of space in permanent depression and accompanied by an expansion of time.

This speed limitation, noted c, could be lifted—thus joining the idea of non-locality — in an unrecognized dimension, where discrete interactions between particles and antiparticles occur, and where our binary system "of universes", in quantum symmetry, would somehow become a border of the multiverse Cosmos.

Can we then argue that photons evolve in a vacuum, understood as the total absence of all things, even though absolute emptiness would have no place in our Universe? Even in a cooled future Universe, almost exclusively composed of mega-massive black holes (MMBH) and excluding any phenomenon of "evaporation" or return to vacuum by annihilation of symmetrical particles, a vacuum of all contents would not exist. It would be more accurate to speak of an energy space in maximum occupation depression.

In a pictorial way, the Universe could be compared to a sponge exposed to the sun: the spongy matter retracts while the cells seem to gain volume. These alveoli represent the depressive space not occupied by matter; the increasingly dense substance represents the matter that is concentrated. The analogy cannot go beyond.

By definition, a space improperly described as void is a field in which baryonic matter appears locally absent. This does not imply that this space is truly devoid of everything. Radiation and scattered elementary particles configure what is called the «space vacuum».

Photons should, in principle, adopt rectilinear trajectories. They do, but these trajectories are affected by the deformations of space induced by gravitational effects. In our eyes, electromagnetic waves (EMW) are all the more deflected by the presence of a body as it is close and massive. Their frequencies increase as the gravitational effects of the approaching body intensify, which captures a fraction of the kinetic energy transported by the EMW. This energy transfer contributes to the energetic depression of interstellar space.

The speed of light—theoretically close to 300,000 km/s in a space assumed to be devoid of any material influence — is necessarily defined in a space that is never totally empty, if only because of the minimal energy represented by the cosmic microwave background. The speed of propagation of EMW thus seems to vary according to the energy density of the medium. Photons appear, for example, to move more slowly in water than in air—only in appearance, as we shall see.

Would the celerity of photons be potentially unlimited in the absence of any interaction with matter? Their speed — defined as the ratio between the distance traveled and the elapsed time—seems to depend on the topology of space. The speed of light is perceived as increased in a low-energy environment. In a complete, purely imaginary vacuum, the speed of photons would therefore be theoretically unlimited.

Such an imaginary void can also serve as an unmaterializable representation of the multiverse Cosmos, understood as a dimensionless reality, carrying an energy devoid of physical presence. However, it remains impossible to conceive a completely fixed time for photons: if this were the case, the speed of light would not be limited to 299,792 km/s. This finite value results from the fact that the EMW are in interaction with the material since the origins. These interactions involve energy exchanges, and more precisely a loss of kinetic energy, for photons considered as virtual particles lacking proper mass.

Imagining photons of infinite speed therefore amounts to excluding any significant relation to time, which would then lose all relevance. Can we even talk about displacement without temporal reference, while in the absence of time measurement, space itself becomes indefinable? The EMW then appear as a primordial kinetic energy, both of infinite celerity and null, due to the lack of measurable displacement. This original energy would have been transformed, by the effect of the Big-bang, into a mediating energy of quantum exchanges.

The photon, for which time — at light speed — does not have the meaning we attribute to it, interacts by energy transfer with charged particles. Energy quantum representing an electric field associated with a magnetic field, it plays the role of mediator in this quantum theory of fields that is electrodynamics. Major advance in the quest for unification, it allows integrating electromagnetism into quantum mechanics.

60

At the scale accessible to our observation and to the laws of general relativity, space-time becomes an essential field of analysis. This scale issue has led to the recognition that in quantum mechanics, the principle of locality—or separability — is partly a matter of interpretation.

The invariant character of the speed of light has as a direct consequence the exclusion of any absolute simultaneity between distant events. Einstein's relativity implies that any dilation of time — that is to say, any relative slowing — is accompanied by a growing depression of the space not occupied by matter. This space is gradually depleted of the kinetic energy of the EMW which, by achieving the "electromagnetic mesh" of mass particles, contribute to the cohesion of matter through charge equilibrium.

At the macroscopic scale, within a body, positive and negative charges compensate each other, with electrons acting as binding agents and ensuring overall neutrality. Under the effect of the gravitation it generates, same body tends to become increasingly massive and, by triggering nuclear fusion phenomena, leads to the formation of brown dwarfs, neutron stars and ultimately black holes. This progressive concentration of matter translates locally into an energetic depression of the fields constituting the so-called "empty" space.

The spin, which represents the intrinsic angular momentum of a particle— as well as the angular momentum of a moving body — confers on the entity in question resistance to any change in its state of motion. This resistance defines inertial mass; the gyroscope illustrates this conservation principle of angular momentum. Gravitational mass is nothing other than the expression of this inertial mass applied to massive bodies.

General relativity implies that same body determines its own space and time according to its inertial mass and the accelerations, positive or negative, it undergoes in a frame of reference that is proper to it. For the distant observer, gravitational effects are manifested by a distortion of space — relative contraction of lengths—correlated to a deformation of time— relative dilation of durations.

There is, however, no subjectivity in the evaluation of time and distances: the ratios between units of measurement specific to each isolated event or system remain constant as long as no relation with an external event is established. Thus, space and time, indissociably linked, shape our Universe

by giving it a constantly evolving relief. This four-dimensional topography
— three for space, one for time—is unique to each observer, related to a
local reference frame that cannot be shared with another distant observer.

The history of our Universe is entirely based on this complex relationship
between space and time, two inseparable notions. It could be described as
follows:

1. If we consider kinetic energy as the first state of matter:
Its ability to disperse in the primordial Universe — dispersion not
necessarily involving expansion—announces the speed of light, still non-
luminous and not measurable at this stage. Without delay, this speed of
dispersion is determined by the level of depression of energy fields,
disturbed by the appearance of objects of increasing mass. The speed of
propagation of this kinetic energy, a component of the current
electromagnetic force, is therefore contextual. Considered impassable, this
speed nevertheless reveals its limits: it cannot manifest itself in a completely
empty space. Yet interstellar space, far from being empty, is structured as a
backdrop to our Universe by photons that generate electromagnetic fields,
interacting with charged particles. In this falsely empty space, the time taken
as a reference in the speed of light is affected by mass effects. It is then
perceived as more or less fleeting, depending on the local intensity of
gravitation.

If light sometimes seems to us to spread more slowly, it is because the
environment crossed differs from that of the observer. In water, for example,
the distance travelled over time remains unchanged, even though shorter
wavelengths appear to propagate more slowly than longer ones. This is
explained by the fact that EMW, by interacting with H O molecules — and
more particularly with encountered electrons — constantly change
direction, lengthening their effective path. This phenomenon gives the
illusion of reduced speed and a loss of intensity.

An infinite speed would necessarily exclude any identifiable referential,
since time and space would lose all meaning. To disregard space-time then
amounts to referring to the multiverse Cosmos as a form of energy that can
only be virtual for the observer that we are.

2. If we consider the potential energy, representative of matter:
Constantly agitated by quantum interactions, fluctuations and incident
62

movements, it generates gravitational effects. These, sometimes perceived as antagonists, modify the energy density of space, slowing down time in the vicinity of massive singularities, to the point of immobilizing it in a Universe imagined at the end of evolution. Inertia, gravitation and fundamental interactions thus impose an unbridgeable limit: everything that has mass is unable to reach the speed of light.

If the masses in presence model space-time, EMW do not possess their own gravitational power but nevertheless undergo, locally, the effects of the deformation of space that shapes their scattering fields. This spatial context, conceived as a set of fields of variable energy density, conditions their relative speed and propagation direction. By interfering with dissociated molecules, atoms and free particles, EMW constitute in a way the moving tissue of space called "empty". Their speed of propagation, regulated by the gravitational obstacles represented by the bodies encountered, finds its limits. Asserting that the speed of light is constant constitutes an acceptable shortcut as long as one limit oneself to a short period of time and local news. On the other hand, if we consider that relativity excludes any idea of observable simultaneity due to disparities and fluctuations imposed on space-time, it becomes problematic to speak of a constant as soon as one is inscribed in the duration or outside a strictly local context.

In short :

- Subject to gravitational effects, EMW are inseparable from a time and space of reference. Their celerity finds its limits there. Moreover, a supposed infinite speed would go against relativity and deprive time and space of any meaning.
- Black holes would ignore time and space. No speed of movement or interaction is observable once the horizon of events has been crossed (accretion zone which should be devoid, in the long term, the MMBH of a cooled Universe about to collapse). If they seem totally integrated into the spatiotemporal fabric of general relativity, it is due to the existence of an observable event horizon. They are, in any case, part of the spatiotemporal continuum of any in situ observer. The event horizon represents an area of material deconstruction. This region, more or less extensive, which marks a border with space-time, is not homogeneous and therefore cannot be assimilated to a spherical surface.

Photons and particles with mass reach quickly exceeded relativistic speeds, depending on their initial speed and the trajectory angle. Time is presumed to stop there once the horizon is crossed. Thus, the image of an object approaching the horizon of events, which may appear frozen to the distant observer, eventually disappears from his field of vision.

We could tell the story of our Universe in terms of mass, considering however that the mass of a particle represents more precisely the amount of energy carried by a permanent assembly of primary waves, grouped here under the term **radiative entanglement** (see: Starting point of space-time or Planck wall, chap. V).

At very low temperatures, close to absolute zero (–273.15 °C), atoms, once trapped in a magnetic field, fuse into a state of total electrical neutrality. No longer being excited, they then access a state known as of lower energy. They seem to occupy the same quantum state, called a **Bose-Einstein condensate**. Atoms then exhibit a behavior analogous to that of photons in a laser beam. By becoming a giant quantum wave of almost zero density, these atoms, correlated to the extreme, pass from a fermionic behavior to a bosonic behavior and behave, in some way, like the photons of electromagnetism. The Bose-Einstein condensate thus confirms the link between a fundamentally undulatory quantum world and our macroscopic reality of corpuscular appearance.

Under these conditions, atoms, by merging into a single giant quantum wave, reveal new properties: superfluidity (frictionless flow) and superconductivity (electrical conductivity without resistance). CBE, considered as a kind of "degraded" state of matter, is distinguished from the plasma state of an ionized gas, characterized by very high energy and strong electrical conductivity.

Radiative entanglement suggests that it might be possible to experimentally confine EMW in a Bose-Einstein condensate. Indeed, in this state of matter, under extreme conditions of density and low temperature, the particles lose their individuality to form an undifferentiated and undivisible whole, evoking a giant exotic elementary particle. This amounts to imagining that one can slow down the speed of light until it stops. One can, however,

64

consider that this apparent phenomenon relates to a particular space-time relativity: the condensate, by compressing space, would slow down time, giving the impression that the speed of light regresses until it cancels itself. The experimental difficulty lies in the fact that projecting a laser beam on such a condensate requires a considerable amount of energy and extreme targeting precision. Therefore, this form of experimental radiative entanglement could only be transient.

Radiative entanglement, by conferring on primitive radiant energy the status of a mass particle, justifies that the Universe is observationally rather corpuscular and intrinsically wave-like in nature. A particle of matter remains fundamentally a wave system. The mass of an elementary particle is equivalent to the sum of energies — assimilable to kinetic energies — waves thus confined, making the elementary particle a quasi-infrangible entity, not representative of an occupied space. It is different for the atom, the molecule or any stellar body: at these scales, time and space account for the exchanges and interactions that make and break down built matter.

The mass that we assimilate to the "weight" of an object on a macroscopic scale represents the combined internal inertia of a complex assembly of wave packets, increased by the associated binding energy.

Mass is thus the rendering that we have of a chosen set of data (spin, charge, color, etc.) representing the intrinsic, inseparable and perennial properties of the elementary particle, considered as the irreducible constituent of matter. In summary, mass is the indicator of a degree of manifest presence associated with the familiar idea of object. That each body be, before reduction, a packet of waves seems founded. However, this is not acceptable in our macroscopic reality, which "breaks" the wave function (Schrödinger equation) and only perceives the massive object, thus justifying the wave-corpuscle duality and the misunderstandings that result from it.

Mass is indeed understood as an intrinsic property of fermions. These elementary mass particles have the ability to form, after decoherence — phenomenon that makes observable what, in a form other than corpuscular, would not be—the constructed matter as we perceive it. But mass can also be considered as the revealer of a context of exchanges and displacements resulting from interactions between these points of energy that are particles. It then becomes extrinsic to the particle. This interpretation led to postulate

65

the existence of discrete fields regulating these exchanges, within which would bathe the so-called matter particles. The Higgs field fits into this hypothesis. Often, when several alternatives arise, we make a clear decision.

In quantum mechanics, we tend to dismiss from the outset alternatives deemed incompatible in order to stick to more or less arbitrary choices, but consistent with physics satisfying our observations. Mass seems inseparable from the particle—therefore intrinsic — but it also reveals a more global context, which would make it an extrinsic property. The notion of mass is decisive in physics: it constitutes the indispensable ingredient for the observation of all things, regardless of the scale considered. Otherwise, we could neither expound on our Universe nor speculate on a multiverse Cosmos outside of our reality.

- **A standard model that challenges**

Our standard model postulates that the radiation travelling in space in all directions is the residual product of the annihilation of mass particles with their antiparticles. The fact that we cannot observe these antiparticles today, except during certain nuclear reactions, would suggest that they were present in lesser quantities at the beginning of the Universe, leaving only a remainder of particles constituting constructed matter. In other words, a concentrated "embryo" of primordial matter, potentially but unequally sharing opposite quantum numbers, would have preceded any form of radiation.

In this conjecture, matter would therefore be at the origin of the OEMs, today representative of the energy of the "vacuum" constituting the interstitial space—between particles, atoms, molecules and stellar objects. The idea of a primitive atom, proposed by Georges Lemaître, or that of a primordial quantum of matter, undoubtedly inspired this now controversial model, which aimed to describe the first manifestations of our Universe. However, this modeling does not allow to go back beyond a certain threshold and excludes any initial explanation, in a reasoning that proves reductive with regard to the cosmological model proposed here, based on a radiative entanglement phase.

66

Is the amount of matter present in the observable Universe stable? If we consider that the energy carried by matter is, ultimately, conserved but deconstructed in the form of black holes — singularities not representative of an occupied space—, this does not seem to be the case. As matter deconstructs, antimatter potentially present in vacuum energy—as evidenced by certain nuclear interactions — should evolve similarly. We would then head towards a Universe emptied of matter, while antimatter, in a dimension that would be proper to it, would deconstruct itself away from our observation. The chirality that distinguished them would dissipate with the cessation of time and the absence of occupied space. Under these conditions, the coalescence made possible of what was matter and antimatter would lead to an ultimate goal: a Universe in which space and time would have lost all meaning.

We can consider another way of conceiving the beginning of our Universe, not from a nucleus or an embryo of primitive matter, but from what represents the energy of vacuum—this interstitial space between particles of matter. What we improperly call vacuum is mostly made up of kinetic energy: the EMW, which can only manifest in the presence of mass particles (and antiparticles).

This vacuum energy is potentially convertible into particle-antiparticle couples, a consequence of a vacuum polarization that induces the idea of quantum symmetry and interfering energy fields. That matter could emerge from the void would justify the existence of a primary phenomenon: a decisive phase of radiative entanglement in the evolution of our Universe. Developed here as representative of the Planck wall, this phase would constitute the missing but decisive link — in the appearance of primitive particles (and antiparticles) endowed with mass. We then better understand the origin of the inevitable wave-corpuscle duality, without this duality implying that these two states can be dissociated when moving from classical relativistic physics to quantum mechanics.

We can thus assume that the essential of matter appeared during this infinitesimal period of radiative entanglement. The matter particles would be the result of a revelation of symmetry by creating couples of primo-particles and primo-antiparticles, from a latent energy without physical representation for us, that is to say, not yet able to fit into a spatiotemporal context. This energy devoid of observable properties cannot be apprehended other than as a virtual concept, designated here under the term, deliberately broad, of multiverse Cosmos.

This phase of radiative entanglement would mark the starting point for the first quantum interactions in a nascent context of time and space. We thus give an intelligible meaning to what we call the Big Bang.

Given the evolution of our Universe, remarkable for its level of entropy and diversity, it is likely that the original "soup" of particles from this phase was not perfectly homogeneous. Everything suggests that it included very early «lumps» explaining the observed anisotropies of the cosmic microwave background, punctuated by vast regions locally poor in matter. Particle fluxes would then have taken place, giving rise to zones of high energy density, at the origin of the first stars. At the same time, regions close to the ground state of quantum fields would have formed, with gravitational wells at their center resulting from the collapse of areas with an over density of massive particles. Recall that, even in the absence of any form of matter, the space vacuum retains a so-called zero-point energy, characterized notably by the presence of virtual particle-antiparticle pairs and the propagation of electromagnetic fields everywhere.

Due to their symmetrical properties—more precisely their opposite quantum numbers—, mass particles and antiparticles would have been led not to share the same 'dimensions' of a space-time within which their interactions would remain discrete. This would explain that our reality occludes antimatter. Particles and antiparticles nevertheless interact during electroweak interactions, in an ephemeral and punctual manner, by creating or annihilating pairs. The unrecognized presence of an antimatter remaining outside our field of observation could thus be at the origin of unexplained gravitational effects, leading to postulate the existence of an unknown material, not observable and not directly detectable: dark matter.

It has been argued that the particles and antiparticles created from vacuum would have been separated too quickly to be able to completely annihilate each other. If this hypothesis cannot be excluded, it does not validate the inflationist theory and does not explain the apparent absence of antimatter. Another hypothesis invokes an initially asymmetric production of particles and antiparticles, but it comes into tension with the idea that physical laws fundamentally proceed from symmetries.

One can consider that this period of radiative entanglement, devoid of significant duration, marks the opening of an extremely accelerated time, unrelated to the current relativistic conception of time and space. Outside the influence of time, in a space then deprived of notable gravitational effects, particles and antiparticles would have remained confined in distinct dimensions of a space-time in gestation. The intensity of non-entangled radiation in primitive mass particles would then have prevented the persistence of this phase of radiative entanglement, explaining its punctual nature and the ephemeral nature of the pair creations observed since.

This approach, admittedly counterintuitive, allows us to no longer question the absence of observable antimatter. It also dispenses with the use of an inflationist theory involving a sudden change in the scale of space-time while maintaining an effective gravitational force. The hypothesis of a mass particle called an inflaton, like that of dark matter or dark energy postulated as fundamental entities, then becomes obsolete. Similarly, the Higgs field can be seen as a default explanation for mass effects that are difficult to justify otherwise.

Phase 1: that of the Big-bang, a non-event

The three most common states of matter are solid, liquid and gaseous, plus a variety of intermediate states. It is appropriate to include the plasma state, less accessible to our direct experience, which requires extreme conditions of pressure and temperature in order to dissociate the constituent particles of the nuclei and to release electrons from the atoms. Even more exotic, we can cite the so-called Bose-Einstein condensate state, in which atoms of different energies, once strongly cooled, collectively behave like waves.

Is this behavior really surprising if one considers that atomic particles can be assimilated to entangled wave packets? Under these conditions, atoms adopt the same superfluid quantum state and take on the appearance of a single giant wave, leaving, for a brief moment, their initial fermion state. This state, obtained at very low temperature, tends to validate the hypothesis of a radiative entanglement phase at the origin of matter.

This first phenomenon is inscribed on the fringes of the commonly accepted history of our Universe and forces us to leave the framework of a physics that we would like strictly in phase with our observable reality. For explanatory purposes, a fermion could be considered as a condensate of photons confined in an optical microcavity — a confinement chamber formed by curved mirrors so that the introduced photons constantly bounce there without ever being absorbed.

But can we really talk about photons at this stage, which would correspond to the opening of the Planck era, when it is a latent energy, devoid of significant wavelength? We are evoking here a prequantum state, cold, without declared symmetry, since matter did not yet exist and energy knew neither space nor time, and therefore no significant kinetics for us. These original photons, thus confined, would present such a close coupling that they would be unable to dissociate. It is under these conditions that they would set themselves up as primary particles of matter, under excessively high temperatures. From these first interactions will be born the current elementary particles (see the development in chapter XIII, with a possible implication of primordial neutrinos). This preliminary, both elusive and violent, would correspond to the very first moments following what we call the post-Big Bang.

Without the presence of elementary particles endowed with mass, constitutive of matter, and without the gravitational effects they induce, what would remain of our Universe? Nothing that allows to confer on space and time the meaning we attribute to them, and no observer to grasp even the question.

If we consider the phase of radiative entanglement as the possible starting point for the formation of all the massive bodies of our Universe, how to reconcile this so-called first phenomenon with a previous physical state? In other words, how can a primary cause—whatever it may be, taken up here under the term Big Bang to emphasize its suddenness and violence — fit into a broader reflection, called here, for convenience, Multiverse Cosmos, without us being able to conceive a link of anteriority? In the absence of established causality, the above can only be hypothetical. This lack of temporal continuity between a 'before' and an 'after' led to the classification of this mysterious Big Bang as a singularity.

Is this not a way of recognizing that we are ontologically incapable of representing to ourselves the appearance of matter, that is to say the birth of our Universe from an absence of time and space—what we could translate by a Nothing devoid of all physicality? It indeed seems inconceivable to us that Nothing, understood in the literal sense of Nothingness, can generate something, except to postulate a potential precedent, taken up here under the term of multiverse Cosmos.

What would precede the Big Bang cannot therefore, it seems, be defined in terms of spatiotemporal dimensions. The Cosmos, qualified here by literary convenience as a multiverse, would then evoke an earlier state of latent potential energy, without physical representation in the classical sense, but which cannot be assimilated to Nothingness. Purely contextual, it offers a default explanation for the appearance of matter and the opening of space-time. Virtual entity and profoundly counterintuitive, the multiverse Cosmos cannot be defined as an infinite set of interconnected universes; it is limited to an exercise in thought, on the fringes of a reality that the observer dresses according to his own ideological constructions (see chapter XXIX on decoherence).

If the particle, as an energy point, is not representative of an occupied space, the internal velocity of the entangled photons in the form of wave packets, by means of a self-sustaining resonance, has nothing comparable with the speed of light defined in Einstein's spacetime. This abstraction of the particle, in which emptiness has no place, would allow us to explain how, at the beginning of our Universe, what today corresponds partly to visible light could become matter, but also antimatter, by radiative entanglement. Symmetry, which excludes the idea of a creation from nothing in the dogmatic sense of the term, is thus set up with the joint appearance of the first particles and antiparticles of matter. Even today, the recovery by material of the OEMs participating in the diffuse background continues, with the electroweak force, combined with the gravitational deformations of space-time.

The particle, considered outside any context likely to interact with it, would manifest neither temperature nor change of state. This is what would make a Universe at the "end of life", reduced to the mere presence of black holes and devoid of EMW, a uniformly flat and cold world. These conditions suggest that the Universe could disappear in the same way as it appeared:

71

without significant temperature, without remarkable symmetry, and outside any spatio-temporal framework.

The primordial kinetic energy, devoid of mass, does not generate any gravitational effect and time cannot, at this stage, imprint its mark. By revealing a symmetry break through the creation of matter and antimatter particles, this energy will open space and establish time.

- **Phase 2, that of a binary system of universes in quantum symmetry**

The primordial kinetic energy is partially converted, by radiative entanglement, into mass energy. Nucleosynthesis then recombination will complete to structure the matter. The proliferation of increasingly massive bodies tends to empty space of what we could call "vacuum energy" — a vacuum that is not, however, void.

The term radiative entanglement used here does not refer to the coupling of distant particles designated by the expression quantum entanglement. He describes, in these lines, constructive interference of primordial waves leading to a change of state that has no equivalent today, if not during ephemeral creations of particle-antiparticle pairs.

These primordial, highly energetic waves, by permanently intertwining in the form of elementary particles, have somehow realized "inextricable knots" of energy, without a physical dimension. Formed at the dawn of the Universe, in an embryonic state, these mass particles will then regroup into increasingly heavier atoms.

This would explain that this disparate matter, which we represent today as a set of elementary entities grouping together constructively, remains, contrary to appearances, fundamentally undulatory before appearing to our eyes as structurally corpuscular..

Why does an atom not collapse on itself under the effect of electromagnetic forces supposed to bring together the atomic nucleus, generally positively charged, and the electron belt of negative charge?

Our macroscopic world prompts us to conceive elementary and composite particles as corpuscles moving in an empty space traversed in all directions by electromagnetic waves. It is to overlook the idea that in quantum mechanics, matter does not really exist in the classical sense.

In the subatomic dimension, nothing is in the state of a compact, massive and heavy energetic entity, such as we perceive it at the contact with constructed matter. This feeling of touch is only an illusion, a simple feeling, because the atoms are far from touching each other. The repulsive electromagnetic force of electrons gravitating around the nucleus keeps atoms at a distance. She thus creates a sensation of contact without allowing us to penetrate the touched object.

Fundamentally, what we consider, at our scale, as an assembly of elementary corpuscles, massive and localizable, is just a complex system of entangled waves—or elementary particles from the radiative entanglement phase — in close and permanent interactions.

It happens that an elementary particle loses this status when creating particle pairs–antiparticle followed by their annihilation, or when it leaves space-time by joining a black hole.

In quantum mechanics, space and time cannot be apprehended as we do in a tangible reality that belongs only to our perception. In the absence of a space-time referential, how to measure quantum interactions in terms of relative velocity and position? This disturbing idea inspired the bewildering idea of uncertainty.

Time is a dimension that has nothing absolute. It is as fortuitous as ephemeral. The chirality that characterizes a binary system of universes in quantum symmetry lies in this «non-smooth» character of time. A single symmetry, the one to which we are attached, is revealed to us.
With a lot of imagination, antimatter could be perceived as the discreet shadow of matter constituting our reality. It would lean against a time that is specific to it and therefore not in phase with ours. This peculiarity would imply the existence of a space that is somewhat parallel — or superposed—to the one to which we are attached and which constitutes our field of observation.

- **Phase 3, precursor of the final collapse**

At this point, the energy is about to lose its mass properties to join the multiverse Cosmos. All kinetic energy carried by the EMW will end up captured by black holes. Time will be suspended when, for lack of significant space, the interactions of deconstructed and gathered matter in this ultimate configuration cease. Dug to the extreme, space will then disappear in the simultaneous collapse of all MMBH, singularities of phase transition where the principle of exclusion is transgressed.

The acceleration of a body supposed to be isolated and initially at rest involves an increase in its mass, indicative of an energy supply. For a body, whatever it is, to be able to approach the speed of light, it should either capture all the matter in the Universe or convert its mass energy into kinetic energy. Deconstructing matter in order to convert it fully into kinetic energy corresponds precisely to the destiny of MMBH in a cooled and "end-of-life" Universe.

EMW, devoid of mass, do not generate gravitational effects. Although they represent energy in motion, they can neither truly accelerate nor slow down. However, the gravitational effects of encountered bodies modify their energy potential by imposing on their movement the curvature and temporality of the space traversed. By giving them a corpuscular aspect, the assimilation of photons to vector particles allows for a better understanding of the role and nature of electromagnetism.

Outside of any energy medium — non-admissible hypothesis—particles could theoretically exceed the speed of 300,000 km/s and move at unlimited speed. This is, however, a purely conceptual case, as it would amount to leaving space-time and rejecting the very idea of referential, as prescribed by general relativity. Moreover, in the absence of units of measurement, how to talk about displacement for those who have no theoretical reference framework?

We can also consider that, for photons, whose speed would potentially be unlimited, time is as if stopped. But to assert that time is stopped is to assume that, somewhere in the Universe, time does not exist. If the photon is only a point in the sense of a localized location, and not an object constituting geometric space, it cannot have dimensions. Not occupying space, he is then out of time.

In a black hole, of unsuspected density and where empty space no longer has its place, the speed of light becomes irrelevant. EMW, having lost their vector status, are confined to such a point that frequencies and wavelengths lose all meaning. This non-occupation of space is a common property of the black hole and the elementary particle, except that in a black hole, matter is deconstructed and any form of radiative entanglement—past phenomenon generating particles — has disappeared. The energy is there in a transitional state that no longer pertains to our space-time.

In the end, nothing really seems to distinguish between an unlimited speed and a zero speed, both of which assume the absence of a space-time reference frame. In both cases, time and space have disappeared.
This could evoke, on the one hand, the cold primordial kinetic energy representative of the Big Bang before the first radiative entanglements of the Planck wall, and, on the other hand, the cold potential energy, deconstructed and deprived of any interaction, of the MMBH. The latter would then restore to the multiverse Cosmos an energy without mass.
Everything would therefore be just a writing game, in which it would suffice to substitute final collapse for the original Big Bang.

We live in an irreducible environment of sounds and lights, two phenomena of profoundly different nature.

Sound is a wave of limited range, which propagates all the more quickly as the particles of the medium traversed are sparse and the molecular bonds are stable and strong. It is often claimed that sound cannot propagate in space. It is forgotten that the so-called 'empty' space contains, in unevenly distributed quantities and at variable densities, clouds of gas — mainly hydrogen — likely to allow the propagation of sound waves. Remember that

75

sound is nothing other than the vibration of molecules within a more or less deformable medium, such as air, water or even metal. The intensity of these molecular tremors depends on the temperature and pressure of the ambient environment. Our Universe thus produces a diffuse background noise, composed of sound frequencies that are mostly inaudible to us.

By nature, these sound waves are distinguished from EMW, although certain frequencies of electromagnetic waves can convey signals — principle of radiocommunication — convertible into sounds, and vice versa. Sound then appears as a mechanical waveform, derived indirectly from electromagnetic waves.

Light, in the broad sense, is an electromagnetic wave covering a wide range of frequencies and intensities, well beyond the only visible spectrum. It results from the inseparable coupling of an omnipresent electric field and a magnetic field, which contribute to giving space its dimension of occupation. Of unlimited range, it interacts with the particles encountered. If it seems to slow down, it is because of the additional paths — not directly perceived — that are imposed on it by the material being crossed, notably by diffraction.

Unlike sound, the speed of OEM is given as invariant: an observer moving in space will not notice any significant variation in the speed of propagation of the light in which he bathes.

This speed is however relativistic, because for two distant observers who would observe each other — assuming shared instantaneous communication — the compared speeds, including that of light, would appear different.

There is every reason to think that the distance traveled/ elapsed time ratio, which defines the speed of light, could be affected by the effects of the «aging» of the Universe. In other words, the increasing depression of space would influence our way of conceiving relativity, and thus the speed of light itself.

The gravitational effects of bodies remain without influence on the intrinsic properties of the elementary particle, which does not occupy space in space. On the other hand, it would be the interactions pertaining to quantum mechanics that would be at the origin of mass effects, by modifying the properties of space-time and inducing relative variations in the trajectory and speed of light rays.

Can we then affirm that the speed of light, corrected for the effects of relativity in time — the Universe not being static — remains an unchanging constant? Could it not be affected, over time, in a doubly relativistic manner, by the global concentration evolution of matter? The constant c would then become a value to be adjusted according to the evolutionary level of depression of the so-called empty space.

General relativity claims that the speed of light is determined and limited by the presence of massive bodies that interfere with it. But can we exclude that the fluctuations of the vacuum energy have no effect on the evolution of what we consider to be an invariant mathematical constant ($c = 299\ 792$ km/s)? The very idea of a reference system, however, implies that two observers obtain the same measure of the speed of light.

All that follows fits into this logic of latent forces, potentially in contrary symmetries, being admitted that:

- A binary system "of universes" in quantum symmetry, reflecting a break in latent symmetry within the Cosmological Equilibrium, has no history with regard to the multiverse Cosmos.
- Cosmological Equilibrium refers to potential energy, that is to say a form of energy devoid of event, without concrete representation for us, but which cannot be confused with an absence of content.

Galaxies and apparent expansion

The galaxies approach each other under the effect of gravitation to then merge. Associated with the kinetic impulse of retrograde dispersion that follows the Big Bang, gravitational effects contribute to giving the impression of an inflationist space. In fact, at observation, galaxies and galactic clusters seem to move away from each other all the more quickly as we place them far in the observable past. But if it was only an optical illusion, how to explain it?

77

To simplify, about two thirds of the observed galaxies are spiral galaxies, which would only represent a quarter of the estimated mass of the observable Universe. The remaining third, mostly composed of elliptical galaxies, would concentrate three quarters of the global mass.

The spiral galaxies seem to be the most recently formed, even if, for the most distant ones, we only observe an image of the past. They are also the most active: gas and diffuse matter abound in them, and stars still form in large numbers. Their central bulge announces the future evolution towards an elliptical galaxy, amputated of its spiral arms.

Elliptical galaxies, with an imposing central black hole, are generally populated by ancient stars and cooled planets. They can also result from collisions between older galaxies, with already blunt rotations. We should therefore observe more young spiral galaxies in the distant, that is to say in the past of the Universe. In fact, it is in these remote regions that dense molecular clouds are detected, heralding protogalaxies, future spiral galaxies.

The oldest galaxies, after having "cleared out" around them, could take the form of dwarf galaxies concealing a supermassive black hole within a residual cloud of gas and dust.

When we observe the distant Universe, we notice a lengthening of the wavelengths, interpreted as a Doppler effect, which could lead to believe that the speed at which the most distant galaxies move away would exceed that of light. It is forgotten that it is the intensity of the OEM fields, modulated by gravitational effects, which gives space its dimension. It then becomes difficult to validate the hypothesis of a superluminal velocity without questioning relativity.

This shift of the light spectrum towards the red could be explained more simply by several closely related phenomena, more marked in the past:

- The forming galaxies produced more stars.
- The current Universe, richer in white and brown dwarfs, neutron stars and black holes, seems generally less luminous.
- The rotation of celestial bodies fades over time.
- The emitted electromagnetic radiation tends to evolve towards red.

- The density of massive bodies increases through gravitational clustering.
- The rate of formation of young galaxies is slowing down.
- The observed distant galaxies have since grown larger and have become mostly colder, elliptical, and less active.

Their present state remains however inaccessible to us. At all scales, it can nevertheless be assumed that the Universe is and will remain globally homogeneous, similar to its observable portion of proximity. Designing a Universe not stemming from a punctual singularity followed by an expansion but created in a relativistic context conforming to its current configuration, already allows solving the question of its homogeneity.

Ultimately, these two types of galaxies must coexist in comparable proportions across the Universe, structured into galactic clusters. The return to the past of distant galaxies also explains the lesser amount of built matter that characterized the young Universe, where matter was more dispersed, giving the illusion of an accelerated expansion. This difficulty of interpretation led to introduce, by default, the cosmological constant and the speculative idea of dark energy. But how could a universe devoid of external benchmarks be defined in terms of inflationary volume? This question leads us to confront at the same time the question of a borderless Universe, that of the link between the infinitely small of quantum mechanics and the infinitely large of relativistic gravitation, and, even more deeply, the very nature of time. Would it be just an artifact of our consciousness, emerging from a reality that still escapes us?

IV Is our universe riddled with tunnels?
(Tunnels that would cross the history of our Universe)

The problem of black holes is perhaps less related to their physical nature than to the very nature of the time to which we relate them, a conception largely conditioned by our experience and by a still partially empirical approach to space-time relativity. Time, as we conceive it, remains inseparable from an anthropocentric perception that tends to fix its linearity and directionality.

If we consider the black hole as an ultimate state towards which converges matter from the first phases of the Universe, the density reached by this matter during its collapse would be such that any internal interaction would become inoperative. For a distant observer, this situation would result in the impression of a stoppage of time. Conversely, in a symmetrical reasoning, the external observer would appear, from the point of view of the black hole, to evolve in an extremely fast present, a direct consequence of the relativity of time specific to each reference. The speed of the flow of time then becomes strictly dependent on the adopted point of view, which constitutes a fundamental principle of general relativity.

The idea of a strictly linear time thus deserves to be qualified. General relativity introduces a form of "elasticity" of time, which excludes any absolute simultaneity in the context of local gravitational fields. Just as one cannot return to a past state, it is not possible to maintain a strictly defined spatial position. Without displacement, without variation in energy, without quantum interactions or intrinsic kinetics of the particles, time loses all operational significance. In this perspective, time would not stop within a black hole, but would tend towards an almost infinite effective value, giving the illusion of an absence of flow.

The density of black holes is often described as infinite. However, such a statement remains conjectural. Can we really speak of compactness for an object that cannot be assimilated to a classical stellar body? If matter is deconstructed in a form without known equivalent in our observable Universe, the tidal effects associated with black holes remain measurable. It is nevertheless remarkable that their attractive power is not proportional to an attributable physical size, which suggests that the black hole does not constitute an evolutive body in the classical sense, but rather a singularity, that is to say, a marginal domain with respect to the usual space-time.

In this perspective, the black hole could be considered as a discontinuity of space-time, a "perforation" that we are forced to characterize indirectly in terms of mass, for lack of being able to apprehend it otherwise. The observation of the most massive black holes tends to support this interpretation: they represent only a minority fraction of the observable black holes of the Universe and could constitute the culmination of so-called primordial black holes, formed at the very first moments of the Universe.

The recent observations made possible by the James-Webb space telescope, revealing very old black holes, weakly surrounded by matter and with apparent mass that is difficult to explain, invite us to reconsider the initial cosmological scenarios. Their early presence questions the validity of a standard cosmological model based on a progressive evolution of structures. It is then conceivable that, during the radiative entanglement phase, the extreme dispersion of elementary particles favored an exceptional density of close interactions, in a still embryonic Universe, devoid of fully constituted spatiotemporal landmarks.

In such a context, the very notion of a spatial void would be premature, as the time scale necessary for its definition remains to come. Primordial black holes could thus have formed very quickly, over an extremely short period of time. Their "mass" deduced today from indirect effects (accretion disks, reverberations, gravitational effects) would then be based on interpretations dependent on the cognitive abilities of the current observer, applied to a reality stemming from an emerging spatiotemporal framework.

The quasars observed in the distant Universe, remarkable for their brightness, could correspond to these primordial black holes in their phase of maximum activity. The images that reach us reflect an ancient state of these objects, which could today have become dimly lit, even totally obscure, and manifest themselves as vast discontinuities in space-time.

At the end of the radiative entanglement phase, the primitive Universe was probably exclusively made up of elementary particles—neutrinos, quarks and leptons — as well as, in equivalent quantities, their antiparticles of opposite charges: antiquarks and positrons. In an extremely short period of time, assembly processes have led to the emergence of composite particles, resulting from the association of several elementary constituents. Thus formed the protons, neutrons and various so-called exotic baryons, characterized by their instability and by a composition exceeding three quarks.

A fundamental property of elementary particles is the existence, for each of them, of an antiparticle associated by an opposite charge symmetry. It is therefore legitimate to postulate that matter and antimatter, initially produced in equivalent proportions, were brought into interaction from the very first moments of the Universe. These interactions have led to mutual annihilation processes, during which particles and antiparticles mainly convert into radiant energy, in the form of photons. This energy, devoid of rest mass and therefore of direct gravitational influence, has spread, contributing to structure what we refer to approximately as the space vacuum..

It should be emphasized, however, that this vacuum is by no means a state of total absence. It remains the seat of quantum fluctuations, manifested by the transient appearance of virtual particle pairs. These phenomena give the vacuum a dynamic structure, which takes on its full meaning through primordial annihilation processes and subsequent gravitational first collapses.

Recent theoretical advances also suggest that the quantum energy "soup" characteristic of the primordial Universe only has a corpuscular character through the interpretation provided by a subsequent macroscopic observer. The entities that we call particles would be more rigorously described as interacting quantum wave fields, in accordance with the formalism of the wave function. Nevertheless, the wave-corpuscle duality still prevails in any attempt at phenomenological description. In the framework of the present development, devoted to primordial black holes, we will therefore maintain an approach based on the operational notion of elementary particles in interaction.

It is also unlikely that the density of interactions, whether they concern entities of identical or opposite symmetry, has been strictly homogeneous within this expanding medium. The temperature prevailing during the radiative entanglement phase should reach values such that the Planck temperature, related to an already structured space-time, cannot constitute a relevant reference.

82

In this extreme quantum regime, succeeding radiative entanglement, the notions of position, velocity, space and time were not yet operationally defined. The residual radiation intended to become observable electromagnetic waves could only acquire determined wavelengths and frequencies from the moment when the first collapses of local particle concentrations began. These processes have allowed the joint emergence of notions of space, time, and physical emptiness.

Prior to the recombination phase, which will see the formation of the first atoms — mainly hydrogen—these concentrations can be described as unstable systems dominated by incomplete random fusion and decay interactions. Initially governed by quantum dynamics, these systems would eventually undergo gravitational collapses, giving rise to so-called primordial black holes. These objects, devoid of significant accretion disks and host galaxies, would have appeared before the necessary conditions for star formation were met..

Difficult to detect and to characterize observationally, these primordial black holes are today manifested by a relatively low activity, although they can still occasionally absorb stellar objects nearby. They would nevertheless have contributed significantly to the cooling of the Universe and to the gradual establishment of an effective space vacuum in a space-time being structured. As such, they would have played a decisive role in cosmological evolution and in the genesis of galactic structures.

Galaxies, resulting from the aggregation of recombined matter and Big Bang residues, have in turn generated, by gravitational collapse, second-generation black holes at their center. They also host many stellar black holes, products of local collapses or remains of supernova events.

The most massive primordial black holes, difficult to highlight despite gravitational lensing effects, are characterized by their isolation and observational discretion. This property is explained in particular by the fact that the tidal forces associated with a black hole are not proportional to its supposed radius. If the gravitational field of a massive body decreases with the square of the distance, the attractive power of a black hole cannot be correctly apprehended by a simple equivalence in solar mass.

The hypothesis of a black hole without mass energy remains controversial. However, it is useful to recall that the constituent elementary particle of
83

matter, the fermion, is modeled as a point without spatial extension. Only through interaction and observation does it acquire physical significance as a massive, localized entity. By analogy, a black hole could be interpreted as a region where space and time are locally erased, although its topology makes this disappearance imperceptible to the external observer, who necessarily attributes it an effective mass.

In this perspective, a black hole could be considered as a discontinuity of space-time, or even as an opening towards the multiverse Cosmos. Resulting from an extreme curvature of space, it would constitute a closed structure outside space-time, attributing to absorbed energy properties—or an absence of properties — analogous to those it possessed before the Planck scale.

One can thus compare black holes to conduits connecting the initial and final phases of the Universe. The so-called "gravitational" effects associated with them would then be of a different nature from those of ordinary massive bodies, which find their origin in underlying electromagnetic interactions related to quantum physics.

In the long run, these structures would close after absorbing the last residues of a cooled Universe, deprived of matter and radiation. In this ultimate state, where the relativistic effects of space-time are smoothed, any matter-antimatter chirality would tend to disappear in the absence of temporality. If the energy associated with antimatter evolves according to a distinct dynamic, this leads to consider the existence of white holes, open on an equally discrete antimatter sector. In the superposition of states, black holes and white holes could then annihilate each other, restoring the cosmological balance broken during the Big Bang.

Mass can be defined as the measure of the inertia of a system, that is to say, its resistance to any variation in its state of motion. It constitutes both an indicator of the potential energy associated with a body and its kinetic energy in dynamic regime. It also characterizes the intensity of gravitational effects experienced and produced, as well as any form of acceleration. An increase in the gravitational field or acceleration leads to a modification of the local reference system, resulting in a contraction of space and an expansion of time for any event considered in situ.

«In all my research, I have never found a material. For me, the term matter implies a packet of energy," said Max Planck, one of the founders of quantum mechanics. This assertion takes on a particular resonance if one considers that, beyond the horizon of events, matter, collapsing upon itself, seems to be converted into a transitory phase assimilable to energy at rest. Such a state, not reflecting any internal interaction or informational exchange, and in which the notion of time loses all operational significance, can hardly be interpreted as an effective occupation of space-time.

In this perspective, matter would return to a massless energy state, realizing an exotic form of cold, radiative plasma, without direct correspondence with a material reality compatible with our observable Universe. The black hole could then be considered as an object of quantum nature, in the same way as the elementary particle, here considered as a packet of entangled waves without defined spatial dimensions. Space-time, in its evolution, would thus constitute the context of a phase transition linking the elementary particle — resulting from the Big Bang — to the singularity of the black hole, the ultimate threshold for the return of mass energy to a primordial state, heralding a final collapse.

Like the interacting elementary particle, an active black hole would not escape the quantum phenomenon of wave packet reduction. The gravitational attractiveness and the presence of an accretion disk suggest some interactions with a space-time from which the black hole would, however, remain distinct. As for an observed particle, initially presumed to be in superposition of states, the observer is led to attribute to the black hole — hidden behind its event horizon — a state corresponding to a macroscopically accessible configuration.

Any measure involving quantum decoherence and the destruction of state superposition, we are thus led to endow black holes with properties that do not necessarily reflect their deep quantum nature, formally described by a wave function. The characteristics attributed to black holes have therefore essentially the function of integrating them into the standard cosmological model, at the cost of a significant part of subjectivity and speculation. This approach is particularly evident in the recurrent use of a definition based on an equivalence in solar masses.

In this regard, several theoretical solutions have been proposed:

- massive black holes in rotation, without electric charge (Kerr).
- massive black holes, not rotating and uncharged (Schwarzschild).
- massive black holes, not rotating and charged (Reissner–Nordström).
- extreme black holes, almost devoid of mass-equivalent, charged and in maximum rotation (associated with Stephen Hawking).
- massive, rotating and charged black holes (Kerr–Newman).

The trajectories of approaching objects and particles, combined with their initial velocity and the apparent speed of rotation of an accretion disk, are in principle the main observational indices of the presence of an active black hole. These elements suggest that the majority of black holes would be in rotation. Such a conclusion, however, raises a difficulty: how to reconcile the idea of rotation with that of an object that cannot be described either in terms of space occupied or in terms of the flow of time?

The hypothesis that a black hole would constitute an escape from space-time for a fraction of the energy it absorbs appears deeply counterintuitive. It almost leads to envisage that it would be the Universe itself which, in a sense, would be in rotation around a singularity escaping any direct observation. In practice, only the apparent area of the event horizon and certain indirect gravitational effects remain accessible to measurement and allow a partial characterization of black holes.

The model of the black hole considered here — devoid of spatial dimension, temporality, intrinsic angular momentum, and electromagnetic charge — does not correspond to that of a massive body in the classical sense. Its attractive power, analogous to the gravitational effects of massive bodies, would result from the fact that it acts as a vacuum pump, gradually stripping space-time of any interaction. This dynamic would be part of a process of returning to a cosmological equilibrium, compatible with the hypothesis of a multiverse-type Cosmos.

A parallel can be drawn with the notion of black energy — formalized by the cosmological constant — characterized by negative pressure and invoked to explain the accelerated expansion of the Universe, although its physical reality remains hypothetical.

Any form of energy crossing the event horizon of a black hole disappears from space-time and loses all temporality. This could explain why, from the internal point of view, past, present and future are confused, with information overlapping without a discernible causal order. The final collapse would already be accomplished there but would remain concealed from the external observer.

Any attempt at introspection of a black hole thus runs up against the fundamental limits of observation: space-time constitutes the binding framework for all measures, and relativity fixes the boundaries. It must be recognized that the scientific dialectic struggles to formalize such radically counterintuitive, largely hypothetical concepts, and that we do not currently have any experimental means allowing us to fully validate their implications.

V Of the difficulty in giving a purpose to our Universe
(But benchmarks are missing)

A fundamental point, often overlooked, is to recognize that any galaxy—and, more generally, any stellar body — can be considered as a reference center of the Universe. Using the notion of a center in the sense of an absolute midpoint gives the Universe a unique character and implicitly joins the idea of an isotropic expansion from a privileged point. Conversely, considering the centre as a simple point of reference implies the existence of a multiplicity of equivalent centres. The center of the Universe is then defined as the place of observation itself, without any position being considered privileged.

In this perspective, the idea according to which a Universe devoid of an absolute center could nevertheless be in global rotation loses all physical significance, both in the literal and conceptual sense. Indeed, a rotation can only be defined with reference to an identifiable axis or plane of rotation, which presupposes the existence of an external or privileged landmark, incompatible with the hypothesis of homogeneity and cosmological isotropy.

However, one might question the possibility that the Universe, without intrinsic rotation, masks increased differential velocities for large-scale structures. Movements not directly observable, possibly oriented in the same direction or according to distributed orientations, could, if their trajectories appeared circular around a common axis, suggest the existence of an effective center of space-relative time in which our Universe evolves. Such an interpretation seems coherent with the idea of a Universe resulting from a singular state, often assimilated to the Big Bang. However, this representation is more a hypothesis reinforced by the illusion of spatial inflation than an established observation. She tries to place cosmological evolution in the context of space-time relativity but remains speculative.

The model of a universe in global expansion, possibly rotating, resulting from an infinite point of energy, is not retained here. It constitutes an alternative construction aimed at exploring the implications of certain observations, without claiming to impose itself as an exhaustive description. The perspective adopted seeks rather to propose an interpretation likely to confer more coherence to a cosmological model which, despite a broad consensus, remains marked by conceptual inadequacies and gray areas.

Potential energy, expressed in mass equivalent, is generally evaluated from cross-observations of gravitational effects. This gravitational interaction opposes any rectilinear trajectory in the Euclidean sense and prevents massive objects from exiting cosmological boundaries. Although the Universe is not infinite in the strict sense, it has no accessible edge. The trajectory or speed of an object changes nothing: any progression remains constrained by the gravitational fields that curve and structure space-time.

Thus, any cosmic movement amounts to a wandering conditioned by the gravitational topology of the Universe. The followed routes result from a complex network of local curvatures — stars, planets, galaxies — and deep singularities that constitute black holes. These trajectories evolve, intersect and transform over time, giving the Universe the appearance of a closed, dynamic system devoid of external escape.

To clarify the notion of no boundary, let us recall that any idea of perimeter presupposes a content and, correlatively, a container. Several consequences follow from this :

- the content of our Universe can only be defined with reference to a symmetry within a binary system of universes in quantum symmetry.
- the multiverse can be considered both as constitutive of our Universe and as representing an infinity of symmetrical pairs of universes, thus assuming a role of container.
- related to our Universe, the multiverse would be simultaneously omnipresent and non-localizable, constituting a virtual entity excluding any direct interaction between systems of distinct universes.

This approach does not contradict the vision of Albert Einstein, who conceived a finite but unbounded space, often represented as spherical or toric, in which the three spatial dimensions are integrated into a fourth temporal dimension. To grasp the intuition, let's consider a spherical space.

89

In plane Euclidean geometry, any closed broken line defines a figure whose sum of angles depends on the number of sides. On a spherical surface, this relation is deeply modified: a triangle consisting of an equatorial side and two half-meridians can cover a hemisphere and present an angular sum of 540°. Figures with a larger number of sides can greatly exceed the corresponding Euclidean values. This property illustrates the difficulty in transposing our classical geometric intuitions to a curved, dynamic space correlated with time.

The Big Bang theory assumes a globally spherical Universe, whose limits remain indefinable. Space-time relativity—time dilation and contraction of space as a function of mass and speed — makes distances unstable and blurred at any scale. It then becomes impossible to determine a shape, volume or global surface of the Universe, as well as to establish an absolute timeline from its origin to a possible end, in the absence of a time unit independent of the spatial context.

This does not mean that the Universe, devoid of "franchisable" limits, is nevertheless expanding. Einstein was long resistant to the idea of an inflation of the Universe. In cosmology, the current "doctrine" advocates an accelerating expanding Universe. Our standard model is based on a general but not unanimous consensus among scientists. It has always been so, the logical tendency being to rally around the opinion of as many people as possible who have extensive knowledge on the subject and not to pay attention to what questions in a marginal way. Critical thinking then partially loses its relevance. This is the whole problem, because this form of membership which seems quite natural, has often shown that it only endorses convictions likely for a certain number to be questioned. The evolution of our knowledge about the Universe is thus marked by a long succession of errors and beliefs that remain for some to this day, in the state of hypotheses. It is enough to consult the scientific works published since Newton, one of the first mathematical physicists of the modern era. Why would it be otherwise today? This observation, which points to inconsistencies or incompatibilities between recent observations and predictions, is moreover echoed in a number of current publications that invite reflection on new theories.

Energy does not fill the space: it structures it. Gravitational space presents itself as a continuum without fixed dimensions, subject to permanent local deformations. Its curvature evolves with cosmic time and could, in the long term, close again during a final collapse. No portion of space can therefore serve as an absolute standard without introducing a fundamental uncertainty.

In the hypothesis of a Universe assimilable to a hypersphere, an anti-universe with negative curvature could correspond to its internal face. The image of a double-sided globe then illustrates a binary system of universes in symmetry. Removing any curvature would be tantamount to annihilating this structure. The most direct trajectory in such a Universe necessarily follows a geodesic. Universe and anti-universe would share similar curvature properties but not superimposable, due to a recursive chirality.

The laws of Euclidean geometry—such as Pythagoras' theorem or the axiom of parallels — adequately describe our local environment but become inoperative in a curved and dynamic space. The Lorentzian geometry, integrating relativity and curvature of space-time, is then indispensable. It allows to link different frameworks, without fully accounting for the non-linear and strongly coupled evolution of energy fields.

The mathematical approach adopted to integrate relativity into the measurement of observed phenomena consists in comparing distinct frames of reference at a given moment, while renouncing the hypothesis of absolute simultaneity. The imprecision of the measurements would not come so much from an instrumental limitation as from the intrinsically non-static character of the physical systems: speeds, masses and positions evolve constantly. In this context, one can question the ability of Lorentz transformations to comprehensively account for complex referentials or interactional conditions, subject to differentiated dynamics within heterogeneous energy fields. Can the same phenomenon, under these conditions, unfold in a strictly identical manner in distinct regions of the Universe, characterized by different energy distributions?

Non-commutative geometry, for its part, favours the description of the state of a system rather than its explicit inscription in classical time and space. However, it remains legitimate to question the truly exhaustive nature of these particularly complex mathematical formulations, since they claim to integrate all the effects associated with a relativity whose recognition remains historically recent. In this regard, it is significant to recall that some equations from Einstein's general relativity do not yet have complete analytical solutions or fully stabilized interpretations.

The geometry of space-time is determined by the distribution of masses and by the fluctuations of energy fields. The classical notion of a moving material point gives way to that of a physical state or quantum field, while the assumption of parallels specific to Euclidean geometry is abandoned. This evolution leads to a joint redefinition of the notions of space, time and symmetry, and requires the adaptation of mathematical tools in order to deal with quantities deemed non-commutative, such as position and momentum in quantum mechanics.

The underlying objective would be to achieve a unified description integrating gravitation, electroweak interaction and strong nuclear interaction. Such a rethought geometry would seek to articulate, within the same theoretical framework, the notion of flexible space, the relativity of spacetime, the non-commutativity related to the factorization order of observables, as well as chirality, which joins the idea that it is not relevant to talk about absolute simultaneity for two distant events. From data expressed in wavelengths for distances and angular measurements integrating the curvature of surfaces, could we reconstruct the past evolution of the Universe in a unified model? The very notion of a «unified model» remains problematic, however, both in its conceptual content and in the intuitive representations it evokes, which are probably not consistent with the physical reality intended.

Chirality, which affects the symmetry between matter and antimatter, makes it difficult to use the concepts of axis of symmetry or orientation, as they are commonly mobilized in mathematics using positive and negative relative numbers defined with respect to a zero origin. During the annihilation of two particles associated by an opposite symmetry, they lose their status as massive wave packets. Energy is conserved, but manifests primarily as electromagnetic radiation, while the excess not carried by this radiation can lead to the creation of other particle-antiparticle pairs. The inverse process, corresponding to the production of such pairs from high-energy photons, can also occur under certain conditions, constituting a transient form of radiative entanglement.

These cycles of creation and annihilation would continue until the ultimate absorption of energy by extreme gravitational objects, such as black holes,

92

the hypothesis of white holes being sometimes evoked in a context of symmetry. During this irreversible process, particles are likely to change nature: thus, the annihilation of an electron with its antiparticle can lead to the formation of a quark-antiquark pair. Such a mechanism would indirectly contribute to the increase of the effective mass of atomic nuclei and, consequently, to the evolution of the concentration of matter in the Universe.

In this perspective, a symbolic writing of the type particle + antimatter electromagnetic radiation + incident production of other pairs of particles, could not be reduced to an equality of particle type + antimatter = 0. This leads to question the relevance of the concept of zero value as a boundary between positive and negative quantities in quantum mechanics. It then becomes conceivable that a coherent model describing the origin, evolution and becoming of the Universe cannot rely on either the strict principle of algebraic equality or the existence of a null value. Should we, in this context, renounce identities such as $+a - a = 0$, or the commutativity of multiplication ($ab = ba$), and even certain elementary relations like $a = \sqrt{(a^2)}$, in order to try to grasp still inaccessible dimensions of physical reality? Matrix algebra, by its multidimensional nature, leads precisely to this type of apparent paradoxes. When its results differ from those of ordinary algebra, they may be interpreted in probabilistic rather than deterministic terms. Since matter and antimatter are considered to be fundamentally correlated without however canceling each other in the classical sense, it becomes legitimate to question the adequacy of the mathematical language used to describe these phenomena.

What is fundamental are the relationships with their information transfers and the mathematical structures that this implies.
It is customary to say that mathematics describes space and time, when in reality, space and time are only manifestations of deeper mathematical structures.
Reality is not fundamentally mathematical and the equations only allow us to glimpse a reality based on an emerging space/time but which has no place in quantum mechanics. This vision, which reflects a confusion between deep reality and observational reality, is far from unanimous.
The whole difficulty in this attempt to go further with a dematerialized approach, out of any spatiotemporal context, lies in the lack of precise formalization, testable predictions, and experimental grounding.
A mathematics "without fundamental space-time" would not start from coordinates, trajectories, or durations but from concepts such as relations,

93

transformations, invariants, and coherence constraints. Imagining a stop of time, although devoid of physical meaning, could also be a way to envision a possible quantum theory of gravitation.

Recall that the wave function is not an object "in space" in the classical sense. We are already approaching it with quantum mechanics (non-locality, superposition), certain formulations of quantum gravity (structures without fundamental spacetime), tensor networks and information approaches.

The physics of the future tends towards a more fundamental but excessively abstract and discreet vision where objects are revealed as secondary while relationships become fundamental.

In quantum mechanics, the effects related to the act of measurement and the observer's methodology imply that the order of measurements influences the result obtained, and thus the interpretation of the phenomenon. Unlike classical relativistic physics, the order of events plays a decisive role here. Since the data is necessarily factorized according to a given order, the result appears to depend on a partial and contextual interpretation, which mathematically translates into a non-commutativity of observables ($AB \neq BA$). Such inequality presupposes that the quantities considered are represented by matrices, each matrix product implicitly integrating a hierarchy in the order of consideration of factors.

In quantum mechanics, reversing the order of events—formulated in terms of transitions or movements — amounts, in a sense, to manipulating the arrow of time. Any measurement performed changes the state of the system and influences subsequent measurements. This mathematical subtlety, at the origin of the principle of uncertainty, can be interpreted as an attempt to extract the particle, as an isolated entity, from the classical categories of time and space. Nevertheless, when it comes to describing or relating interactions between particles, the reference to time and space remains inevitable.

The difficulty is accentuated by the absence of a fully completed mathematical formalism to describe what could be described as feedback loops, in which effects influence causes. How, then, to conceive a time that would exist or not depending on whether the particle is engaged in an interaction or considered outside of any interaction? It is in this context that Planck's constant fits, which one can wonder if its value should not tend towards zero, and if it's supposed invariance is really founded. More generally, can we identify anything truly constant, immutable or static in a Universe characterized by the permanent evolution of its structures and effective laws? A set of constants and physical parameters constitutes the foundation of the contemporary cosmological model and provides reference values considered fundamental in astrophysics. Their epistemological status, their real scope and their supposedly invariant character deserve however to be examined critically.

The cosmological constant and the Hubble constant are primarily mathematical formulations intended to account for certain large-scale observations. They notably allow to introduce, in the equations of the standard cosmological model, a term interpreted as an accelerated expansion force, attributed to a hypothetical entity commonly referred to as «dark energy». The physical nature of this energy remains unknown, and its assimilation to vacuum energy as defined in quantum field theory remains problematic, as the conjectures used are different.

The fine structure constant, a dimensionless number, establishes a relationship between the elementary charge and the intensity of electromagnetic interaction. It is not based on a primary explanatory theoretical foundation but constitutes an essential empirical parameter for interpreting many observations in quantum mechanics. Its value, however, explicitly depends on other constants, in particular the speed of light and Planck's constant, which raises the question of its autonomy and its fundamental character.

The gravitational constant, on the other hand, fixes the coefficient of proportionality between masses and distances in the law of gravitation. It is based on measurements expressed in units of mass, length and time, assumed to be invariant with respect to each other. This hypothesis implicitly implies the existence of constant conversion factors between these quantities, whose arbitrariness and physical validity can be questioned, especially from a relativistic point of view where these units are not necessarily absolute.

Planck units aim to define theoretical minimum scales of mass, length and time, as well as a lower limit of temperature, deemed non-franchisable. They constitute a powerful basic tool in a mathematical approach to fundamental physics. However, the orders of magnitude they imply appear highly disproportionate compared to the scales accessible experimentally, which makes their direct integration into a physics oriented towards observation and application particularly delicate.

95

The Planck constant plays a central role in quantifying physical phenomena, by relating energy to frequency. It thus allows assigning a granularity to energy exchanges. Paradoxically, elementary particles are often considered to lack their own physical dimension. This conceptual tension invites us to question the deep meaning attributed to this constant and the interpretation in which it is used.

The expression of the speed of light in kilometers per second implicitly assumes that units of length and duration are invariant and insensitive to relativistic effects or to the global dynamics of space-time. Such a hypothesis amounts to considering that the evolution of the energy density of the Universe, including in what is called vacuum, would have no impact on the propagation of electromagnetic waves. This statement deserves to be questioned. In general relativity, the speed of light intervenes as a fundamental constant; any variation thereof would imply a profound revision of Einstein's theory, likely to lead to a new interpretation of gravitation and its role in cosmic evolution.

Without questioning the principle of the relativity of space and time, one cannot exclude, on a speculative level, that the speed of light may have varied over the course of cosmic history and may continue to evolve. Such a hypothesis would have direct consequences on the measurement of cosmological distances, in particular for objects observed at great distances, thus at remote times. The very idea of an expansion of the Universe should then be reconsidered, which could call into question the standard model of the Big Bang as well as the scenarios envisaged for the future of the Universe.

Moreover, nothing requires that the so-called fundamental constants be universal in the strict sense, that is to say applicable without modification to any other universes in the theories of the multiverse. Are these constants really universal, or do they reflect a physics built above all to describe the particular conditions of our observable reality? Most of these parameters are related to each other by common relations, so that a variation in one of them would necessarily lead to correlated changes in the others. Their interpretation also relies on a cosmological model that is still incomplete, marked by numerous uncertainties, unverified hypotheses and internal tensions.

96

It is frequently argued that any modification of the physical constants would lead to a radically different universe, even unstable, chaotic or incompatible with the very existence of complex structures. This claim, although widely spread, is not rigorously demonstrated. If the constants are considered invariant, it is always in relation to an implicitly chosen global context. Yet this context, in an expanding and constantly evolving Universe, is never strictly fixed. Time and distances are locally and globally heterogeneous.

Even for dimensionless constants, asserting their invariance does not mean that their value is absolute. The digital quantities that define them are based on standard units of mass, length and time, established in a local and contemporary environment. Considering general relativity, these units cannot be considered as fundamentally absolute in relation to each other.

If we adopt a conception of the cosmos that includes the hypothesis of a multiverse, our Universe ceases to be a unique case. In this perspective, it becomes relevant to consider these constants not as strict invariants, but as correlated variables likely to evolve non-linearly over cosmic time. Other universes, characterized by different parameters, could follow distinct evolutionary trajectories, without necessarily being compatible with the emergence of observers. Such an approach nevertheless comes into tension with the anthropic principle, defended by certain scientific currents and by many philosophical or religious traditions.

These reflections, although particularly demanding, emphasize above all the need to evolve our modes of analysis and to refine the theoretical tools that are physics, chemistry, and mathematics. In an ideally comprehensive approach, understanding a single event should lead to tracing all the conditions that preceded it. In practice, our understanding remains largely confined to an immediate present, which limits our ability to fully grasp the interactions between the infinitely large and the infinitesimal, as well as their inscription in cosmic time.

The history of humanity has been marked by major transitions—from prehistory to digital and nuclear technologies. Applied to the knowledge of the Universe, this analogy suggests that we may only be at a rudimentary stage of understanding. However, the perspectives opened by artificial intelligence and advanced computational tools suggest new theoretical exploration capabilities, not without raising, in turn, the question of their influence on our ways of thinking and interpreting.

Locate to describe

The notion of position becomes problematic when applied to an unbounded space that is constantly evolving. Any measure, as soon as it is carried out, immediately loses its relevance, unless it tends towards limit values - almost zero or almost infinite. To overcome this difficulty, one approach consists in discretizing space into a multitude of elementary volumes, defined as as small as possible, in order to introduce operational value units. Each volume is then assumed to correspond to a minimum amount of energy, identified as a quantum. Below this minimum unit, no continuous description seems accessible; we then enter a domain that can be described as a discrete state or a dimension not directly observable, opening onto a regime still largely unknown, sometimes associated with the hypothesis of a cosmos of the multiverse type.

Renormalization appears as a mathematical process intended to avoid the appearance of uninterpretable values—often assimilated to infinite divergences. This method, frequently criticized, amounts in practice to reformulating a borderless universe as an effectively circumscribed system, within which no magnitude can diverge. Renormalization thus constitutes an effective mathematical artifice, whose legitimacy as a description of a physical reality remains however discussed. Certain formal structures thus allow to approach phenomena that otherwise escape any direct modeling. In this perspective, Paul Dirac defended the idea that a mathematical formulation could be considered valid, even in the absence of a fully satisfactory physical interpretation. Mathematics indeed offers a language that is both rigorous and deeply abstract. Nevertheless, an excessive conciliation between formal pragmatism and abstraction could lead to losing sight of a reality that is already difficult to grasp. This invites us to consider exploring new paths, even if it means questioning the foundations considered as established.

The proposal of a "pixelization" of space fits into this logic, by attributing a discontinuous structure to radiation and thus joining the notion of quantification. Photons, mediators of electromagnetic interaction, make it possible to account for energy variations that would otherwise be described as undulations propagating in space. If matter was formed from the radiation

98

of a primordial universe, a corpuscular description leads to consider it as an assembly of irreducible entities. This is how fermions were defined as elementary particles of matter. Such an approach, however, amounts to superimposing two representations: that of a universe dominated by matter and that of a universe fundamentally composed of radiation.

We can then assume that non-directly observable interactions, occurring in a context of symmetry, lead to perceive electromagnetic waves in the form of interference patterns, characterized by alternating intensity maxima and minima. The ability of these waves to entangle themselves beyond certain energy thresholds, giving rise to particles of matter, as well as their tendency to be absorbed by extreme gravitational objects, explains why they are also apprehended in the form of quanta, as photons. This duality of observation is not limited to electromagnetic waves: material objects can also manifest wave-like properties. Wave–corpuscle complementarity thus played a determining role in the major advances in quantum physics.

This complementarity suggests a form of equivalence between the movement of particles and the propagation of waves. From the elementary particle to the most massive astrophysical structures, what is referred to as matter could, without altering its fundamental nature, be described as a set of processes of dematerialization and rematerialization. Such a perspective implies to conceive all physical phenomena as information transfers resulting from field interactions, essentially electromagnetic. This includes both the orbital properties and angular moments attributed to particles, as well as the movements of atoms within molecules, the organization of stellar systems, galaxies and their clusters, or even the fluxes of photons characterized by frequencies and amplitudes of coupled electric and magnetic fields.

Electronic transitions between discrete energy levels, often described as orbital "jumps", are directly related to this wave-like dynamics specific to electromagnetism. When an electron absorbs the energy of a photon and releases from its atom, it adopts a behavior analogous to that of a diffracted photon. Interference experiments, such as those involving individually sent photons through slit devices, show that interference patterns emerge even in the absence of a continuous beam. Each photon can then be interpreted as a punctual manifestation of an electromagnetic field having interacted with numerous systems before the act of measuring. Observation requires an additional interaction, during which the photon loses its global wave

99

character to manifest locally in a corpuscular form, limited to the duration of the observation.

The wave-particle duality thus leads to considering macroscopic matter as resulting from complex intertwining of waves. The notion of massive particle then appears as an operative construction related to the cognitive abilities of the observer, adapted to scales where quantum effects are not directly perceptible. In the absence of any measurement interaction, an electron, like any other particle or composite system, could be described as fundamentally undulatory. It is the interaction with the electromagnetic waves associated with the act of observation that induces corpuscular behavior in accordance with our modes of perception.

What is conventionally referred to as a particle or an object located in space would thus correspond to a set of packets of entangled waves, engaged in interactions of charges generating magnetic fields. These effects, negligible at the atomic scale, could nevertheless participate in the emergence of gravitational interactions on broader scales. In this perspective, matter would not constitute a fundamental reality, but an appearance conditioned by the position of the observer within the studied system. Our perception, essentially empirical, leads us to materialize what our senses apprehend, while the particle could be described as a packet of waves in superposition of states, only one of which becomes physically accessible during observation.

Electrons can thus be assimilated to packets of confined standing waves, whose interactions within the atom are perceived as a probabilistic halo around a central nucleus. This nucleus itself constitutes a confinement region for other standing wave packets, associated with quarks. The relative stability of composite particles results from the collective organization of these wave states, independently of a classical temporality. Protons, and to a lesser extent neutrons, play a regulatory role in charge interactions, their durability being closely linked to the presence of electrons, which adjust their states to ensure the overall equilibrium of the atom. This charge neutrality results mainly from continuous energy transfers between electrons and electromagnetic radiation, which explains the great diversity of possible molecular structures.

Experiments also show that individually projected fermions produce interference patterns similar to those observed with photons. A strictly corpuscular description, however, does not allow to account for the apparently continuous and progressive character of the transformations of matter, inherent in its deep undulatory nature. This limitation is not, however, prohibitive, insofar as the mathematical tools of quantum physics favor a discrete and unsmoothed interpretation of energy exchanges.

In radiative entanglement phenomena, primordial wave amplitudes, devoid of any defined spatial extension, can combine in such a way that they become inseparable. They then evolve into confined wave packets; at the border of a space-time whose properties emerge from their interactions. The particle of matter thus appears as a stationary system of self-sustaining waves, interfering continuously, while collectively interacting with open electromagnetic fields and with other particles.

In these confined wave systems, polarization can only be described in circular terms, without reference to a physical plane or axis of rotation in the classical sense. Spin then constitutes a formal representation of these intrinsic internal dynamics, irreducible to spatial rotation. Devoid of spatial occupation in the classical sense, the elementary particle can cross energy barriers that extended molecular structures cannot. Constructed matter thus results from the collective interaction of these wave packets, whose emergent properties depend on their modes of interaction, rather than their individuality.

Curiously, regardless of the direction considered, the observation of distant galaxies indicates that they seem to move away faster than nearby galaxies. This observation could, at first glance, suggest a violation of the speed limit set by the speed of light. However, it is also possible to interpret this phenomenon as **an apparent effect related to perspective**: if the Universe was created simultaneously in its entirety, rather than from a singular point, the perceived distance of galaxies could result from the way we receive electromagnetic signals (EMW) emitted in the past. In this context, the Universe, considered as a whole and in the absence of an external reference frame, could appear as a singularity devoid of quantifiable dimension, without measurable intrinsic expansion.

101

The analogy with a mist cloud condensing into raindrops illustrates a phenomenon of densification without real expansion of space: initially distributed in a fixed volume, the fluid gathers locally, forming more concentrated structures. **Similarly, the gravitational aggregation of matter and the redshift of electromagnetic wavelengths (not just visible light) are two interconnected phenomena but perceived differently depending on the scale of observation.** In the distant regions of the Universe, corresponding to its past, space could be in a state of local depression or reduced density, which can be wrongly interpreted as an accelerated expansion.

Indeed, distant galaxies seem to move away faster than nearby galaxies. This observation led to the hypothesis of an accelerated expansion of the Universe, but this interpretation raises different problems. By considering the Big Bang as the starting point of the Universe and without including the phase of radiative entanglement theorized here, the primordial energy would have been released at an extremely high speed. This speed of diffusion would then have decreased gradually until becoming negligible in the current near Universe. This initial energy distribution corresponds to the classical bell shape describing the genesis of the observable Universe. Thus, observations from the far reaches of the Universe correspond to states of the past, where time and space differed from those of today. The current real velocities of distant galaxies, which escape our observation, could be comparable to those of nearby galaxies.

The hypothesis of an expanding Universe also raises the question of the center of expansion. If the Universe does not have a definite center, the interpretation of an exponential expansion of space could result from a confusion between measures of the present and past states. Gravitational evolution concentrating matter creates an illusion of expansion, as galaxies follow trajectories dictated by local dynamics and gravitation. As G. Lemaitre had suggested, the perception of a rapid distancing from distant galaxies can be a consequence of the concentrational evolution of matter rather than a real expansion.

During the early phases of the Universe, after the radiative entanglement period, matter was scattered and poorly differentiated. The local gravitational effects were weak and the curvatures of space negligible, making space/time relativity unpronounced. The objects observed today at a great distance, coming from this period, therefore seem to move away

102

linearly and quickly, while the close objects appear slower. The opposite occurs in the vicinity of massive objects, such as black holes, where intense gravity induces an apparent slowing of time for approaching objects. These objects undergo extreme tidal forces, resulting in their spaghettisation and the gradual loss of properties such as mass, spin, and charge, to become only latent energy. These ultimate singularities could then represent the terminal phase of cosmic evolution, prior to a possible collapse of the Universe, forgetful of time, space, and observer.

The Universe, although dynamic, tends to present a thermal and material homogeneity on a large scale, which can reduce the role of the cosmological constant. The distinction between a positive (Einstein) or negative constant (string theories) then becomes secondary. Originally, matter and antimatter could constitute a binary system of virtual energy, evolving into symmetrical states and now manifesting as electromagnetic signals from multiple and radio concentric sources. The EMW we observe result only from their interactions with matter and each other, in the form of constructive or destructive interference. These interactions give rise to phenomena of diffraction, refraction, absorption and dispersion, and define, with gravitation, the observable space/time.

Electromagnetic waves as we observe them represent only a limited part of phenomena whose real complexity exceeds our detection capabilities. What we perceive as undulations in an energy field could actually be structured hierarchically: each wave front could be composed of secondary fronts, themselves subdivided into tertiary fronts, and so on. This fractal organization apparent at the microscopic scale suggests a fully curved Universe, where the "perimeter" can tend towards infinity while the "volume" remains finite.

Although the Universe has no identifiable physical edge, there is also no boundary for EMW. These waves are only observable when they interfere with each other by superposition or interact with matter via diffraction, refraction, absorption or dispersion. It is these interactions, combined with gravitational effects, that define the observable space-time and give it its dimensions, even if the global scale remains non-quantifiable. In the absence of matter or gravitational curvature, EMW would cease to behave as interfering waves and would be "marginalized" with respect to space-time. In this scenario, they could leave space-time and, without charge interactions, their electromagnetic fields would stop manifesting, thus

103

returning to a ground state, non-undulatory, comparable to the virtual energy before the Big Bang. This energy would then return to a latent state, defining what we call here, the multiverse Cosmos.

The EMW remain, however, in constant interaction, their frequencies, amplitudes and wavelengths evolving continuously. A photon can be interpreted as an elusive point of interference, which provides a foundation for wave/corpuscle duality. Similarly, the particles constituting matter can be described as corpuscles — points arbitrarily located in space and capable of interaction — or as waves, energy flows revealing ongoing interactions. The phenomena we observe are the result of interference and energy or information exchanges resulting from the initial latent energy activated during the Big Bang, an initiating process of space-time.

Finally, it is conceivable that the collapse of other quantum symmetry binomials in the Universe could be linked to this initial singularity of the Big Bang, which thus finds itself "out of play" with respect to the multiverse Cosmos, without revealing any symmetry at the pre-quantum state. Therefore, it remains difficult to determine whether there is a succession of distinct universes or whether each cosmic event corresponds to a specific manifestation of the same energy continuum.

Electromagnetic waves can be considered as the inheritance of the initial latent energy that defines the multiverse Cosmos. They intervene as essential mediators in the process of evolution of our Universe, which can be ordered according to the following steps:

1. **Big Bang**: initial singularity corresponding to a disruption of the cosmological equilibrium, revealing a chiral symmetry and the emergence of a time without macroscopic significance.
2. **Radiative entanglement and decoupling of primordial radiation**: formation of current electromagnetic radiation, with its full range of wavelengths. Appearance of elementary particles then composite particles. Space/time opening.
3. **Recombination:** grouping of composite particles into atomic nuclei and association of electrons, giving atoms a neutral charge ensuring their relative stability.

4. **Primordial and stellar nucleosynthesis**: formation of heavy nuclei and progressive densification of matter.
5. **Gravitational gathering and space depression**: progressive conversion of kinetic energy into potential mass energy, formation of galaxies and clustering.
6. **Energy accumulation in supermassive black holes**: global cooling of the Universe and relative decrease in entropy.
7. **Final collapse of black holes**: return to a cosmological equilibrium, reintegration of energy into the multiverse Cosmos and disappearance of space/time specific to our Universe.

Quantum mechanics allow the creation and point annihilation of particle/antiparticle virtual pairs from vacuum energy, defined here as a fluctuating field of energy, composed of EMW and carrier of free particles (quarks, electrons, neutrinos) as well as atoms and molecules detached from any lasting interaction. In this quantum void, particles and antiparticles can interact temporarily, giving the impression of appearing spontaneously, before being reabsorbed or generating new lower-energy particles. These fluctuations modify the quantum grid of space/time.

In this model, a **chirality between symmetries** prevents the majority of interactions between particles of opposite symmetry on a large scale, making these phenomena largely undetectable. In the long term, particles and antiparticles would end up confined and destructured in black holes, representing the concentration of universal energy. The final collapse translates the ultimate annihilation of matter and antimatter components, leading to a return to the multiverse ground state.

According to Eddington, the Universe could have been born from a minimal fluctuation breaking cosmic symmetry. The multiverse Cosmos would thus represent a latent symmetry, constituting both the starting point and the culmination of our Universe.

The idea of a non-expansionist Universe suggests global homogeneity at a given moment (however, can one speak of absolute simultaneity without contravening relativity?). The galaxies seem to move away in all directions, but without a single center or initial point being identified. If the Universe emerged simultaneously from multiple "points", it cannot be measured or quantified in terms of volume or absolute distance. The notion of space expansion then becomes ambiguous, in the absence of an extrinsic

105

reference frame. How to imagine an expanding Universe if one considers that it cannot be measurable or quantifiable, due to the lack of an index or extrinsic unit of measurement likely to serve as a reference?

Observations of very distant galaxies confirm this perspective. They appear young and relatively poor in heavy elements like carbon, the thermonuclear product of massive stars that have evolved and exploded into a supernova. Having not been able to be made in the first moments of the Universe, this carbon is therefore less present in the young stars that we scrutinize in the most distant past. On the other hand, in closer and therefore more recent galaxies, carbon is abundant, reflecting the gradual evolution of the chemical composition of the Universe. This distribution indicates that looking into the distance is equivalent to observing a distant past, populated by young stars relatively poor in carbon, without hope of glimpsing a distant news which, unlike the image received, would be richer in this element.

VI <u>The Universe plays boules</u>
(An unpredictable game on an uncertain terrain)

Each species conceived and developed in symbiosis with a contact environment. Darwin's theory of evolution leads man, as the dominant species, to consider himself a major event, predestined and unavoidable in the Universe. But one can also think more simply that the living is nothing other than the product of a mineral, solid, liquid, gaseous environment having reached a stage of evolution conducive to the emergence of a particular macromolecule, described as biological. In these chromosomes carrying genes, will duplicate, self-program and evolve the information initially of a viral nature, which will develop life. The latter recorded in any cell in the form of a synthesis element at the basis of organic chemistry is DNA.

To achieve the more or less perennial architecture of matter, atoms generally share one or more electrons through a so-called covalent bond. They can also exert between them, a low intensity electrical interaction, necessary to approach the thermodynamic balance. The hydrogen atom (the most widespread atom in the Universe) has the particularity of binding stably with certain electronegative atoms such as oxygen, nitrogen, and fluorine. This hydrogen bond under favorable temperature conditions allows for the creation of intermolecular links between hydrogen and 3 other elements in quantity in the universe. These are oxygen, nitrogen and carbon which has the distinction of allowing a wide variety of molecular bonds. Yet it is precisely these constituents with other elements that are rarer but essential to cells such as phosphorus and sulfur that make living organisms. Hydrogen bonds are at the origin of these molecular structures in the form of double helices that is DNA. This hydrogen bond, particularly through oxygen, the water molecule (which accounts for 66% of the human body) would therefore be decisive in the genesis of life.

One cannot say that in a given environment, choices are left to genetics. In this logic, if the advent of the living in planetary evolution is indeed in the order of things, notably with photosynthesis and the carbon cycle, the destiny of man whatever he does would be mapped out in advance. It is the radiation born from the Big-bang that after radiative entanglements and information sharing, allowed the symbiotic formation of the first viruses and single-celled host organisms. It should be noted that one sperm alone would represent, in genetic form, several hundred megabytes of data. It is a

107

summary of technology that is far from being within our reach. It is not unreasonable to assume that particular quantum phenomena are the basis of the chemistry of life. This long process led to the presence of man on earth. Unfortunately for us, it is this same radiation that largely causes the aging of our cells. But it is to be feared that man's worst enemy is in him. His oversized ego encourages him to want to rule everything, if necessary through constraint and take ownership of everything without sharing. Unless it is life itself in its most rudimentary form, parasitic inside our organisms; a viral form against which man would one day be powerless to react. A very derisory end for a humanity that since Einstein in particular, should learn to review its behaviors, by relativizing everything and not only time and space! Without forgetting that our planet does not have, in the Universe, the particular status that we attribute to it and that all the stellar bodies which gravitate around us are swords of Damocles on our heads. The scenarios that will lead humanity to its end are not lacking. In any case, for each of us, taken individually, the tomorrows unfortunately have no great future.

But how to explain that in the immensity and uniformity of the Universe, life could have made its cradle of planets like earth? The evolution of matter gives the impression that it is becoming more complex in order to come together better. Thus, in a punctual and rather marginal way, has been able to develop on planets predisposed by their biotope, a biodiversity of which we are part. The hydrogen atom is at the base of organic molecules. This explains that the seabed with the presence of water (H_2O) facilitated the emergence of the first forms of life in a rudimentary state then unicellular. Some of these first organisms left the marine environment out of necessity or opportunity for an atmosphere now composed of 21% dioxygen (O_2). The radiation, reduced to the appropriate wavelengths due to the presence of an atmospheric layer, provided these proto-organisms with the energy required for their evolution, playing in a way the role of catalyst. To the reign of the plant thus installed, came to be superimposed an animal life which was not slow to diversify and for certain species no longer satisfied with a strictly vegetable diet. Thus, the food chain was built. No doubt a constant need for more complexification but also for sustainability - the inevitable evolution of life - has led part of this animal population, freed from the aquatic environment, to be taken from a population of herbivores. The instinctive behavior of these predators will evolve to become more and more conscious and reasoned.

108

The human organism is ultimately an assembly of hydrogen (10%) and oxygen (65%) with a carbon (19%) in strong connection with the first 2 components that make water. Added to this is nitrogen (3%), which helps promote lasting covalency between these various components. The man who is at the top of this food chain and distinguishes himself from it by developed cognitive abilities, has arrived at the current stage of this evolution, to be able to make thoughtful and organized actions. These have the particularity, although insignificant at the scale of the Universe, to go against the almost programmed if not 'normal' evolution of the latter. This performance of the living could be interpreted as the finality, the ultimate goal of everything that makes evolution and why not, the raison d'etre of our Universe. But let's rest our feet on the ground! Everything shows that life will disappear as it appeared. It is written in the genesis of our planet: life remains a parenthesis.

Before the Big Bang, no observable event would allow us to deduce the subsequent characteristics of our universe. The analogy with living systems, used here, has the sole purpose of facilitating the understanding of an initial singularity and should not be interpreted as a biological or teleological causality. The notions of chance, inexplicable or singularity often represent heuristic responses to phenomena whose exact cause remains undetermined. However, the idea that the universe emerges "from nothing" that is physical, does not necessarily imply that it follows absolute nothingness.

This reflection is part of the hypothesis of a potentially infinite succession of binary systems of universes in quantum symmetry. These universes are independent, non-quantifiable in number and have characteristics close to those of our universe, while being able to differ in their dynamic development. A variable chirality could suggest processes of deconstruction or transformation of these universes, more or less rapid depending on the initial conditions.

Before addressing these hypotheses further, it is necessary to examine supermassive black holes, astrophysical objects whose density and compactness rival those of elementary particles. These structures could represent both an end point and, possibly, a starting point of cosmic dynamics.

The first so-called primordial black holes would have formed rapidly from the gravitational collapse of extremely dense hydrogen clouds present in the

109

primitive universe. Some of these primordial black holes could still be detectable as extremely bright and active objects, notably via quasars, but the majority probably remain out of reach of current instruments.

Black holes can merge through gravitational interactions, giving rise to objects of increasing mass-equivalent (see chapter XXII on the definition of mass equivalent with regard to black holes). Although these events are not spectacularly visible in the classical sense, they represent one of the most energetically extreme astrophysical phenomena. The generated energy density locally modifies space-time significantly and generates very high-energy radiation fluxes, notably gamma-ray bursts and X-rays. These fluxes, particularly penetrating, cross space without substantial interaction with matter, except for the production of electron-positron pairs and possibly neutrinos and antineutrinos (see chapter XI).

Given the difficulty of direct observation, it is likely that the actual number of black holes is much higher than currently reported. Interactions between galaxies, such as collisions, can project certain black holes, whether supermassive or stellar, into the intergalactic medium. This migration is not exceptional and could explain the presence of many undetected supermassive black holes in intergalactic space. Although theoretical limits, such as the Oppenheimer-Volkoff limit, prevent the lasting existence of black holes below a critical mass-equivalent, it remains possible that they reach masses much larger than those initially estimated.

A black hole does not directly emit detectable radiation. However, it can be observed indirectly by the light emitted by its accretion disk or via gravitational lensing effects. The perceived light intensity depends on the rotation of the disc and the observation angle. These observations are strongly modulated by the gravitational perturbations of the surrounding space, making the interpretation of the received images particularly complex.

It is possible to consider, hypothetically, the internal state of a black hole mega massif (MMBH) in a Universe that has reached an advanced stage of cooling. Such a universe would be globally homogeneous, weakly structured and devoid of the atomic and particulate diversity that characterizes the previous phases of cosmic evolution. The content of a MMBH could then be described as an extremely compact, quasi-uniform system dominated by maximum energy concentration, without measurable

110

internal dynamics and modes of oscillation comparable to those of ordinary matter.

At a sufficiently advanced stage of the evolution of the Universe, any supply of matter to black hole accretion disks would cease. In the absence of a power supply, the phenomena associated with these discs—intense electromagnetic radiation, relativistic particle emissions and energy losses — would gradually disappear. In this context, supermassive black holes, considered as macroscopic quantum objects, would no longer produce observable thermal radiation. The emissions conventionally attributed to black holes would cease not due to an intrinsic modification of these objects, but due to the disappearance of the accretion zone which usually makes their presence detectable.

The gravitational boundary associated with the black hole, commonly referred to as the event horizon, defines the limit beyond which no information can be transmitted to an outside observer. However, this effective limit depends on the dynamic parameters of the incident particles, notably their energy, angular momentum, and trajectory. The notion of a Hawking radiation, often associated with an own emission from the black hole, can then be reinterpreted as an indirect manifestation of energy processes located in the peripheral region of the object, rather than as an intrinsic thermodynamic property of the black hole itself. In this reading, the evaporation of black holes would be more a description of energy exchanges in the accretion zone than a fundamental mechanism affecting the internal structure of the black hole.

The radiative mantle surrounding a powered black hole constitutes a region of high entropy and effective high temperature. The observed emissions — mainly in the X-ray and gamma-ray domains—as well as the relativistic jets expelled along the axes of rotation result from complex electromagnetic and

111

magnetohydrodynamic processes. These phenomena allow the evacuation of an excess of angular and gravitational energy, without however involving a thermal dissipation of the black hole itself.

This interpretation, which deviates from the canonical formulation proposed by Stephen Hawking, is part of a projection towards a future state of the deeply cooled Universe, distinct from the current observable Universe. In such an extreme regime, the classical principles of information conservation and thermodynamic equivalence could lose their operational relevance. The matter absorbed by black holes would be gradually compacted in a cold and transient state, intermediate between pure gravitational energy and primordial plasma residue, before an ultimate collapse on itself.

In this perspective, it becomes conceivable—although speculative — to associate the collective collapse of the MMBH constituting a Universe at the end of evolution with a cosmological event of the Big Bang type of second generation. This hypothesis suggests a dynamic continuity between phases of expansion and contraction, without strictly belonging to a universe cycle, as developed in certain ancient cosmological traditions.

The existence of antimatter is today firmly established on a theoretical and experimental level, although its direct observation remains limited to the fleeting production of antiparticles. The matter-antimatter annihilation processes are well known and lead to the emission of very high energy electromagnetic radiation. Conversely, the creation of antimatter requires a considerable energy input and is only achievable in infinitesimal quantities. Experimental devices, notably those developed at CERN, allow temporarily confining antiparticles in electromagnetic traps at very low temperatures, highlighting the artificial and unnatural nature of these conditions.

The fundamental question remains that of the fate of antimatter, which is supposed to have been produced in equivalent quantity to matter during the first moments of the Universe. If it has not been entirely annihilated, a hypothesis consists in considering that it evolves in a distinct space-time domain, not directly accessible to our observations. Antimatter could thus be distributed in superposition with ordinary matter, while remaining invisible to classical instruments. Such a distribution would allow the interpretation of certain gravitational anomalies observed on a large scale, without resorting to the hypothesis of a distinct dark matter.

This unobservable antimatter would then indirectly participate in what is referred to as vacuum energy. The gravitational effects measured at the scale of galaxies or clusters could thus differ from those observed at more local scales, such as planetary systems, due to this hidden component.

The cosmological model known as the Dirac-Milne universe fits into this logic by postulating a global symmetry between matter and antimatter, the latter being gravitationally repelled by ordinary matter. The existence of dark matter or dark energy then becomes superfluous, and the Universe can be described without a major inflation phase, with a cosmological horizon without clear boundaries. This approach meets the idea developed here of a structuring of cosmic reality into binary systems of the Universe in quantum symmetry, evolving largely independently.

How to explain more precisely this notion of a "borderless" Universe? Perhaps starting from the idea that it is difficult for us to imagine a Universe having the shape of a polyhedron, a cylinder, a cone, a torus, a Klein bottle or any other complex geometric form. Arbitrarily, we exclude a Universe whose edges would present a negative curvature.

The configuration that is both the simplest and the most consistent with the idea of a Universe (whether it is expanding or not) born from a singularity, remains the sphere. We see this geometric figure of perfect symmetry as possessing a unique center and a volume circumscribed by an equally perfect curved area. If the sphere is the object with the lowest area/volume ratio, determining its area or volume as locating its center makes any calculation incomplete or incomplete for 2 reasons:

- First difficulty: the number $\prod$ that allows to define the ratio between the radius (r) of a sphere (distance between surface and presumed center) on one hand and its area ($4\prod r^2$) or its volume ($4/3\prod r^3$), is an irrational number, transcending which includes an infinite number of decimals (3,141592653589.......).
- Second difficulty: space/time relativity makes the Universe a kind of entity with uncertain contours, while curved, whose content presents fluctuations in energy density that make measurements imprecise.

Imperfect distance assessments, a factor that, regardless of the degree of precision sought, does not provide a definitive measure! How under these conditions, position a center equidistant from an insufficiently determined perimeter? And how could this perimeter, being a presumed borderless Universe, be considered a traceable limit? Imagine for this, a form of sphere called relativistic space/time, remarkable mainly by two mass components which are matter and antimatter. Do we have this sphere of multiple centers, not positionable and undefined edges? This demonstrates the precariousness and incompleteness of even our most advanced mathematics.

This parallel with the sphere is just one more mathematical artifice to transpose into our reality, phenomena that refuse to integrate.

The matter/antimatter symmetry which relates to elementary particles of the same nature, is distinguished from the electrical charge distribution symmetry attributed to particles of different properties and which confers a certain stability by charge neutrality to the atom. That the electrons remain at a good distance from the nucleus, could be explained - if we get to the bottom of things - by the fact that the electron cloud of the atom is susceptible, as we have seen, to be considered not as a flow of matter particles but as a packet of entangled waves. These cohesion-forming waves are then assimilable to an electrically charged event horizon. Similarly, we can consider that the atomic nucleus realizes an equivalent charged system, of a fundamentally undulatory nature. As energy carriers (see chapter XVIII), EMW achieve load neutrality of the built material. Everything suggests that the universe is globally charge-neutral.

Antimatter would be defined as the other side, the hidden energetic reflection of a 'palpable' reality made of this material that is familiar to us.

The particle perceived as an undivided entity would only be a wave packet but we can hardly consider it as such. Should we imagine antimatter as a true copy of constructed matter (molecules, stellar objects...) knowing that the antiparticle, too, is just a wave packet whose properties would be imperfectly symmetrical to those of its dedicated particle? We make matter a tangible as well as subjective reality. This reality belongs only to us, of which we are moreover an incarnation as a living organism. It is a surface reality, an interpretation of what our senses deliver to us in a logic that stems from learning knowledge and satisfying needs dictated by an unspeakable precariousness. It is to be feared that we are not at this day, in capacity of apprehending and understanding a more complex reality which escapes our sight but also our intellection. Obviously, observing constructed antimatter is not within our reach today.

We know that EMW not captured by the material continue to interfere with each other. In phases, they add up and produce a wave of greater amplitude. In total phase opposition, the wavelengths harmonize and no emission peak is detectable. Between these 2 extreme cases, depending on their particularities of emission and path, the waves interfere with each other in a more or less "constructive or destructive" way. That particles and antiparticles completely annihilate each other (without incidentally producing new particles), under conditions of unprovoked destructive interference, would suppose that the waves, confined in packets, which are associated with them:

- Are of similar intensity (same field orientation, same amplitude, same frequency). Which would involve a common sharing of time and space.
- Propagate in the same field of interaction. Which is not the case, antimatter remaining without observable effects
- Share a common, imaginary time for the observer that we are. This last condition will only be fully fulfilled at the stage of the final collapse when the MMBH have gathered all the energy that our Universe carries.

Not satisfied, these drastic conditions represent what makes the chirality of symmetry.

The notion of forces in presence makes it possible to provide an observation framework for energy transmission and transformation.

This symbolism was born from the idea that energy "stricto sensu" has no definable material reality. Protean, it becomes difficult to explain. However, in an antithetic logic, we could say that energy represents the movements and interactions of everything that contributes to giving a dimension to a doubly relativistic Space/time due to quantum symmetry.

What would become of space if time did not exist and vice versa. One then imagines an environment where nothing happens, deprived of what makes energy, and therefore an impossibility of space. Revealing a symmetry break, time is the representation related to our symmetry, that we make of a certain chirality between quantum symmetries.

115

VII **The "Risen" Universe**
(Not to be taken literally)

Quantum symmetry posits that the fundamental properties of matter apply to antimatter as well. Attempting to describe a binary system of universes in quantum symmetry by assimilating the 'anti-universe' to a simple inverted image of our universe (like a photographic negative) seems insufficient to account for the real complexity of these structures.

As will be detailed in chapter X, a universe tends to cool down by reducing the energy occupation of space. In the long term, our universe, in its terminal phase, would essentially be reduced to isolated black holes, corresponding to a state of maximum energy depression. In this configuration, massive black holes (MMBH) appear far apart from each other. This apparent distance would indeed be significant if the surrounding space retained a notable energy density, but at this stage the space is almost empty and no longer provides effective separation.

During this ultimate phase, an event would occur, escaping any spatiotemporal description, leading to the simultaneous collapse of all MMBH at a non-localizable point. At this stage, space, called «empty», is almost nonexistent and time is practically stopped. The multiverse does not retain any memory of these binary systems of universes in quantum symmetry (see chapter X), it is improbable to find traces of a previous universe in the cosmic microwave background, contrary to the hypothesis proposed by Roger Penrose. Observing the cosmic microwave background involves receiving photons from both distant and nearby regions, some of which have been refracted or deflected by gravitational deformations of space, complicating any direct interpretation.

If the Big Bang is defined as a "primordial" singularity, the final collapse could be considered a "terminal" singularity. Now, a singularity, by definition, is a unique event, independent of any previous or future context. One could therefore also consider that the true singularity is not a specific event, but the universe itself, emerging from «nowhere» and destined to return there. This scenario, although speculative, has the advantage of consistency and explanatory simplicity in the perspective of a plausible multiverse.

The moment of the Big Bang is characterized by a density and an energy such that the quanta could not be distinguished individually. In this initial state, the universe is homogeneous and smooth: it is not possible to define wavelengths or talk about particles, and the notion of time remains potential, not updated.

Quickly, with the first radiative entanglements, nascent matter absorbs some of the intensity of diffuse kinetic radiation (later designated as EMW). In a universe in loss of energy continuity, the notion of corpuscular photon becomes relevant. Wave phenomena do not have precise spatial coordinates and are described as energy fields difficult to quantify in terms of space occupancy. On the other hand, a particle can be represented as a point moving in space, which implies linking space and time to define its trajectory. The wave-corpuscle duality therefore constitutes a reflection tool allowing to represent mathematically the space-time of the universe.

As the universe cools down, the frequencies of radiation decrease, with high frequencies disappearing mainly at the end of the radiative entanglement phase. The energy divides during photon splitting, enhancing the corpuscular representation. In a cooled universe, the energy carried by photons, less disturbed by now weak energy fields, decreases in frequency and amplitude. The initial corpuscular vision becomes less appropriate. If one maintains a wave-like representation, the wavelengths are stretched to the point of becoming negligible, and the energetic relief of the empty space gradually fades.

Illustrations

The illustrations that follow only dress up with images, the ideas included in the text but are not really transposable as they.

Standard model elementary particle table

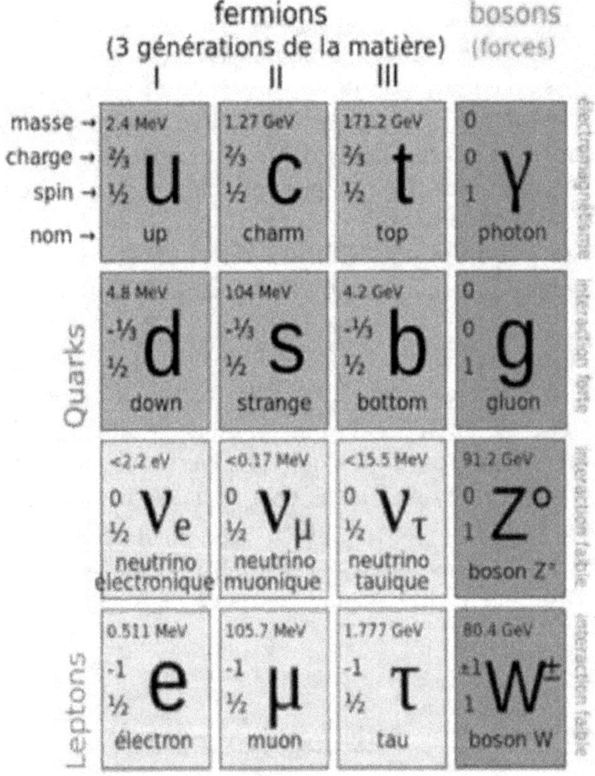

2

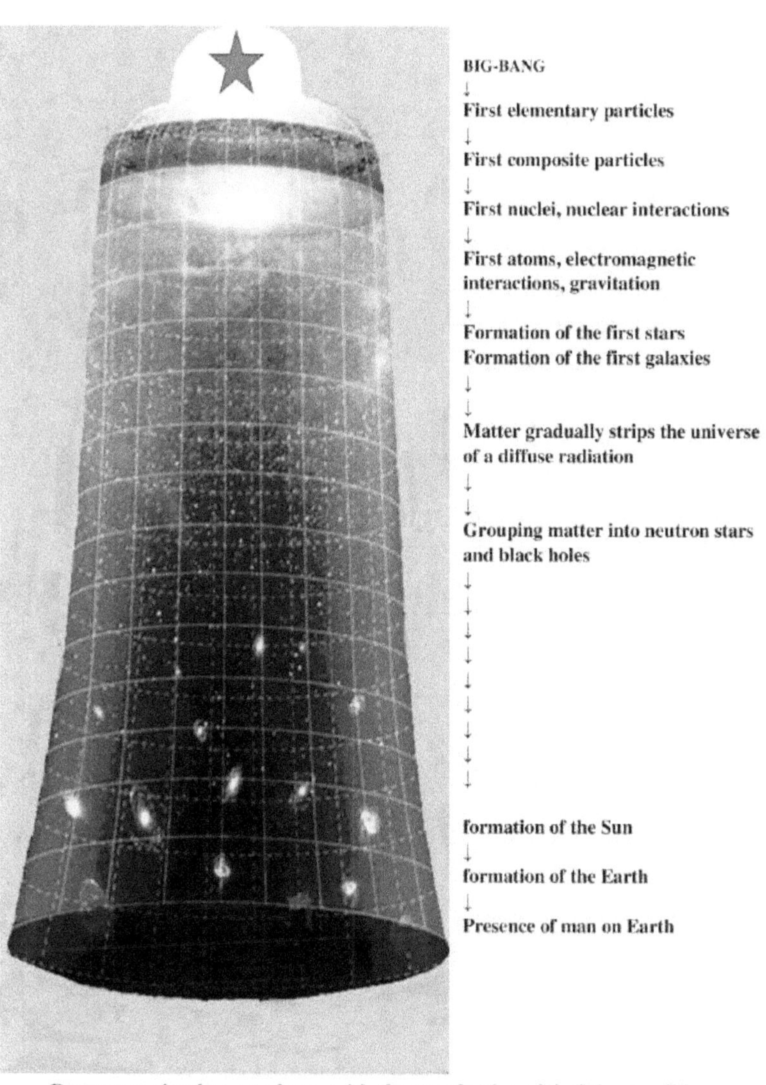

Representation in accordance with the standard model of the possible evolution of an imagined expanding universe

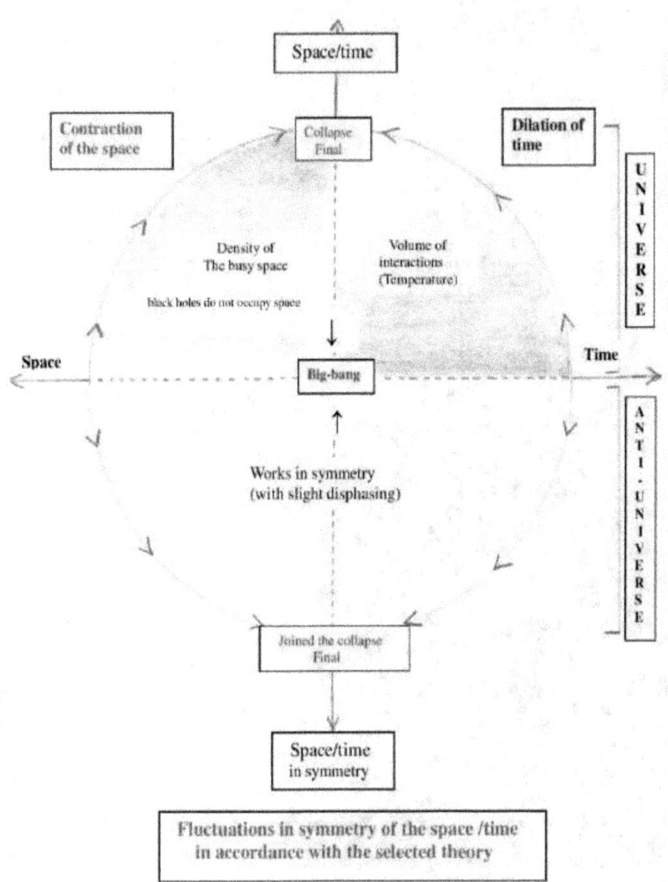

Space/time
Contraction
of the space
Collapse
Final
Dilation of
time
UNIVERSE
Density of
The busy space
Volume of
interactions
(Temperature)
black holes do not occupy space
Space
Big-bang
Time
ANTI - UNIVERSE
Works in symmetry
(with slight disphasing)
Joined the collapse
Final
Space/time
in symmetry
Fluctuations in symmetry of the space /time
in accordance with the selected theory

120

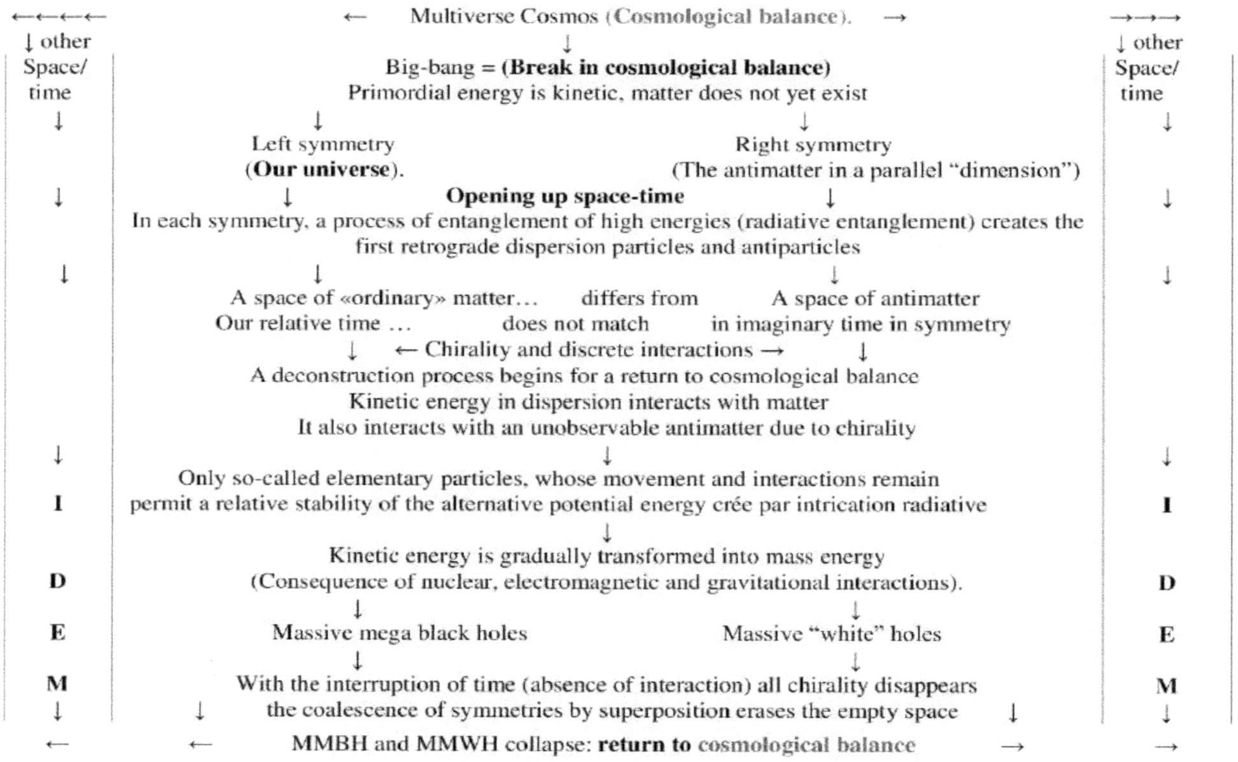

121

←←←← ← Multiverse Cosmos (Cosmological balance). →

↓ other
Space/
time
↓

Big-bang = **(Break in cosmological balance)**
Primordial energy is kinetic, matter does not yet exist
↓

Left symmetry   Right symmetry
(Our universe).   (The antimatter in a parallel "dimension")
↓   **Opening up space-time**   ↓
In each symmetry, a process of entanglement of high energies (radiative entanglement) creates the
first retrograde dispersion particles and antiparticles
↓   ↓
A space of «ordinary» matter…   differs from   A space of antimatter
Our relative time …   does not match   in imaginary time in symmetry
↓   ← Chirality and discrete interactions →   ↓
A deconstruction process begins for a return to cosmological balance
Kinetic energy in dispersion interacts with matter
It also interacts with an unobservable antimatter due to chirality
↓

Only so-called elementary particles, whose movement and interactions remain
I permit a relative stability of the alternative potential energy crée par intrication radiative **I**
↓

Kinetic energy is gradually transformed into mass energy
D (Consequence of nuclear, electromagnetic and gravitational interactions). **D**
↓

E Massive mega black holes   Massive "white" holes **E**
↓

M With the interruption of time (absence of interaction) all chirality disappears **M**
↓ the coalescence of symmetries by superposition erases the empty space ↓
← ← MMBH and MMWH collapse: **return to** cosmological balance →

↓ other
Space/
time
↓

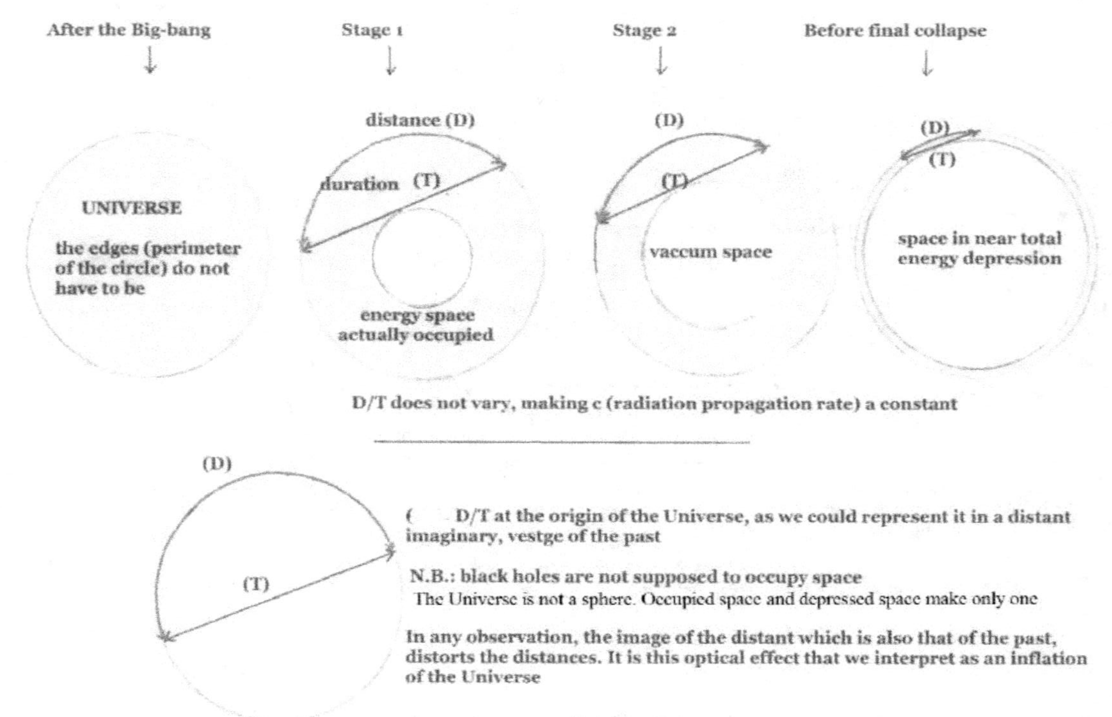

Time, Space and Relativity

**Electromagnetic interactions
as a force representative of
strong interactions**

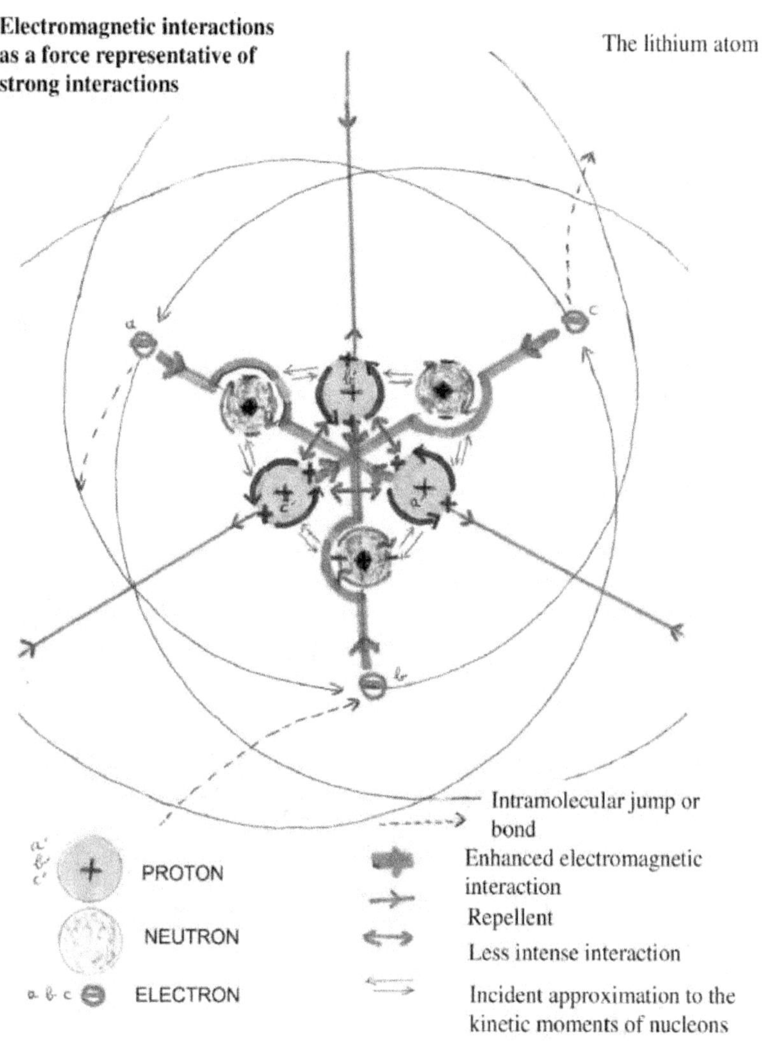

SPACE/TIME   AND   SYMMETRY OF UNIVERSE

THE CURRENT PROBLEM
IN AN EXPANDING UNIVERSE
SYMMETRY IS ABSENT

↓

68% DARK ENERGY

HYPOTHETIQUE (!)
(to justify the apparent expansion of the
Universe)

27% BLACK MATTER
HYPOTHESIZED
(to justify all observed gravitational effects)

5% KNOWN ENERGY/MATTER

THAT IS 100% OF CONTENT BY
OF OUR OBSERVABLE UNIVERSE

WITH RETROGRADE DISPERSION
IN A DEPRESSED UNIVERSE
THE SYMETRIE IS CONSIDERED
FOR ITS GRAVITATIONAL EFFECTS

↓

the expansion of the Universe is considered a

optical effect /

Actually, due to the energy
depression of space

**GRAVITA-
TIONAL
OBSERVED
EFFECTS**

50% ENERGY/ANTIMATTER

(shared gravitational effects)

**50% ENERGY/MATTER   AFTER
RE-EVALUATION OF THE MASSES
IN PRESENCE**

THAT IS 100% OF ENERGY CONTENT
OF A UNIVERSE IN SYMMETRY
included electromagnetic fields

VIII <u>A singularity that would have nothing singular</u>
(And who, out of time, would conjugate in the plural)

The notion of energy remains fundamentally defined from its observable manifestations: movement, transformation of matter, heat transfer or radiation. Apart from these measurable phenomena, energy does not have an independent intrinsic definition of its effects. Any attempt to characterize it otherwise thus runs up against the inevitable dependence on observation and on any measure taken.

At any scale considered, from the elementary particle to cosmological structures, energy manifestations seem to be part of a global process of relaxation towards a state of equilibrium in cosmology. The hypothesis adopted here is that any admissible physical process participates, directly or indirectly, in the progressive resorption of the initial imbalance introduced during the primordial cosmological event commonly referred to as the Big Bang. Any phenomenon that would not contribute, in the long term, to this dynamic of rebalancing would then be excluded from the field of physically feasible processes. Weak interactions can in this perspective be considered as one of the fundamental expressions of this dynamic of irreversible evolution.

At the formal level, the energy E is quantified differently depending on the nature of the physical medium considered.

For particles endowed with mass — mainly fermions—the total energy is written:

$$E = mc^2 + K$$

where m denotes the rest mass (understood as intrinsic mass independent of motion), cla propagation speed of electromagnetic waves in a vacuum, and K the kinetic energy associated with the variation of motion in a given frame.

For phenomena where mass is not revealed—in particular in the case of electromagnetic waves ensuring the interactions and cohesion of matter — energy is described by the Planck relation:

$$E = hf$$

where h is Planck's constant and f the frequency of the radiation.
In the first case, regarding potential mass energy, the constant is the photon displacement velocity ($\approx$ 299792 km/s) squared.

In the second case, in a more arbitrary but judiciously chosen way, the constant is given by Planck's formula (± 6.63 x 10^{-34} j/s). The Planck constant is supposed to represent a certain observed relationship between the wave frequency and the energy carried by this same wave. It implies that the energy of a particle cannot be measured below a certain threshold thus defined.
It may nevertheless be asked whether the constants f and c could not vary significantly over time (see chapter XVIII).
These 2 equations mean that a mass, representative of a given amount of energy, can be translated in terms of wave frequencies with reference to the Planck constant. **This equivalence would validate the idea of particles considered as the product of radiative entanglements.**
M x c^2 = h x f only makes sense by referring to a contextual environment made of space (km traveled) and time (seconds elapsed). Space/time is an essential framework of analysis for the observer that we are. Describing over time without referring to space seems impossible and vice versa. This explains why we are not able to describe an antimatter that does not share the time that is ours and occupies a dimension of space somewhat parallel.

To return to Einstein's famous formula, E= mc^2 and without wanting to commit further in the field of mathematics, how to explain that the photon, a particle devoid of mass, is nevertheless a carrier of energy. In fact, this equation thus formulated, responds to the particular case of a mass particle considered at rest, without movement, that is to say, out of any gravitational and displacement reference frame. However, this case is purely theoretical. In reality, E=mc^2 is a simplified formula of the equation $E^2=m^2c^4+p^2c^2$ in which:
E: is the energy carried by the particle or body considered
M: is the rest mass or intrinsic mass when the momentum
(p) is assumed to equal 0.
c: is the speed of light
p: is the amount of movement that any particle holds.

For a massless particle (m = 0), which is the case of photons representative of EMW, we would obtain with the simplified formula:
E = 0 x c^2 let be E = 0 whereas starting from the general formula we obtain:

126

$E^2 = 0 \times c^4 + p^2c^2$ let be **E=pc** (p being supposed to represent the oscillations of electric and magnetic fields occurring perpendicular to each other).

This result is in line with the idea that the energy of photons, in other words of EMW, lies in their propagation speed alone, translated into wave frequency and amplitude. The EMW thus bring additional energy (kinetic) to the material particles without transport of matter. The energy/mass of the latter varies with the contribution of intrinsic movement that is thus conferred on them. If the photon represents a unit of energy measurement associated with OEM, the mass particle as an entangled wave packet can be understood as a concentration of potential photons confined in a form of contact unification or circular polarization in a closed structure. The mass particle would have the particularity of not being more representative of occupied space than a point which by definition has no dimension. In other words, the intrinsic mass could be defined as the level of radiative entanglement characterizing any particle of matter (see wave packet in chapter V).

An observed particle reveals a state reduced to the only properties that an observer is able to more or less presuppose before measurement. These properties, which are actually prescribed by the choice of measurement tools and observation methods, belong to a reductive vision specific to the observer. This is called wave packet reduction. It leads in a simplistic but logical way to define the particle or any observed system, in terms of mass, electric charge, spin, color notably.

It must also be considered that some of these properties, which are closely correlated with each other and interdependent, cannot be attributed to certain types of particles. Indeed, photons have no mass, neutrinos would have no electric charge, free electrons do not have spin, leptons would not possess color charge (this term is not to be taken, here, in the literal sense). What makes the properties of a particle, would it not rather be, its potential to interact with any other particle capable of doing so with it? However, this potential can only be determined randomly or statistically, given that quantum mechanics which continues to build and deconstruct is essentially unpredictable because it is insufficiently understood.

Mass confers an operative reality on the observed phenomena. All observation is necessarily referred to matter, and both the observer and the observed system share a common dependence on a given material state.

Electromagnetic waves, devoid of mass and charge, do not exhibit differentiated symmetry and would interact equivalently with matter and antimatter. The energy they transport simultaneously participates in the structuring and destructuring processes of the Universe.

At equivalent energy, a formal correspondence exists between mc^2 and hf, suggesting a partial substitutability between baryonic material and radiation. However, in the extreme regimes constituted by cosmological singularities — whether it is the initial state of the Big Bang or a final state of a cooled and annihilated Universe — the constants this h, intrinsically linked to the notion of time, lose their descriptive relevance. In the absence of temporality, these states escape any conventional physical representation and justify the use of the term singularity.

The extremely concentrated energy during the Big Bang could not be governed by physical laws that were originally non-existent and only apply to the Universe after the emergence of time. The theories developed since relativity and quantum mechanics remain valid in their field of application, that is to say beyond the Planck wall. Planck units then serve as a formal tool to delimit the field of the describable, while suggesting the existence of an earlier regime of a virtual nature, relying on quantum symmetries inaccessible to observation but constitutive of the observable Universe.

Primordial phase: Big Bang and emergence of physical structures

The Big Bang can be described as a **limit-event**, or more precisely as **a non-event in the dynamic sense**, insofar as it corresponds neither to a measurable phase transition nor to an evolutionary process with a definable duration. It constitutes the **simultaneous establishment of space and time**, made effective by the appearance of the first physical interactions. Before this opening, the very notions of change, causality, or temperature are not operative.

The primordial energy associated with this initial state has no differentiated symmetry. The wave frequencies are not yet distinguishable, their distribution being comparable to a smoothed and undifferentiated state. The subsequent modalities of conservation, transformation and structuring of this energy will determine the entire evolutionary history of the Universe.

The disruption of cosmological equilibrium introduced by the Big Bang generates an extremely high energy field, deployed in a space still devoid of gravitational structure. This energy, of incommensurable intensity, can be interpreted as an initial impulse announcing the emergence of electromagnetism, in the form of very high-energy radiation, comparable to gamma rays of previously unknown frequencies. As the Universe unfolds, differentiated wavelengths appear, paving the way for radiative entanglement phenomena and, subsequently, primordial nucleosynthesis.

The primordial energy cannot be described in terms of temperature, which is a thermodynamic quantity related to entropy and assuming an already established temporality. Time becomes relevant with the appearance of the first radiative entanglements, from which gradually distinguish quarks, electrons, neutrinos, as well as other elementary particles now disappeared or even hypothetical. Among this initial diversity, only particles whose properties and interactions allow a relative stability of the potential mass energy will remain.

Not every conceivable process is physically allowed. The interactions and types of particles must meet the constraints imposed by the initial symmetry break induced by the Big Bang. The complexity of the observable Universe would thus result from a fine adjustment of fundamental parameters, not arbitrary, but intrinsically linked to the initial conditions, and in a way pre-inscribed in the original dynamics.

This constraint is found in the description of atomic systems, where the electron field is limited to a discrete set of allowed states. These conditions, far from being contingent, guarantee the relative stability of matter, a necessary condition for its subsequent evolution towards a state of global deconstruction, interpreted here as a gradual return to cosmological equilibrium. This dynamic could culminate in a state of extreme energy concentration, in a form excluding any manifestation other than that associated with black holes.

Heat constitutes an indicator allowing the interpretation of certain primordial phenomena in a Universe undergoing ionization, that is to say presenting the first charge manifestations. The primordial energy, initially
129

diffuse, both latent and kinetic, without mass or defined charge, is perturbed by the appearance of radiative entanglements that generate symmetrically particles of matter and antimatter. The progressive occupation of space by massive particles gives the Universe a growing inertia, while freeing the so-called "empty" space.

Heat is produced by the agitation of matter: radiation, whatever its intensity, does not generate heat in the absence of material support. Thus establishes an irreversible evolution during which potential energy, associated with gravitational effects, gradually replaces the primordial kinetic energy.

The first interactions are accompanied by a sudden rise, followed by a gradual decrease in temperature. The extremely intense radiation interacts with itself and with the nascent matter, breaking the initial uniformity and generating local fluctuations of intensity. These fluctuations result in the appearance of frequencies associated with very short wavelengths, revealing an incipient granularity of the primordial plasma and a scission of mass energy into two symmetrical states.

At this point, space-time is operational. The temperature continues to decrease, while the wavelengths increase inversely at their frequencies. The two symmetrical states of matter then begin to interact within an excited plasma, where electrical currents and magnetic effects are gradually distinguished. The gamma rays observed today constitute an indirect remanence of these initial states, resulting mainly from collisions between astrophysical objects with a very high energy density.

This ionized plasma constitutes the cradle of an embryonic material. The first particles from this state include:

- residual neutrinos and antineutrinos (see Chapter XIII), playing a regulatory role in energy exchanges during nuclear reactions.
- quarks and antiquarks, fundamental constituents of atomic nuclei.
- the electrons and positrons, ensuring the balance of charge around the nuclei.

The primordial plasma associated with the Big Bang should not be confused with plasmas accessible experimentally. It is an extreme transient state, distinct from both the fusion plasma and the one associated with black holes.

These plasma states can be interpreted as transition phases marking the beginning, evolution and end of the Universe.

The primordial plasma can be described as an unstable energy configuration, devoid of significant temperature and excluding any classical corpuscular description. The irregularities that develop there, in the form of primordial radiation, interact with each other and gradually constitute the known electromagnetic waves. Radiative entanglements then transform this energy configuration into a plasma soup populated with charged primo-particles, embryos of current neutrinos and, indirectly, massive particles.

When the temperature reaches a maximum and then decreases, particles of opposite charges appear, with identical spins, in a context of broken symmetry. After crossing the Planck wall, electromagnetic interactions become operative: opposing charges attract each other, identical charges repel each other, while certain neutral particles contribute to the overall stabilization of the system.

With the continuation of cooling, space and time take on a fully defined meaning. Primo-particles evolve into quarks, electrons, and neutrinos. The formation of nucleons precedes that of atomic nuclei, then ions. Primordial nucleosynthesis marks the exit from the plasma regime, later relayed by stellar nucleosynthesis.

Electrons interact with newly formed nuclei, ensuring the electromagnetic equilibrium of the atom. The grouping of atoms allows the formation of clouds of hydrogen and helium, then molecules, stellar objects and galaxies. The large-scale structures thus formed question the hypothesis of a perfect homogeneity of the Universe.

Aware of the speculative nature of this description, it should be remembered that the standard cosmological model, although incomplete and debated, provides a coherent interpretative framework for describing the cosmic environment of which we are part. An exotic plasma state, devoid of significant temperature, could reappear in a cooled Universe on the way to collapse, notably within massive black holes.

In the current era, protons, neutrons and electrons constitute the stabilized configurations of constructed matter. Neutrinos, discovered later, would

131

play a regulatory role in energy exchanges, facilitating certain quantum interactions without disturbing the electrical neutrality of atoms. Photons remain the fundamental vectors of energy transfers, ensuring the dynamic balance of forces within atoms and molecules.

Electromagnetic interactions at the subatomic scale produce attraction effects analogous, in their global manifestation, to those attributed to gravitation at the macroscopic scale. This analogy suggests that gravitation could result, on a large scale, from the superposition of quantum interactions contributing locally to charge neutrality.

The neutron can be interpreted as the product of a proton transformation by capturing an electron and an antineutrino. Unstable in the free state, neutrons provide a stabilization role within the atomic nucleus, compensating for electrostatic repulsion between protons. Electrons, by their charge and binding properties, are decisive in the construction of built matter.

Electron–photon interactions and matter stability

When an electron absorbs a photon, it acquires additional energy in the form of kinetic energy, which translates, by mass-equivalence–energy, into an increase in its effective mass. This energy acquisition changes the quantum state of the electron and manifests as a transition to a higher energy level. The electron is then statistically more distant from the atomic nucleus with which it is associated, and can, in certain configurations, become shared between several atomic nuclei. This sharing constitutes the fundamental mechanism of molecular bonds, in which one or more electrons simultaneously participate in the cohesion of several atoms.

Conversely, when an electron emits a photon, it loses a fraction of its kinetic energy. This energy loss results in a transition to a lower-energy quantum state, corresponding to an orbit closer to the nucleus. These absorption and emission processes illustrate that atomic and molecular interactions essentially relate to energy transfers between differentiated wave states, organized around a global balance of charge and energy.

The stationary state of a particle or an atom is generally defined, in an idealized approach, as an equilibrium state with no observable interaction. In practice, this state corresponds rather to a stable dynamic situation, within which exchanges of information—in the quantum sense — continue continuously. Thus, an electron describes probabilistic trajectories characterized by a variable spatial distribution and velocity, continuously exchanging information with the nucleus to which it is associated as well as with neighboring atoms involved in binding forces.

The relative stability of molecules is based on this continuous exchange dynamic. At the atomic scale, electrons interact with each other while maintaining average distances compatible with the system's equilibrium. They collectively form an electronic cloud, a distributed and correlated structure, coupled with a nucleus consisting of quarks globally carrying an opposite charge. This cloud does not correspond to classical individualized trajectories, but to a superposition of entangled quantum states.

When atoms come together, their electron clouds interact through electromagnetic influence, allowing a partial sharing of valence electrons. This sharing is at the origin of chemical bonds and conditions the formation of stable molecular structures. Electromagnetic waves play a central role in these processes, ensuring the transfer of energy necessary for electronic transitions, whether by absorption or emission of photons.

In summary, photons, by modulating the energy, speed and spatial distribution of electrons, make molecular bonds possible and give to the built matter a relatively perennial structure. From the perspective developed here, the hypothesis according to which photons could interact in a discreet way with antimatter is not without consequence on their interactions with matter and could subtly influence the mechanisms of energy exchange and stability of atomic systems.

Self-programmed deconstruction of the binary system of universes in quantum symmetry

133

Electromagnetic waves, in the phase of dispersion, absorption, and refraction, gradually lose amplitude. At the beginning of our Universe, these waves, characterized by very high amplitudes and significant frequencies, continued, for an extremely short period of time and at a decreasing rate, to entangle and interfere, favoring the formation of elementary particles. These particles then assembled to form mainly light hydrogen atoms.

A fraction of the primordial kinetic energy was thus mobilized and integrated into matter. The global temperature of the Universe is decreasing continuously. EMW manifest a generalized elongation of their wavelengths, and their light spectrum shifts towards the red. This observation has inspired a now-abandoned hypothesis, known as the "aging of photons", developed in particular to support the idea of cosmic expansion without explicitly resorting to space dynamics.

In an increasingly cooled Universe, these free photons, having become energy-poor, would develop extremely extended radio wavelengths, until the space, dilated to the extreme, seemed emptied of any significant occupation. However, in the hypothesis adopted here, the total energy contained in a universe does not undergo any real loss. At the final stage of its evolution, this energy would be fully recorded in MMBH.

The totality of energy contained in a binary system of universes in quantum symmetry, at all stages of its evolution, could be expressed essentially as the sum of:

- the potential energy mc^2 associated with the inert masses present in each symmetry.
- the kinetic energy $\pm\frac{1}{2}mv^2$ corresponding to the movements of all matter and antimatter.
- the kinetic energy hf transmitted by electromagnetic radiation from the so-called 'empty' space, interfering between the two symmetries.

The masses would thus gradually increase in density until, in a Universe at the end of its life reduced to the sole presence of MMBH, destructured

134

matter and antimatter only show residual agitation. The constant c, representing the speed of light, would then tend to become a constant devoid of effective physical range. Similarly, the average level of EMW frequencies would slowly but inexorably evolve towards ever lower values, so that Planck's constant would ultimately reveal a value that has become non-significant in this limit context.

Wave and corpuscle, radiation and matter thus constitute two representations of the same amount of energy. At the origin of its quantum constituents, matter retains the imprint of a broken symmetry. This symmetry would be double. It would involve, on the one hand, charge interactions between particles and antiparticles of the same nature, and on the other hand, charge interactions, specific to each symmetry, between particles of a different nature.

Thus, antielectrons would carry a positive charge, inverse to that of nuclei made of antiquarks, which have elementary charges of opposite signs but an overall negative charge. Antimatter would be, like matter, globally neutral. This neutrality of load, shared at all levels of organization, would allow by gradual coalescence of loads a return to a cosmological equilibrium initially broken.

In the case of the electron, it can be described as a beam of waves associated with a plural orbit, or more precisely with a fluctuating ellipse, maintained at a distance from the atomic nucleus. The orbit, supposed to represent the trajectory of a particle such as an electron, is essentially a heuristic representation of an atom in the equilibrium state. Nothing indicates that the electron, as a localized particle, physically revolves around the nucleus.

The electron is bound to the nucleus according to a configuration determined by the overall charge balance of the atom, which itself depends on the more or less stable molecular context in which it falls. The atom then becomes a distinct and identifiable system, manifesting a physical presence that can be located in the space and time of relativity. Conversely, the notion of a particle orbiting around the nucleus disappears as soon as the electron is considered a delocalized quantum object, unsusceptible to precise positioning.

The 3-D space represents an obligatory positioning framework that we cannot conceptually dissociate from observation time, however short it may
135

be, for any observed subject. This time that passes in an environment where entropy keeps changing the situation, makes us feel the need to give a precise position to what would be fundamentally undulatory in nature and therefore not truly localizable. We are therefore led to consider the observed particle as if it were a macroscopic object with classical behavior, whereas in all cases, such a position can only be relative.

However, the requirement of charge neutrality for the atom imposes the presence of a number of electrons equal to the number of protons in the nucleus, these electrons sharing a synergy in the form of orbitals. These cannot be definitively attributed to them but result from the relative instability of the charge equilibria ensuring molecular assembly. The electron horizon of the atom, generally negatively charged, adjusts its intensity by distributing itself over orbitals adapted to the mass and configuration of each atom.

The nucleus, whose charge is not uniformly distributed, adjusts its proton-neutron structure in coherence with the angular momentum of the electron horizon. A fine equilibrium is thus established, governed mainly by electromagnetism, resulting from permanent exchanges between the nucleus, the electron cloud, and neighboring atoms.

By simple charge effect, the electron, considered as a packet of entangled waves without proper spatial dimension, should be attracted to the proton of the nucleus that corresponds to it. Now, proton and electron being both entangled wave packets without spatial extension, the electron's perigee point—whose velocity in vacuum is close to that of light — can only be located as close as possible to the nucleus, without direct interaction with it. In other words, the electronic nucleus and orbitals interfere but remain spatially disjoint. If this were not the case, matter would be unstable and the Universe could not structure itself sustainably.

The electron would leave the nucleus if it were devoid of mass, like the photon, or would collide with it if it had excess mass, unless it constantly adjusted its speed and orbital perigee. Its charge is in adequacy with its mass; however low it may be (about 1/1850 of that of the proton). Minute variations in mass, related to its speed and trajectory, allow it to maintain the necessary distance from the nucleus. These energy adjustments, ensured by the absorption or emission of photons, also allow the electron to change atoms. No nucleus is definitively assigned to it. By jumping from one orbital
136

to another or by changing atomic partners, the electron ceases to belong to the atom considered as a discrete entity. It structures matter by moving within a molecular field in perpetual evolution, its speed varying continuously according to this environment.

The atomic nucleus can be considered as a minimalist center of gravity. It consists of quarks grouped into protons and neutrons. Protons carry the positive charge necessary for electrostatic equilibrium with electrons, while neutrons, which can be assimilated to protons neutralized by modifying their internal structure, ensure the compactness of the nucleus without affecting the overall neutrality of the atom.

The photon, a so-called virtual particle due to the absence of mass at rest, makes it possible to account for the energy losses and gains of electrons by transporting quanta of energy between them.

It can be assumed that an old phase of shedding at the subatomic scale explains the current scarcity of heavy quarks (C, T, S, B), which would have fragmented into lighter quarks (U and D). Similarly, generations of fermions more massive than current neutrinos and electrons could have abundantly populated the young Universe before disappearing.

Permanent exchanges are established, dictated by the need to preserve the fragile balance of matter. Protons and electrons, of opposing charges and remarkably stable outside nuclear interactions, currently achieve a transient equilibrium state of matter. This phase would only constitute a preliminary step in the cosmic evolution, ultimately leading to a return to global equilibrium and a final collapse.

Imagining that protons retain an indelible memory of their history since the origin would suppose that time has a meaning at the scale of elementary particles. However, there is nothing to indicate that this concept is relevant at this fundamental level. Their history could not in any way go beyond what is referred to as Planck's wall, corresponding to the first constituent radiative entanglement of matter.

Therefore, being ourselves made up of these particles assembled into atoms and molecules, can we really formulate otherwise than by assumptions what would be the origin of our own emergence? Claiming to go beyond this singularity, improperly qualified as a primary cause, remains highly speculative. Yet it is indeed this attempt that underlies our notions of infinity, eternity, or even the recourse to metaphysical entities intended to fill the gaps in our understanding.

If protons and electrons have a stability and a lifespan possibly comparable to that of the Universe, it is not the same for the atom. This is the result of a succession of fusion and fission nuclear reactions from primordial hydrogen. The ionized hydrogen atoms, the starting point for the structuration of matter, have gathered into molecular clouds called H II regions. By densifying and absorbing the energy carried by the EMW, these clouds gave birth to stars.

The latter produced, by secondary nucleosynthesis, heavier elements, which were partly dispersed during stellar collapses (supernovae, hypernovae), at the origin of planets, neutron stars, black holes and galaxies. The first massive groupings of matter, comparable to giant protogalaxies, would thus have formed from particularly dense clouds of ionized hydrogen.

The primordial Universe would present an initially poorly differentiated topology, in the absence of marked gravitational effects. Currents of matter at relativistic speeds would have gradually distinguished dense regions, causing a local rise in temperature and initiating the formation of these protogalaxies. In return, vast regions poor in material would have developed, drawing on a large scale the spider-web structure observed today. These areas would not be completely empty and could host extremely massive isolated black holes, remnants of ancient galaxies engulfed by their central black hole.

The properties of space in the primordial Universe thus suggest that these first disappeared galaxies reached dimensions beyond those formed later, in a space already depleted of gas and free particles. Gigantic galaxies, equipped with very active central black holes and directly powered by abundant hydrogen, could thus have formed quickly.

At some point in its existence, an atom has a more or less massive nucleus, which directly determines its degree of stability. Atomic stability depends primarily on the composition of the nucleus. The most stable atoms are those whose atomic mass—defined as the sum of the masses of protons and neutrons — remains relatively low. Few isotopes are stable beyond an
138

atomic mass of 209, and all isotopes above 238 are unstable. Moreover, when the number of protons exceeds 82, nuclear interactions are no longer sufficient to ensure the integrity of the nucleus. The same goes for nuclei of atomic mass 43 and 61, for which no stable isotope is known.

Within these limits, isotopes are generally stable when the number of protons and neutrons is comparable, with a few exceptions, such as beryllium-8, which nevertheless plays a transitory role in stellar nucleosynthesis of heavier elements, particularly carbon. Moreover, beyond 137 protons in a nucleus, it seems that the latter is no longer able to maintain stable electron orbitals for more than 137 electrons, as binding interactions become insufficient. The balance of matter thus appears as inherently fragile.

Atoms exhibit increased stability when their electron orbitals are filled, as is the case with deuterium or lead-208. Similarly, atomic nuclei are relatively perennial when the "layers" of nucleons are saturated, which reinforces the binding energy (examples: helium-4 and lead-208, the latter being doubly stable). It should be noted, however, that the notion of quantum layer used here must not be understood in the classic sense of superimposed spatial strata.

The neutron has a lasting existence only when it is confined within an atomic nucleus, in the presence of protons. This leads to consider that protons need to associate with these electrically neutral composite particles to regroup in a stable, durable way and to allow the increase of atomic mass. With the exception of the most common hydrogen isotope, consisting of a single proton and a single electron, atomic nuclei cannot form stably without neutrons.

Then arises the question of the very short average life span of the free neutron — of about fifteen minutes—compared to the extreme longevity of the proton. The neutron is a composite particle made up of quarks and generally electrically neutral, unlike the proton, which carries a positive charge. While the antiproton has a negative charge, opposite to that of the proton, the antineutron exhibits the same electrical neutrality as the neutron. The quantum correlation associated with symmetry, distinct from non-local quantum entanglement between particles of the same symmetry, would therefore not be of the same nature for the proton-antiproton pair as for the neutron-antineutron pair.

139

One might wonder if this symmetry difference would not help explain the instability of the free neutron, that is to say, not confined in an atomic nucleus. However, this hypothesis, based on a synergy related to quantum symmetry, remains to be elucidated.

It is on this basis that matter structures and densifies, leading to the formation of massive stars, then neutron stars, before finally deconstructing in the form of stellar black holes. The atom keeps changing its configuration, mainly according to two mechanisms.

The first is nucleosynthesis, under the effect of high temperatures and pressures prevailing at the heart of massive stars. Following the radiative entanglement phase, nucleosynthesis corresponds to the formation of atoms heavier than hydrogen. It operates mainly in the central regions of the stars. During explosive events, this nucleosynthesis can produce particularly heavy nuclei, such as iron-56, one of the most massive stable metal atoms. Some of these heavy elements, notably carbon, nitrogen and oxygen, are then projected into space and join the diffuse particles of cosmic radiation, transported by electromagnetic interactions.

Moreover, these violent events — supernovae and novae—which mark the end of life of massive stars, fragment atoms and scatter their electronic procession. The lighter atoms thus produced, mainly hydrogen and helium, contribute to the formation of new interstellar clouds.

The second mechanism is photodisintegration, which occurs at temperatures of several billion degrees within extremely massive stars. In this process, the heaviest nuclei, subjected to an intense flux of very high-energy photons, are fragmented into lighter nuclei.

Over these cycles dominated by nucleosynthesis, light atoms tend to become increasingly rare. Like the free particles of cosmic radiation, all the matter constituting stellar bodies would eventually join the black holes of a gradually cooling Universe.

Mass variations are made by successive energy adjustments. Particles devoid of electrical charge but possessing a non-zero mass, electronic neutrinos — sometimes noted v_e— participate in this delicate balance, by contributing to the redistribution of energy during nuclear reactions involving quark transfers between neutrons and protons. In this logic of interaction vectors, uncharged and massless bosons, such as gluons for the strong interaction and photons for electromagnetic interaction, play the role of mediators by framing these exchanges. Other bosons, Wet Z, with large masses and zero or transient charges, also occur in the atomic nucleus, accounting for the shedding and rebalancing processes associated with electroweak interaction.

Space and time quickly lose their operatory character in quantum physics, where interactions between particles cannot be described satisfactorily in terms of trajectories or displacements. The difficulty lies in the fact that we spontaneously assimilate these interactions to exchanges, a notion which classically implies a distance covered and a duration of transmission of information. For charge transfers, this difficulty is circumvented by the introduction of the photon as a vector of electromagnetic force. Virtual particles, in the sense of not directly observable, such as gluons, as well as W et Z bosons, are introduced to account for respectively strong and weak interactions.

By "dressing up" quantum interactions in this way, this formalism stages exchange vectors — the bosons — without which, from the point of view of an observer subject to space and time constraints, energy transfers would remain inconceivable.

In a hypothesis of perfect symmetry between matter and antimatter, present in equal shares, it would be possible to dispense with the introduction of such a large number of bosons to ensure energy transfers between particles, as well as between particles and antiparticles. Photons would remain the vectors of vacuum energy and intervene in unobservable charge interactions between symmetries, as well as in those participating in the charge equilibrium of matter. Electromagnetic waves, specific to each symmetry and evolving in distinct "dimensions", would then play the role of discreet mediators of these exchanges. Such an archetype meets that proposed by the theory of supersymmetry, according to which each particle has an associated partner particle, in particular its antiparticle.

These potential interactions between particles and antiparticles could be represented, heuristically, by one-dimensional objects assimilable to threads, thus joining some aspects of the very hypothetical string theory.

141

Inscribed in a context of multiverse Cosmos and chirality — considered here as an unrecognized cause — this representation echoes the unifying theme of M-theory.

In a more global approach, bosons can be understood as discrete exchange agents operating at the "osmotic" boundary between two quantum symmetries. The particles, associated with these bosons, then acquire, from the observer's point of view, the ability to overcome distances. The theory of supersymmetry, on the other hand, aims to attribute to each elementary particle of matter intrinsic properties analogous to those of the corresponding bosons. In this perspective, the bosons introduced by the Standard Model would appear as an artifice of thought, necessary to describe phenomena that we do not yet know how to explain otherwise.

Nuclear fusion and fission constantly modify our fields of observation at all scales. These interactions, described as weak, constitute spectacular manifestations of accidental but necessary events, intended to correct certain imbalances in the pre-plotted evolution of the Universe. Favorable to the cohesion and gathering of matter, could they not result from discrete exchanges between quantum symmetries, leading to a progressive rarefaction of heavy elements in their free and diffuse state?

One way to represent the forces at work in this binary system of universes in symmetry is to assign them a physical appearance compatible with our mode of representation essentially cognitive. The chosen logical artifice consists of describing the interactions between packets of entangled waves — referred to here as particles — using familiar concepts such as mass, charge, color, displacement or intrinsic movements. From these criteria, the elementary constituents of matter are classified into quarks, leptons, and bosons.

This nomenclature responds to an analytical vision of the Universe, but quantum physics suggests that particles and antiparticles, sharing the two symmetrical states of a same binary system of universes, may not possess the phenomenological existence we attribute to them. This is not inconsistent when it is admitted that particles and antiparticles annihilate each other when they meet. In this perspective, everything becomes potentially virtual, the energy they represent being destined to return to a multiverse Cosmos devoid of proper physical reality.

What is energy?

This question alone summarizes an essential part of the problem of fundamental physics. Despite its central use, energy remains difficult to define ontologically. It manifests empirically in different forms, mainly through electromagnetic radiation and fields, as well as by matter and the gravitational fields associated with it. This descriptive observation, if it allows classifying the observable manifestations of energy, however, does not shed light on its fundamental nature. Therefore, can we consider another orientation allowing us to deepen our understanding of this principle which seems at the same time to structure, govern evolution and, possibly, condition the ultimate transformation of the Universe?

Major advances in this field frequently result from thought exercises that deviate from more intuitive ideological schemes. These approaches, often counterintuitive, lead to question the usual notions of space and time, or even to suspend their use. On the scale of our ordinary experience, physical reality is described using quantities such as distance, position, displacement, duration and energy exchanges, that is to say through the categories of space and time. Thus, in the well-known expression $E=mc^2$, the term c^2 refers to the speed of propagation of light, defined as a ratio between a distance traveled and an elapsed time.

However, at the level of quantum mechanics, and more particularly in a context of quantum entanglement, these notions seem to lose their classical meaning. The non-locality observed in entangled systems suggests that space and time are no longer relevant parameters to describe certain physical correlations. Quantum entanglement characterizes a system composed of two or more particles sharing a common state, provided that their fundamental properties—those that define their nature as elementary particles — are not modified by an external interaction. The entangled particles would not exchange in this way, neither energy nor information in the classical sense, but form a global system whose state is correlated in a predetermined manner. This correlation can give the illusion of an instantaneous interaction, independent of any speed constraint, which seems to place these phenomena in tension with the principles of general relativity.

143

At the macroscopic and supra-atomic scale, the description of the Universe is traditionally based on three fundamental principles, which however seem to be tested in the quantum context:

- **the principle of locality,** according to which two spatially separated objects cannot interact instantaneously, any influence being limited by the speed of light.
- **the principle of particularism,** which assigns to each physical entity well-defined intrinsic properties.
- **the principle of causality,** according to which any effect results from a previous cause inscribed in a temporal succession.

The existence of systems of entangled particles suggests a form of non-separability of their components, although they are perceived as spatially distant. This non-separability seems, at a minimum, to call into question the principle of locality, giving the impression that space and time no longer play their usual structuring role.

An alternative interpretation is to assume that the observed correlations originate from a set of common properties acquired during the genesis of the system. These initial properties, inherited from a unique prior state, would be conserved and shared by the entangled particles, regardless of their subsequent spatial separation. This hypothesis implies the existence of non-local discrete variables, called hidden variables, which are not explicitly considered in the wave function. Although strongly counterintuitive, this approach allows preserving the principles of locality and causality considering that the apparent violations of Bell's inequalities—highlighted by the EPR paradox — reflect above all the informational limits of the observer, unable to access the set of quantum parameters of the system.

The use of hidden variables thus constitutes a tool intended to overcome the impossibility of fully integrating space and time into the quantum description. It leads to accept a form of apparent indeterminism, notably illustrated by the impossibility of simultaneously measuring the position and velocity of a particle with arbitrary precision. This indeterminism reflects less an absence of physical determination than an incompleteness of accessible information. An entangled quantum system cannot then be satisfactorily described using the classical categories of space and time, if it is admitted that the intrinsic properties of particles are not fundamentally defined by their spatial location or their temporal evolution.

144

In this perspective, hidden variables play the role of a bridge between classical physics and quantum mechanics, attenuating quantum indeterminism and inviting to rethink the notion of space-time in a less restrictive way. The charged elementary particle could thus be considered as a fundamental actor, still poorly identified, in the emergence of the gravitational phenomenon, by a local modification of the structure of space.

This approach ultimately leads to questioning the global vision we have of the Universe. Whether it is the infinitely small or the infinitely large, space and time seem to lose the universal and invariant character that we attribute to them at our scale. One can then formulate the hypothesis that energy, in its most fundamental form, would not intrinsically depend on space and time. Although this idea is deeply destabilizing, it might be more relevant to consider that in quantum mechanics, space and time are no longer dissociable entities, contrary to what our macroscopic experience suggests.

At the scale of interactions between elementary particles, the notions of location, displacement, and velocity lose their operational character when they are considered together. The arrow of time no longer appears as a predefined parameter, space seems to lose its classical metric structure, and information seems to be freed from the constraints imposed by the speed of light. This observation, difficult to reconcile with the principle of causality as it underpins our classical logic, fuels the idea that quantum mechanics does not fall within a determinism in the traditional sense.

The phenomena of **state superposition, quantum entanglement, tunneling and delocalization** profoundly deviate from the thought patterns derived from classical experience. Their counterintuitive character suggests that the usual conceptual frameworks, based on a continuous and local spatiotemporal representation, are either insufficient or inadequate for describing fundamental processes. These phenomena indeed lead to consider that elementary particles, as they are described by quantum mechanics, are not intrinsically defined by space and time coordinates, even if the collective extension of their interactions contributes to make them emerge, at the macroscopic scale, a structured space-time. This then manifests itself in the form of a set of massive objects interacting mainly through electromagnetic and gravitational effects.

To accept this conclusion, however, would be to postulate a fundamental incompatibility between quantum mechanics of elementary particles, still largely interpretative, and classical or relativistic physics, more accessible experimentally, which describes the macroscopic world. An alternative hypothesis is to consider that space-time is not fundamental in itself but would represent an **emergent property from quantum physics**. In this perspective, our difficulty in conceiving the transition between the infinitely small and the observable material world — which tends towards the infinitely large — would be ontological: our cognition, shaped by macroscopic experience, would not be suitable for the direct representation of this emergence process. This limitation would partly explain the obstacles encountered in developing a unified model valid at all scales of observation.

The notions of time and space are thus part of an essentially empirical understanding, depending on the observer's point of view. We can then formulate the hypothesis that time and space emerge from the underlying quantum dynamics. Recent developments in quantum physics indeed suggest that the entity designated as an elementary particle has neither a determinate spatial position nor intrinsic temporality. Cause and effect relations then only make sense when they are transposed into a macroscopic context, where space and time appear as effective parameters for observation and measurement. The notion of space-time, as a mathematical structure unifying space and time, thus retains all its relevance, but only relative to the scale considered. It constitutes the indispensable framework on which our lived reality is built, given our limited observation and analysis capacities.

In this perspective, space-time could be interpreted as the awareness, by an observer, of his ability to intervene locally on an environment perceived as coherent and stable within a given time interval. This awareness necessarily implies that any act of measurement induces a partial and reductive description of the observed phenomenon. This process is referred to as **quantum decoherence.** It is not a defect of the theory, but the explicit recognition of the cognitive and instrumental limits that constrain our apprehension of phenomena likely to be at the origin of the Universe, to govern its evolution and to determine its future. Furthermore, the tools and experiments necessary for the exploration of these fields are becoming increasingly complex and difficult to design and implement.

From this point of view, the hypothesis of a virtual energy at the fundamental level, although apparently irrational, deserves to be considered. The use of abstract constructions is already widely accepted in other fields

146

of thought, such as metaphysics, existentialism or belief systems. Applied to physics, such a concept could lead to a joint re-evaluation of the standard particle model and the resulting cosmological model.

Any observation is based on the implicit reference to a given time interval and region of space. Now, in quantum mechanics, particles seem to free themselves from distances in the classical sense. Asserting that a particle can simultaneously occupy several positions, or that a wave can propagate without apparent speed limitation — as the tunnel effect or the Hartman effect suggests — amounts to acknowledging that these entities do not comply with spatial constraints-ordinary temporal. This situation comes into tension with general relativity, in which the notion of simultaneity does not make sense for spatially separated events. When distances lose their operational significance and no direct observation can be firmly established, elementary particles acquire a status close to that of virtual entities.

To say that a particle is potentially present at any point means that it can change state without this change being able to be explained locally. This property corresponds to the principle of **superposition of possible quantum states**, which allows transitions between symmetrical configurations in a context of chirality not directly observable, without the need to introduce hypothetical superluminal particles such as tachyons. Although experimental physics, particularly within CERN, allows the creation of exotic particles and atoms, these entities generally have an extremely short lifespan. Their instability reflects the absence of a lasting structural role in the observable Universe.

Finally, the inability to directly identify discrete interactions linking certain quantum symmetries leads to a probabilistic description of the observed phenomena. Without access to all the underlying parameters, physical quantities can only be determined statistically. Uncertainty does not then appear as a necessary ontological property of nature, but as the consequence of incomplete information on the fundamental mechanisms governing quantum interactions.

Universe cooled down

147

In the hypothesis of a Universe reaching an extremely evolved thermodynamic state, dominated by MMBH, the notion of speed of light would lose its operational relevance. The electromagnetic waves would have been completely absorbed by a deeply transformed material, concentrated at the heart of these structures, which would no longer present measurable thermal energy. The degrees of freedom associated with atomic oscillations — vibrations, electronic transitions or collective excitations—would have disappeared, the former atomic constituents being supposed to have lost all individuality by merging into an undifferentiated global state. The Universe would then reach a critical state extending that which preceded electromagnetic decoupling.

In such a cooled Universe, black holes would be electrically neutral and would no longer fit into a space with a structuring background energy. The space itself, emptied of its effective energy contents, could be described as a region in global energy depression, rendering inoperative classical physical laws based on space-time dynamics. This situation leads us to question the universal and immutable nature of physical laws: are they valid in any cosmic context, or only in a restricted domain of scales and conditions? Do their complexity exceed our reflective and observational abilities?

It indeed appears that certain fundamental interactions still elude us, notably those likely to involve non-local variables or mechanisms related to hidden variables. In this perspective, the standard cosmological model proves to be incomplete and insufficiently coherent, calling for a deep extension or revision. Several major limitations can be identified:

- it neither describes nor explains the Big Bang as a fundamental physical event.
- it relies on the concept of space-time derived from relativity without questioning the very origin of space and time.
- it does not account for the initial appearance of the material.
- it is unable to fully integrate gravitation into the other fundamental interactions.
- it predicts the existence of antimatter without explaining its almost total absence in the observable Universe.
- it posits the existence of a dominant dark matter without being able to specify its nature.

- it attributes the accelerated expansion of the Universe to a hypothetical dark energy, introduced by the ad hoc addition of a cosmological constant (Λ), without direct detection.
- it remains imprecise regarding the possible cosmic end scenarios.

The final state envisaged here, corresponding to the set of MMBH, could be described as a **collective singularity**, comparable to a non-localizable gathering of energy devoid of measurable intensity, but characterized by an extreme density. Under these conditions, particles — if at all they have retained their particularism — could be considered as at rest, the wave-particle duality ceasing to be relevant. The Universe and a possible "anti-Union", confined in parallel configurations of opposite symmetries, would eventually annihilate each other, restoring an undifferentiated energy to a Cosmos envisaged as multiverse.

The fact that these interactions between symmetries are imperfectly described does not necessarily reflect an incoherence of nature, but rather the limits of our observation and investigation capacities when the change of scale modifies the very nature of the phenomena studied. Discrete interactions, today inaccessible, could play a determining role in the unstable balances that mark cosmic evolution. Their consideration would allow establishing a subtle link between the infinitely small and the infinitely large, two notions that only make sense in reference to the idea of a multiverse Cosmos. In this perspective, elementary particles of matter, as we classically describe them, would only constitute an effective and intelligible manifestation of more fundamental interactions between quantum symmetries. They would be the observable expression of an underlying structure of discrete interactions, inaccessible directly to measurement.

Thus, which is sometimes interpreted as a violation of the universality of flavor — notably for certain fermions and more specifically leptons, who share similar interaction properties while differing in their mass — could be the observable extension of deep interactions not detectable at our scale.

Planck's wall marks the limit of the pre-quantum era of a universe still un-ionized and practically unrepresentable. What precedes this threshold cannot be described either by quantum physics or by general relativity, just like the discrete interactions likely to characterize the boundary between two quantum symmetries. Before the appearance of the first particles, as in the very heart of them, the usual categories of reality cease to be operative: phenomena become quasi-virtual in a regime dominated by unrecognized interactions.

Before the rupture of symmetries leading to the Big Bang, as after the complete annihilation of matter during a final collapse, no event would be recorded by the so-called multiverse Cosmos. Time then appears as a relational construction, inseparable from space and calibrated by the human observer from the fluctuations of matter perceived as conflicting.

A theory today abandoned postulated that after an expansion phase, the Universe would enter into contraction to result in a Big Crunch. This hypothesis finds a certain consistency if one considers that the material, after a dispersion phase, could be recomposed in the form of MMBH. These latter, in their apparent flight towards infinity, could be interpreted as the beginnings of a new cosmic episode, analogous to a Big Bang, followed by a redistribution of matter through the reconstitution of black holes.

In this perspective, the idea put forward by Stephen Hawking, according to which a black hole could constitute a passage towards another Universe, is not necessarily pure speculation. If the confused symmetries during the final collapse generate a new cosmic state, it is however not possible to affirm that these events are causally related. It also remains uncertain whether such a mode of transition can be envisaged as anything other than a notional limit, devoid of operational significance for any observer or "traveler". It is not certain, however, that this mode of transport would be appreciated by the traveler who would use it without certainty as to its destination.

Thermodynamics provides a framework for estimating the orders of magnitude of the energy states of atoms, considering that these states result from underlying quantum phenomena. **Absolute zero** corresponds theoretically to a state in which an interacting particle system reaches a minimum mean entropy. It cannot therefore be assimilated to a total absence of temperature, but rather to a limit situation in which thermal agitation is reduced to the minimum compatible with quantum principles. In a gas close to this state, the particles are strongly slowed down, the whole tends towards a stable configuration, and the kinetic energy available for the interaction and chemical bonding processes is extremely low.

However, a gas whose average entropy corresponds to absolute zero does not exclude the existence of local fluctuations, or more precisely, differentiated levels of entropy within the system. In a universe in permanent evolution, it is reasonable to suppose that this so-called absolute zero does not possess a strictly invariant character and could, in certain extreme quantum configurations, be redefined. Absolute zero thus remains a theoretical minimum temperature, associated with a limit state where material interactions are almost non-existent. This «void», although devoid of ordinary matter in interaction, is however not free of energy. The fluctuations of quantum vacuum indeed allow for the transient creation of particle-antiparticle pairs from radiant energy. This vacuum dynamic is probably dependent on the global evolution of the Universe.

If it is assumed that the potential energy of vacuum decreases over cosmic time, then the effective temperature associated with this vacuum should evolve in a correlated manner. The characteristic temperature of the cosmic background, today in the order of a few degrees Kelvin, has not stopped decreasing since the initial phases of the Universe. Observations and experiments related to Bose-Einstein condensates confirm that at temperatures extremely close to absolute zero, the atoms constituting matter collectively tend towards their ground state of lowest energy, even towards a regime where thermal energy becomes negligible. In this speculative perspective, a Universe reaching rigorously 0 kelvin would correspond to a global collapse, marked by a generalized collapse of gravitational structures, including black holes.

151

In a Universe in the terminal phase of collapse, any region with significantly high temperature would have disappeared. The vacuum temperature, previously a local indicator of the entropy level, would no longer be a measurable quantity. The classical laws of thermodynamics would then cease to be operative. It also becomes legitimate to question the relevance of the very notion of temperature for objects such as black holes, particularly beyond their accretion disc. An elementary particle, considered as an entity without its own spatial extension, cannot be described either as hot or cold, nor can it be said to be luminous or dark. Nevertheless, quantities such as temperature or light intensity remain, on our scale, accessible indicators of the degree of entropy, interaction and instability of a physical system.

The notion of **critical threshold**, associated with that of thermo-activity, constitutes a relevant tool for describing the concentration camp evolution of the Universe and considering the conditions of its final outcome.

- The constituent particles of matter can be considered as resulting from a primordial process of entanglement of very high frequency waves. The level of energy required for this determining radiative entanglement would have been reached within an initial plasma, isotropic but unstable, announcing a symmetry break. In a non-measurable time interval, the temperature of this ionized plasma would have reached a maximum before starting a gradual decay, allowing the formation of the first gaseous molecular structures.
- A gas cloud, by densifying under the effect of gravitation, can cross a critical threshold called Jean's mass, a necessary condition for collapse leading to the formation of a star. This process involves a rise in the cloud's internal temperature. Since the lifetime of a star is inversely correlated with its mass, there exists an upper limit to the stellar mass, estimated at about 150 times that of the Sun, beyond which internal pressures make the structure unstable.
- After the exhaustion of its nuclear fuel, a star can evolve into a white dwarf state, persisting as long as its mass remains below a critical threshold of about 1.4 times the solar mass, known as the Chandrasekhar limit. At this stage, the residual temperature of the star remains high, but not in keeping with that of previous phases.
- If the mass of a white dwarf increases by accretion, the object can transform into a neutron star, an extremely dense residue resulting from the gravitational collapse of the core of certain massive stars. Under these conditions, the pressure is such that electrons and

protons combine to form neutrons. The strong interactions that ensure the cohesion of neutrons persist, while global electrical neutrality prevents any electromagnetic interaction of repulsion or attraction between them. The density then reaches an extreme regime, close to collapse, in which the space between particles is strongly contracted and relativistic effects become dominant. When the mass exceeds about 3.2 solar masses, collapse becomes irreversible and leads to the formation of a black hole. Such objects can also result from the fusion of white dwarfs or neutron stars.

- The ultimate critical threshold of density, considered fatal to the Universe, would be reached when almost all primordial energy is concentrated in the form of TNMM, in a Universe characterized by an extreme depression of space and a maximum slowing of time. In this regime, the notion of temperature loses its usual meaning. This «cosmological» zero should not be confused with the thermodynamic absolute zero (-273°), which actually corresponds to a non-zero temperature associated with a minimal residual agitation.

- The final collapse would then gather together, in a virtual singularity devoid of spatiotemporal coordinates, all the energy of an eventually binary cosmic system of universes. Such a configuration cannot be described in classical thermodynamic terms, the very notion of temperature becoming inapplicable to it.

There is another way of associating the notion of **critical threshold** with that of **thermo-activity** in the genesis of the Universe, if one posits that the multiverse Cosmos constitutes an abstract entity, devoid of direct thermodynamic description. The **absolute cosmological zero** would then correspond to a state excluding any electromagnetic manifestation. Such an absence of electromagnetic phenomena then raises the question of the physical meaning of conductivity in a regime where neither free charges nor electric fields are definable.

By coherence, a primordial Universe, anterior to the wall of Planck and devoid of identifiable electromagnetic effects, should present these same characteristics: a non-significant temperature and a conductivity without object. However, a minute elevation above this cosmological threshold would have been enough to introduce a break in this ideal state. The

153

appearance of imperfect conductivity would then have manifested itself in an opaque primordial plasma, through the emergence of electrical phenomena revealing the first radiative entanglements. This transition would have led to a gradual decrease in the opacity of the Universe, allowing the release of photons in a new context, called recombination, covering a wide spectrum of wavelengths.

The correlative increase in resistivity and temperature would have favored the differentiation, within each quantum symmetry, of primitive forms of energy associated with opposite charges. In an initially quasi-homogeneous plasma, locally more energetic regions—hot spots — would have multiplied. This thermal agitation would have modified the dynamics of the ionized plasma. Energy transfers, particularly in the form of electron fluxes, would have generated magnetic fields whose intensity, orientation and polarity testify to the propagation of electrical phenomena. These fields become all the more marked as the conductivity of the medium deteriorates. The electromagnetic waves that remain today in a space largely occupied by matter can thus be interpreted as the remains of a primordial energy that has remained without mass, whose intensity has gradually diminished due to its continuous interactions with matter.

The subsequent evolution of the Universe then seems dominated by mass density and the intensity of gravitational interactions.

Classic stars

The gravitational collapse of gas clouds resulting from nucleosynthesis leads to the activation of nuclear fusion reactions and, secondarily, fission. This process generates stars covering a wide spectrum of masses, from solar-type stars to super giants, some of which will end their evolution with supernova-type explosions.

Neutron stars

At the end of these explosive events, there frequently remains a compact residue composed essentially of neutrons, formed by electron capture by protons. The density achieved prohibits any continuation of nuclear reactions, temporarily stabilizing the object in the form of a neutron star.

Black holes

When the density increases beyond a critical threshold, by accretion or fusion, neutrons can lose their effective individuality, their fundamental

154

constituents ceasing to manifest their usual properties. The object then changes its nature and can evolve into a stellar black hole. Such objects can also result from binary systems of compact stars or form the nucleus of some galaxies at an advanced stage of their formation.

Return to the fundamental state

This evolution leads to consider that what was initially constituted by charged particles will regain, within the black hole, a status close to that of coherent wave packets. The black hole could then be interpreted as a global quantum entity, analogous to an elementary particle, in which any notion of empty space becomes inoperative. This analogy would give the black hole a status on the fringes of classical space-time. In a Universe dominated by such objects, the notions of chirality, matter-antimatter separation and temporal causality would lose their relevance. The once distinct symmetries could coalesce, returning the whole energy once interacting to a latent state of energy, associated with the multiverse Cosmos.

Thermodynamic synthesis

The minimum conceivable temperature thus remains an essentially theoretical quantity. It would correspond to the state of a system in which matter, and more fundamentally the particles described by quantum mechanics, would be frozen, without interaction and without temporality, in an immutable space allowing no significant observation. Nothing imposes that the threshold of 273.15 °C, extrapolated from the minimum entropy of ideal gases, indeed represents the lowest possible temperature. Heat is only a macroscopic indicator of energy transfers between systems. In theory, zero entropy would imply the total absence of these transfers. In such a regime, molecules, atoms and particles could return, through a process akin to 'quantum evaporation', to a primordial state without differentiated properties, without interaction and without displacement.

155

In a Universe where the energy of vacuum would be inactive, where photons would be immobile and where matter would manifest no interaction, space and time would cease to have an operational significance. Outside any spatio-temporal context, the very notion of temperature becomes indefinable, all the more so since no observer could refer to it.

Symmetrically, the maximum conceivable temperature is also purely theoretical. Planck's temperature would correspond to the thermal state of the Universe at the time of the joint emergence of space and time, immediately after the Big Bang event. It can be described as that of a system subjected to such extreme quantum agitation that it would emit gamma radiation at frequencies likely to overlap. Such a state would escape any description based on wave packet reduction. The energy involved would likely lead to a collapse of the system upon itself, producing a singularity located on the edge of spacetime. This singularity could be assimilated to a black hole, considered here as a transition step towards an ultimate collapse.

In this approach, the notion of temperature loses all relevance when applied to black holes, highlighting the intrinsic limits of our reflection capacities in the face of the unobservable. These objects, improperly described as «black», conceal their dynamics behind a boundary without return for the accreted energy. From then on, any projection on the ultimate future of the Universe is necessarily a theoretical choice. The scenario proposed here is part of this assumed speculative approach, without claiming to avoid it.

X <u>E = m c2 in light</u>
(An equation that highlights but does not illuminate everything)

We consider the relationship $E=mc^2$, where E denotes energy, m la mass at rest — understood as the corrected mass of energy contributions related to motion changes — and c the propagation speed of photons in the gravitational "medium" that defines the reference space. This relationship implies that a very large amount of energy is required to produce a relatively small mass: as an order of magnitude, 1 gram of matter would correspond to about 25 million kilowatt-hours. In this perspective, black holes can be interpreted as energy concentrations of extreme magnitude.

Under certain conditions, matter releases a fraction of this energy in nuclear reactions, either by fission—splitting an atomic nucleus — or by fusion—assembling light nuclei into heavier ones. These processes can be accompanied by signals compatible with the transient appearance of antiparticles. **When particles and their corresponding antiparticles are brought together in this context, their annihilation leads to the integral conversion of their mass into high-energy radiation, mainly in the form of gamma or X photons. In this annihilation, the state symmetry disappears and the mass of both partners is entirely transformed into energy. This mechanism constitutes the inverse of the material creation process by radiative conversion and quantum entanglement.** It essentially produces a flow of photons corresponding to electromagnetic radiation, accompanied, in accordance with the principles of conservation of energy and charge, by secondary particles of lower energy.

At the cosmological scale, however, it is postulated that the reverse process — characterized by an irreversible concentration of matter — globally dominates evolutionary trends. This regrouping dynamic would have been particularly active during the initial phases of the Universe.

In particle physics, fermions interact through integer spin particles, such as photons and, more generally, gauge bosons described by the standard model. These bosons can be interpreted as mediators of interactions between quantum states, including between symmetries that are not directly accessible to observation. In a symmetrical extension of the formalism, the existence of antiparticles implies that of antinucleons and, by extension, more complex sets. It is then conceivable that antimolecules are not strictly

157

superimposable to their material counterparts, due to a chirality likely to affect their atomic organization.

The distinction between particles and antiparticles is not only based on the opposite electric charge, but also on the imprint of a chirality associated with a so-called imaginary temporal dimension. This imaginary time, devoid of temporal direction in the classical sense, does not fit into the relative time we experience. Like imaginary numbers, it escapes direct representation on an ordered scale. This unrecognized temporal dimension can be interpreted as a temporal disparity between two otherwise symmetrical quantum states. Such a spatiotemporal chirality would make the system formed by these quantum symmetries both discrete and metastable and guide its evolution.

The spin of quarks and leptons appears as a property difficult to interpret from our macroscopic intuitions. The analogy of an intrinsic rotation — often associated with magnetic moment or electrical charge — constitutes a heuristic image but does not correspond to real mechanical motion. The internal dynamics associated with the entangled waves that constitute the particle, and which determine its spin and helicity, cannot be described within the framework of classical relativistic space-time. Spin can then be interpreted as the effective manifestation of internal confined wave movements in the form of stationary packets, giving the particle an intrinsic angular momentum.

The photon, a spin 1 particle lacking mass and charge, has polarizing properties that allow it to ensure energy transfer between charged particles. It thus plays the role of state vector in electromagnetic interactions, both between particles of matter and between particles and antiparticles. Photons allow these energy exchanges in systems where fermions, due to the exclusion principle, cannot simultaneously occupy the same quantum state. Most fermions having a half-integer spin, they must, for equal energy, be distinguished by opposite spin orientations, a necessary condition for the stability of matter.

Spin is not a classical geometry. We can formulate the hypothesis that a property of spin still incompletely understood contributes, through its dynamic effects, to the appearance of electric charge for the majority of particles. Without this charge, the constitution of complex material structures would be impossible, as illustrated by the case of neutrinos. This fundamental property of spin is also sufficient to justify the existence of antiparticles of identical spin, in accordance with the principle of quantum symmetry. In composite particles, the combinations of spins make the overall magnetic properties less directly observable than those of the elementary constituents.

Although not mechanical, spin defines intrinsic angular moments and, consequently, the magnetic fields associated with particles. It plays a central role in understanding chemical bonds, both between atoms and between molecules, which do not possess spin in the strict sense attributed to elementary particles. In this perspective, spin and electric charge occupy a place in quantum mechanics analogous to that of body rotation and gravitation in the macroscopic world.

In quantum mechanics, the analysis of phenomena is based on concepts of effective mass and probability of presence, rather than on defined trajectories and precise locations. This probability density of presence collapses under the gaze of the observer in a retained context of space/time from which he cannot escape. This idea, which may seem fallacious, is similar to that of quantum entanglement in the case of particles that, although delocalized, form an intrinsically linked system, independently of any spatial consideration. In quantum physics, the context of causality must then be replaced by that of duplicity or quantum entanglement, without one being able to mention a shared space/time referential. A probabilistic approach makes it difficult to establish a continuous transition between the quantum description and our macroscopic reality. The change of scale requires abandoning schemes based on movement tracking and strict spatial location.

To this difficulty is added the practical impossibility of directly distinguishing antimatter, insofar as we ourselves belong to one of the branches of a global symmetry. If this symmetry was not discreetly distributed according to differentiated times and spaces that can be described as parallels, the observable Universe would probably not allow the emergence of observers capable of debating it. The only accessible clues of this antimatter then lie, on the one hand, in certain still unexplained gravitational effects (see chapter XIV on dark matter) — often associated with dark matter — and, on the other hand, in punctual manifestations observed during specific nuclear reactions.

General relativity means that the description of physical phenomena requires any measure of duration to be related to distance measurements. Time and space are inseparable and jointly defined by the structure of space-time. However, this conjugation does not necessarily imply a strictly parallel evolution of these two quantities if one adopts the hypothesis that, on a cosmological scale, time tends to expand while space becomes energetically scarce, notably by a gradual decrease in the vacuum energy. In this perspective, relativity itself becomes evolutionary, dependent on the global state of the Universe. We can also postulate that two fundamental symmetries interact in a shared temporal context that does not correspond to our observable time, and in spatial dimensions whose imbrication remains inaccessible to direct observation.

During the first billions of years of cosmic history—the year here being considered as a relative and not an absolute unit of time — the first black holes, associated with extremely dense accretion disks, would have shown a particularly high efficiency in the capture of energy and matter. The quasars, mainly observable at long distances and therefore at remote times, testify to the existence of young and very active galaxies in the primordial Universe. These systems gathered, around a central black hole in high activity, significant quantities of material still largely diffuse, composed mainly of gas and stellar bodies in the process of formation.

The clouds of neutral hydrogen were then heated and ionized before collapsing under the effect of gravitation at the approach of the central black hole, which gave these galaxies an exceptional brightness. The quasars observed today would thus correspond to an intense phase of accretion of a matter occupying space more uniformly, then much less structured than at the present time. These clouds, consisting mainly of hydrogen and secondarily of helium, also fueled the formation of the first stars. Heavy elements from stellar nucleosynthesis were still scarce, except for those produced during extreme energy phenomena such as supernovae, neutron star collisions or absorption by black holes. By crossing a space still rich in free particles, young stars and various objects, these quasars could quickly accumulate considerable amounts of matter, which would explain their significant size and the relative poverty of their immediate environment in stars.

Since that time, the average temperature of the Universe has been decreasing continuously, although this cooling tends to slow down over time. The

160

quasars, today observed as distant remnants, provide an image altered by the remoteness of galaxies forming in the young Universe. After having concentrated a large part of their baryonic matter in the form of stars and gases, these galaxies gradually lost activity, evolving towards elliptical galaxies characterized by a strong gravitational influence but poor in gas. A significant part of their initial matter would now be concentrated at the heart of supermassive black holes. Elliptical galaxies thus appear relatively devoid of hydrogen-rich regions, unlike spiral or irregular galaxies, which have not yet converted all their gas into stars. The galaxies observed in the distant Universe correspond more to spiral galaxies in previous states of evolution. However, the very notion of past remains difficult to specify, insofar as the observable fraction of the Universe could represent only a limited part of its totality. The exact age of the Universe remains uncertain, and any estimate of its future duration is still largely speculative.

If we consider the evolution of the Universe as a whole, several hypotheses can be formulated:

- The constant c, the propagation speed of light interactions, would not be an absolute constant but an evolutionary quantity, defined in a complex and changing gravitational frame of reference, characterized by an increasingly depressionary space. It would thus appear correlated with the age of the Universe.
- The mass would move within each quantum symmetry, depending on the dynamic state of the considered systems.
- Total energy is quantitatively conserved, although its forms of expression are transformed. The latent primordial energy would be converted into mass energy, each atom being able to be considered as a dynamic micro-system animated by vibrations, oscillations, transitions and exchanges described through the intermediary of elementary particles. However, the conservation of energy does not seem to imply a global conservation of movements, if one considers the capture of angular momentum by black holes and the hypothesis of an absence of internal dynamics observable within them. At all scales, the Universe remains fundamentally non-static. Space-time can then be understood as the framework in which any variation of motion tends to converge kinetic energy and mass energy into an undifferentiated state. Energy would remain globally conserved,

161

whether it manifests itself in thermal, mechanical, radiative form or, at the extreme, in a cold form, of negligible mass and possibly plasmatic, confined to the heart of black holes.

A composite particle presents emergent properties distinct from those of the elementary particles that constitute it. Similarly, a molecule has its own characteristics that cannot be directly deduced from those of the atoms that compose it. On a larger scale, stellar objects exert gravitational effects that cannot be directly related, in a first approximation, to elementary interactions between particles. This discontinuity illustrates the difficulty of relating quantum mechanics to macroscopic phenomena. It is reinforced by the wave-corpuscle dualism, which imposes depending on the situations a description sometimes corpuscular, sometimes undulatory.

A massive body can be considered as a tangle of entangled wave packets evolving in space and time. Conversely, an isolated elementary particle is reduced to an entity made up of indissociable waves, non-localizable and escaping relative time as we conceive it. The particle then appears as potentially present at any point. For an observer devoid of this capacity for ubiquity, it can only be defined where a measure allows it to be singularized, that is to say in a particular state predicted from the set of possible states.

Bringing matter, whatever the scale considered, back to wave functions would make it possible to overcome an approach that is too compartmentalized. Such an approach would involve a deep questioning of the foundations of the standard cosmological model. Theoretically conceivable, it nevertheless poses major conceptual difficulties. By partially renouncing a strictly corpuscular vision of energy, it would become possible to more finely apprehend the superposition of states in quantum mechanics. This superposition, which is not limited to two discrete states, implies that particles and antiparticles would theoretically share the same set of potential quantum states. They could however only imperfectly put these states in common, due to a chirality associated with shifted temporalities.

Quantum superposition, closely related to the notion of wave function, implies that a particle can be associated simultaneously with several positions and several states. This idea clashes with the intuition resulting from our macroscopic experiment, based on the observation of trajectories. It becomes more coherent if we consider the particle not as a localized corpuscle, but as a wave packet evolving in a complete mathematical space, such as the Hilbert space. A relativity freed from direct reference to mass or corpuscle could integrate the idea of simultaneous distant present.

A particle could then be described as a node of vibrating waves, without effectively occupying a measurable space. The notions of locality and separability would essentially concern the analysis of the observer and measurement devices, in accordance with the intuition developed by Niels Bohr. Spatial location and inscription in duration are essential to our representation of the observable world but become unsuitable when it comes to describing the infinitely small or the infinitely large. In quantum mechanics, it would be necessary to disregard, at least partially, the classical frameworks that are space and time, although these notions, non-absolute, ground our relativistic logic applied to the macroscopic world. Relativity only fully applies from scales where physical quantities are measurable; it follows that at the quantum scale phenomena are described in a probabilistic way, while at the macroscopic scale they become relativistic and predictable.

The fundamental question that arises remains this one: where does time begin and where does it end?

- We can assume that it originates with the Universe, in the extension of the Big Bang.
- It can be considered that it ends with ultimate states of matter concentration, such as those associated with supermassive black holes, heralding a final collapse.
- We can also consider that, for an elementary particle, time is quasi-stationary.
- It remains, however, that without reference to time, it would be impossible to describe the very interactions that ground the existence of these particles.

If we admit that time has no meaning of its own for elementary particles, several fundamental questions arise in quantum mechanics: how to distinguish causes and effects, how to define a sense and a speed of movement, and how to assign a unique location to a quantum entity? Since

163

the founding works of quantum physics, notably those of Albert Einstein, Niels Bohr and their contemporaries, these questions have deeply divided the scientific community. It then appears necessary to extract oneself from the observable environment in which we are immersed corporeally and cognitively, namely that of space and time. Now this mathematical representation in four dimensions constitutes precisely the natural context of structured matter at the macroscopic scale, which explains the intrinsic difficulty of the exercise.

This constraint leads to describe quantum phenomena essentially in terms of probabilities and presence distributions, which amounts to formulating predictions based on hypotheses rather than on deterministic trajectories. In this perspective, one can consider that time is a quantity dependent on the scale of observation, emerging significantly only from the atomic level. It then acquires physical relevance with electromagnetic interactions and, more generally, fundamental interactions that modify the structure of organized matter.

The major difficulty consists in reconciling a classical relativistic physics, formulated in terms of space and time, with a quantum mechanics whose constituents, strongly dematerialized, seem freed from any spatio-temporal reference. At this scale, the very order of events becomes ambiguous, which led to Max Born's statement of the principle of quantum indeterminism. Classical physics, although counterintuitive when extended into the quantum domain, nevertheless remains related to the latter. Faced with persistent paradoxes and hypotheses long considered unacceptable, scientific reflection has gradually moved towards increasing levels of abstraction. In this context, the notion of virtuality, although confusing, opens the way to new interpretations of fundamental phenomena.

Wave-corpuscle duality illustrates the difficulty of thinking about the physics of the infinitely small, whether it is mass particles described as packets of waves or electromagnetic waves considered as fields. It then becomes legitimate to ask whether notions such as mass, density or even charge —closely related to a corpuscular representation — do not above all constitute operative constructions. In this perspective, mass and charge could be interpreted as mathematical tools allowing to quantify and locate a certain amount of energy, according to the relation resulting from special relativity, $m=E/c^2$.

The photon, in this approach, should not be conceived as a point object of infinitesimal dimension, but as a quantification of purely kinetic energy, mainly observable through the energy transfers it operates with the electron cloud of atoms. The attribution of a corpuscular character to the photon nevertheless remains a useful artifice for relating certain experimental phenomena to measurements expressed in frequencies and wave amplitudes. These frequencies would result from interactions accumulated over time between the elementary particles and the residual component of primordial radiation that has escaped the radiative entanglement phases. Wave frequencies, dependent on the properties of the propagation medium, would thus evolve continuously through energy and information exchanges with matter. Although particles, considered as packets of waves, may exhibit diffraction phenomena analogous to those of light, their lack of a defined spatial dimension prevents them from propagating in an open field in a manner strictly equivalent to electromagnetic waves..

The notion of mass then intervenes as a means to integrate these wave packets in a perceptive way, by assigning them characteristics compatible with our sensory experience of the world—movement, compactness, temperature or spatial extension. Our cognitive abilities are indeed shaped to primarily interpret the macroscopic environment useful for our survival, which limits our access to the ultimate foundations of physical reality, despite the advancement of knowledge and technologies.

The electron illustrates this difficulty: it does not constitute a point located in space, although we often describe it as such. The mass attributed to it essentially serves to quantify, through a mathematical formalism, the potential energy associated with its state and the kinetic energy related to its interactions. If mass is understood as the effective signature of a wave packet, then it would be more appropriate to describe the electron cloud as a flux or an energy field structured by the presence of a central field associated with the atomic nucleus. These two fields, correlated and interdependent, prescribe the existence of opposite charges. The notion of charge, intimately linked to that of energy-mass, could thus result from the projection, in our physical environment, of a more fundamental quantum symmetry still unobserved.

So-called massive particles cannot reach the speed of light. This limitation does not concern so much their internal wave components as the collective organization of the entangled waves that constitute them, which interact in a confined manner, analogous to a system without possibility of escape. The internal dynamics of these waves do not fall under classical relativity and are not defined in relation to the limiting speed of light propagation.

Describing the electron as an isolable particle logically leads to adopting a similar approach for electromagnetic waves. The photon, which has become a corpuscle by convenience, then represents the minimum unit of energy likely to be exchanged with an electron. In this perspective, the electron cloud of an atom can be considered as a collective bundle of wave packets orbiting around a more stable aggregate of complex wave packets forming the nucleus. The electron field thus confines the atomic nucleus in a manner analogous to a confinement device. To account for the observed equilibrium of this structure, the notion of opposite charges and polarities is an effective operational explanation, validated by numerous experimental applications.

A parallel can be drawn with gravitation, which attracts bodies and can keep them in orbit. At the macroscopic scale, however, gravitation seems independent of load effects. This does not exclude that electromagnetism and gravitation can share a common foundation. An astral body can indeed be considered as a complex open system of electromagnetic interactions. Electromagnetic waves follow the curvature of space-time and contribute to modifying it during their interactions with matter. At our level, we interpret these effects as an expansion of the so-called empty space, while observing that the more a body is massive, the more it distorts space-time. This interpretation results from the fact that we assimilate the time associated with deep space observed to our local time. Gravitational effects could thus emerge from a set of electromagnetic interactions between charged particles. Even when bodies are globally neutral, the fields they produce are not strictly so. The electromagnetic fields associated with mass and motion energy would then constitute the macroscopic manifestation of these interactions, contributing to the observed gravitational effects. In this hypothesis, gravitation would have a fundamentally quantum origin (see chapter XVIII).

The central question remains that of the stability of the atomic structure: how can a system of packets of electronic waves in a closed oscillatory regime confine another wave system sharing similar properties within the nucleus? The introduction of electrical charges associated with these wave packets provides a coherent explanatory framework, in terms of balance of forces, and allows modeling the emergence of space and time at scales

166

higher than atomic. The notions of mass and particle thus serve to mathematically formalize what could be, at a more fundamental level, only a set of entangled waves interacting with the electromagnetic fields that structure space.

If any form of energy corresponds to a state selected from an infinity of possible states, it becomes conceivable that there exists an infinity of potential universe states. The cosmos could then be considered as a virtual mosaic of universe systems in quantum symmetry. Each act of observation implicitly imposes conditions that select certain quantum states compatible with our observation conditions and with the representation we have of our Universe. Ultimately, we only access a partial reality, most often indirectly, shaped by our condition as a living organism seeking to understand a truth whose access remains fundamentally limited.

XI <u>A mysterious absence: The Antimatter</u>
(Really missing or simply hidden from our eyes?)

The heart of the stars is the seat of intense physical activity, indirectly accessible through the observation of the heat produced and the electromagnetic radiation emitted. Temperature and luminosity thus constitute two observable manifestations of the same fundamental process, generally interpreted as resulting from nuclear reactions. These take mainly the form of fusion of light nuclei—essentially hydrogen and helium inherited from the early phases of cosmic evolution — but may also include, depending on the contexts, mechanisms involving heavier nuclei.

These nuclear reactions can be considered, in a broader quantum perspective, as the consequence of close, discrete and highly structured interactions between symmetrical quantum states. However, their effects are only accessible to us through their projection in our own symmetry, that of the particles of matter, as well as by the punctual and transient appearance of experimentally detectable antiparticles.

In this context, several elements can lead to consider the hypothesis of a system of universes organized in quantum symmetry.

- The observed asymmetry between matter and antimatter in the visible Universe generally leads to reject the idea of a global symmetry between these two forms. However, this rejection raises more fundamental questions than it solves, particularly regarding the origin and persistence of this dissymmetry.
- The hypothetical existence of a quantum symmetry capable of neutralizing the usual notions of time and space would allow for conceiving the emergence and disappearance of universes within a more general context, virtual, where the concepts of finitude and temporality would lose their relevance. Such an approach excludes the idea of a nothingness conceived as absolute absence.
- A binary system of universes in quantum symmetry would allow the introduction of a shared universal temporality, or an imaginary temporality, distinct from the relative time specific to our observable Universe.
- Apparent deficits in matter (attributed to dark matter) and energy (attributed to dark energy) could invite a re-examination of the

168

potential role of antimatter, provided that current accepted estimates and their interpretations are revised.

- Thermodynamic quantities such as temperature, pressure and energy intensity, which condition the entropy of a system, characterize all forms of energy. These parameters could constitute the guiding principle of a coherent reconstruction of cosmic history if one adopts the schema of a multiverse-type Cosmos.
- Finally, the space-time that structures our observable Universe cannot, in the current state of its evolution, be confused with that corresponding to its opposite quantum symmetry. These space-times, although inseparable, could evolve in shifted, overlapping or parallel dimensions, and remain in permanent interaction in a form of dynamic equilibrium.

In quantum mechanics, time is not a directly measurable quantity in the classical sense. It could rather be reduced to a continuous succession of possible state redistributions. The phenomena of quantum superposition and entanglement can then be interpreted as the observable expression of information exchanges without clearly defined classical causality, within systems consisting of more or less entangled wave packets and associated electromagnetic modes. Elementary particles thus share common and persistent fundamental properties, which confer on their interactions an intrinsically non-local character and hardly compatible with the emergence of a time in the macroscopic sense.

Under these conditions, any attempt at spatiotemporal localization based on the wave function can only be formulated in probabilistic terms. This approach could help explain the absence of direct observable interactions between matter and antimatter, as they do not necessarily evolve in the same space-time. The question then arises as to whether antimatter belongs to a dimension distinct from that constituting our physical reality. Such a hypothesis remains difficult to access for an observer composed of matter and inscribed in a specific spatiotemporal framework. The famous remark attributed to Einstein — according to which what man ignores does not exist for him — illustrates this cognitive limit.

What escapes direct observation can nevertheless be considered by extrapolation, provided that the hypotheses formulated remain compatible with a set of robust empirical observations. This also implies not to question, beyond the necessary, general relativity and the architecture of the standard

169

model of particles, which constitute to date indispensable foundations for any subsequent theoretical exploration. These conditions must remain points of support, even when they are extended by deliberately audacious assumptions.

In the hypothesis of a binary system of universes in quantum symmetry, the two corresponding states would not strictly superpose each other, but would partially and differentially overlap, according to a multiplicity of referentials associated with particles and antiparticles. These energy strata in quantum symmetry would interpenetrate without being entirely confused at the current state of cosmic evolution. Particles and antiparticles would then share complementary states in an undiscernible conjunctural temporality, evolving in a quasi-parallel manner, with slight divergences attributable notably to chirality effects.

Is the idea that antiparticles can belong to a dimension distinct from that of our reality more counterintuitive than other concepts now accepted, such as decoherence, non-locality, quantum correlations, the emergence of a Universe from an initial state without classical structure, a finite space without boundaries, string theories, dark matter, dark energy or even quantum gravity approaches? Although the cosmology outlined here remains speculative, it proposes an interpretative framework likely to respond to certain inconsistencies of a standard model continuously adjusted. It should be remembered that this model is based not only on increasingly precise instrumental observations, but also on mathematical constructions, thought experiments and postulates relating to scales inaccessible to direct observation.

At these scales, experimental research reaches its limits and mathematical formalisms become increasingly abstract and complex. Theoretical physics is then engaged in a process of feedback, where each response raises new questions and where certain advances challenge previous achievements. It remains legitimate to question the relevance of the terms used in our equations when they are applied to still unrecognized areas of physical reality. These constructions are based on a perception of reality shaped by our observer status, even though we know that this reality is only an interpretation conditioned by our perceptual and cognitive limitations.

Finally, conceiving a parallel dimension devoid of direct physical representation and nevertheless in symmetry with our reality poses a major challenge. Our thought is structured by a relativistic environment based on time and space as we experience them. Describing the dynamics of the infinitely small and that of a universe in its quantum symmetry is undoubtedly beyond our current capabilities. An analog approach can then serve as heuristic support. As a model of thought, we can consider that massive elementary particles, assimilated to wave packets without defined spatial extension, may change, during their evolution, from a particle state to an antiparticle state. Such a possibility, assumed to be inherent in any massive particle, would correspond to a superposition of symmetrical states that we are not able to discriminate.

These quantum states would then be potentially alternative, without implying that a particle is simultaneously its own antiparticle. An extreme energy density, characterized by the absence of interstitial space and therefore of temporality, could alone lead to the fusion of these symmetrical states. The final collapse of extreme compact objects in a cooled Universe could be compatible with such a boundary condition.

If we assume that a binary system of universes in quantum symmetry evolves towards a final state equivalent to its initial state, this process can be interpreted as a global return to an earlier configuration, without it being necessary to invoke a return to the past in the relativistic sense. Such a hypothesis leads to consider that these two states share a common form of temporality, distinct from the relativistic time specific to our observable Universe. This temporality, described here as imaginary or absolute, would distinguish neither past nor future and would therefore not be ordered according to a causal chronology of events. It could be described as a succession of permanent presents, which is consistent with the role attributed to so-called virtual forces and particles in quantum physics.

The existence of a shared universal time could shed light on the particular behaviour of certain weakly massive or practically massless particles, such as photons and neutrinos. Devoid of significant electrical charge and rest mass, these quantum objects appear to appear and disappear without exhibiting stable observable symmetry. Virtual particles, understood here as energetic entities intervening in explanatory formalisms without being directly observable, could then be interpreted as mediators of unrecognized connections between two symmetrical states of the Universe.

171

The dominant interpretation in cosmology postulates that matter and antimatter would have interacted massively during the early phases following the Big Bang, leading to an almost complete annihilation of antimatter and thus explaining its apparent rarity in the current Universe. However, such a hypothesis necessarily implies an initial asymmetry between matter and antimatter, which remains problematic. The hypothesis developed here, based on a chiral symmetry between universes, offers an alternative to this difficulty without assuming an original predominance of matter. This dissenting approach does not resolve all issues, but it takes into account that our observation capabilities remain inherently limited, despite the constant progress of instruments and methods.

It is thus conceivable that antimatter is not absent, but simply inaccessible to direct observation. The matter constituting our observable reality would then act like a screen, allowing only its own manifestations to be perceived. In this perspective, the gravitational effects attributed to dark matter could be interpreted as indirect effects of antimatter concentrations. These concentrations would intervene in the gravitational dynamics of massive structures, notably galaxies and galactic clusters, by measurably modifying the orbital trajectories of the most massive bodies, such as stellar black holes and galactic systems.

Antimatter manifests in a punctual manner during certain nuclear reactions involving ordinary matter, but it does not reveal autonomous nuclear interactions, nor directly observable strong or electromagnetic interactions. In contrast, gravitational effects not fully explained could indicate its discrete presence. As with matter, antiparticles of the same charge would repel each other, which would limit their observable clustering. When it manifests, antimatter appears in the form of symmetrical antiparticles of the corresponding particles, of the same masses and spins, but of opposite quantum numbers.

General relativity describes the effects of densified baryonic matter on spacetime geometry. It is plausible that antimatter obeys the same laws, but in a chiral symmetrical space-time of that associated with matter. Such a hypothesis would allow understanding why antimatter, although having measurable gravitational effects, remains largely unobservable. In theoretical astrophysics, the introduction of hidden dimensions is not unusual; postulating distinct dimensions associated with antimatter is therefore no more daring than some extensions of the standard model, such

172

as the additional dimensions of string theories or type formulations of Kaluza-Klein.

Photons, open to both symmetries, could play the role of energy vectors ensuring discrete exchanges between matter and antimatter. Without mass and charge, they could circulate between quantum symmetries in the form of limited and indirect interactions.

Electromagnetic waves interact with each other and with matter through diffraction, refraction, absorption, and emission. The absence of analogous phenomena observed with antimatter could be explained if it evolves in a distinct dimension, made transparent to our detection devices. A quantum entanglement of photons, considered as the corpuscular representations of electromagnetic waves, could allow interactions interpreted as non-local, in accordance with Bell's theorem, but operating in a context inaccessible to direct observation. In this case, gravitational effects would remain the only indirect indications of the presence of antimatter, which would have led to the idea of dark matter as a distinct entity.

Our inability to perceive the interactions between electromagnetic waves and an unobservable antimatter is part of a more general limitation of our access to physical reality, limited to a finite volume of the observable Universe. This restriction imposes caution in the interpretation of observational absences.

One may then wonder if the observers that we are would not, unbeknownst to us, be involved in both symmetries. Each symmetry would interact with itself, while indirectly reacting to the other through echo effects. In ordinary matter, atoms exchange electrons to form molecules through chemical bonding. Analogous processes involving antielectrons and antiatoms are conceivable, without constituting perfect symmetry in the classical geometric sense. Antimatter particles, although not directly observable, would have the same masses and spins as their material counterparts, and would exert gravitational effects on it. The Universe could thus be described by two coupled Riemannian metrics.

This porosity between quantum symmetries would allow the coexistence of complementary wave functions, for which the probability amplitude would lose its usual meaning. The distinction between waves and corpuscles would then fall under observational modes applied to common underlying

173

mathematical processes. Corpuscles are associated with traceable positions and defined trajectories, while waves are non-locatable and propagate over extended fronts. The wave function allows to overcome this duality by introducing a probabilistic description, based on the superposition of potential states.

The wave function, conceived as a probability cloud, does not imply a fundamental inconsistency, but translates a statistical description of quantum phenomena. It is then legitimate to ask whether the presence of unobservable antiparticles, but potentially in interaction with their material partners, could not explain the apparently random nature of the observed trajectories, especially in experiments such as that of Young's fissures. The antiparticle, although not detected, could indirectly influence the trajectory of the associated particle, thus contributing to the measured uncertainty and statistical formalism of quantum mechanics (Schrôdinger equation).

Quantum superposition can be interpreted as a construction of the mind, intended to account for discrete effects attributable to antiparticles that escape any direct observation, except during indirect manifestations associated with certain nuclear reactions. The difficulty increases as soon as one considers the implicit interactions between the observed system, the observer and the measuring devices. If it is assumed that the antiparticle intervenes, directly or indirectly, in determining the trajectory of a particle — whether strictly matched to it or not — then the idea of a fundamentally random or intrinsically indeterminate quantum trajectory becomes debatable. Such a hypothesis could help mitigate a persistent incompatibility between quantum mechanics and gravitation, by reintroducing an unrecognized form of correlation between seemingly independent phenomena.

A heuristic way, although necessarily imperfect, to illustrate the principle of quantum symmetry consists in resorting to an optical analogy. *Consider two slides representing the same landscape in black and white, without intermediate levels of gray. The first consists of black areas on a white background, the second of white areas on a black background, each being an exact inversion of the other. If these two slides are projected simultaneously using a single projector and superimposed on the same screen, the resulting image appears uniformly black, the landscape becoming indistinguishable. If two separate projectors are now used to project the two slides onto a single screen, the illuminated areas of one*

174

overlay the dark areas of the other, this time producing an evenly bright screen. In both configurations, the superposition of inverted images prevents any reconstruction of the initial landscape, as long as the mechanism of superposition remains hidden from the observer. This analogy suggests that perceived reality can be a partial projection, resulting from the superposition of symmetrical structures one of which remains inaccessible.

In this perspective, matter and antimatter can be described as an asymmetric superposition of constantly evolving multiscalar fields. Such a configuration would prevent their effective meeting before the end of the process of cosmic deconstruction of our Universe. Unlike the particle of matter, whose existence is indirectly revealed by interactions corresponding to an effectively measured state, the antiparticle, evolving in a dimension that would be proper to it, remains inaccessible to any form of observation, even indirect. When it nevertheless manifests—generally in the transient form of particle pairs–antiparticle — it reveals only a possible state, strictly symmetrical to that of its associated particle, and this in an extremely fleeting manner.

If we assume that antimatter was produced in equal quantities with matter during the initial phases of cosmic evolution, it would then occupy a central place in the global understanding of the Universe. For the observer made up of matter, antimatter appears absent or indifferent to ordinary matter and the background electromagnetic radiation that makes the Universe observable. This apparent absence could however result not from an effective disappearance, but from a structural inaccessibility. Antimatter would thus constitute a fundamental but veiled element of cosmic reality.

The observer, as a consubstantial entity with matter, is forced to consider it as the only physically accessible reality. This situation naturally leads to interpret the discretion of antimatter as the consequence of an almost total annihilation. A CP asymmetry is sometimes invoked to explain the observed predominance of matter over antimatter, but this explanation remains insufficient to account for such a marked deficit. Everything indicates rather that antimatter only manifests itself through quantum states that are difficult to grasp, comparable to wave packets whose evolution would not be strictly symmetrical to that of matter on the atomic scale.

Assigning the antiparticle an electrical charge opposite to that of the corresponding particle is a convenient simplification for formalizing matter-antimatter symmetry. This representation is consistent with the observation of annihilation and pair creation phenomena, particularly during certain nuclear reactions, such as beta decay, or in extreme astrophysical environments, near neutron stars, of black holes, or under the influence of intense magnetic fields. The particles of matter cannot annihilate each other, for lack of adequate symmetry of their properties or quantum numbers; it is probably the same for antiparticles between them. Similarly, a particle and an antiparticle could not annihilate unless they rigorously shared the same mass and spin characteristics, despite opposite quantum numbers.

Annihilation, by removing the involved particles from the quantum system under consideration, does not destroy the energy associated with these particles. It is conserved and re-emitted in the form of quanta of radiation, mainly photons, likely to exhibit several polarization states. This process emphasizes that the apparent disappearance of matter and antimatter from the 'quantum landscape' is merely a transformation of energy's modes of existence, not an effective loss of physical information.

The question of quantum entanglement between particles of the same nature remains delicate. An interpretative hypothesis consists in considering that this phenomenon might not be fundamental in itself but represent the indirect expression of deeper quantum correlations linking particles of opposite symmetry, in particular matter and antimatter. Such a perspective would allow to assign an extended physical status to antimatter, beyond its sole manifestation during processes of pair creation or annihilation.

Several arguments are put forward in this regard.

- Firstly, the photon, a particle without an electric charge and without mass at rest, can be described as an excitation of the electromagnetic field. During certain division or conversion processes, the photons produced exhibit correlated quantum states, which are likely to remain entangled as long as no interaction causes decoherence.

- Secondly, the electronic neutrino, almost devoid of mass and charge, could also manifest similar behaviors. In the hypothesis where it would be a Majorana particle, that is to say identical to its antiparticle, it would constitute a limit case reinforcing the idea of an entanglement related to deep internal symmetries.

- Thirdly, the electron, although endowed with a non-zero mass, sees some of its dynamic properties (energy, pulse, effective velocity) depend directly on its electromagnetic interactions, mediated by photons. The fact that electrons can be entangled suggests that these correlations could be inherited, at least in part, from the photon-electron interactions themselves.

Quantum entanglement can thus be interpreted as a non-local interaction that does not fit into ordinary temporal causality. It would constitute the projection, in a relativistic space-time where simultaneity has no absolute meaning, of correlations established in a more fundamental level of description, which can be described as strictly quantum. This level would not be directly accessible to observation, notably because of decoherence mechanisms that mask the superposition of quantum states in favor of effective states compatible with a classical description.

Thus, in the absence of any measure — that is to say independently of an observation device — a particle would not possess a physical reality determined in the classical sense, but only a set of quantum potentialities. The observed reality would then emerge as an interpretative result related to the act of measurement, which operates a transposition of quantum phenomena into a spatiotemporal framework compatible with macroscopic relativistic physics. From this point of view, the superposition of states and the quantum correlations it implies confer on quantum mechanics a singular status, largely freed from usual spatial and temporal references.

It is notable that quantum entanglement does not seem to manifest in a non-local manner for certain categories of particles, notably quarks, whose strong interactions lead to permanent confinement within hadrons. Under these conditions, the individuality of the elementary constituents is somehow lost. Similarly, for composite systems—atoms or molecules — dominant internal correlations confer on the system global properties that render inoperative the notion of non-local entanglement between its components, in the sense that it applies to isolated elementary particles.

177

The profoundly counter-intuitive nature of quantum entanglement explains the persistent difficulty in proposing a coherent representation. Yet this phenomenon constitutes a major lever for new interpretations of quantum physics. Entangled photons, for example, form a global system whose properties are not reducible to those of its constituents taken separately. Although their states remain correlated, they remain spatially separated and are never simultaneously observable in their entirety. This intrinsic limitation of the field of observation could mask interactions involving entangled photons out of experimental scope, as well as processes involving antimatter, assumed not to share the same observable spatiotemporal framework as matter.

In this perspective, certain phenomena — such as photoelectric effects, pair creation or diffusion processes — could involve extensive quantum correlations between distant photons and degrees of freedom associated with antimatter, remaining inaccessible to direct observation. The apparent asymmetry between matter and antimatter, or global chirality, could thus result from an intrinsically restricted field of observation, unable to simultaneously encompass all the symmetries involved.

Even if Bell-type correlations can be highlighted for spin states between atoms under highly controlled experimental conditions—particularly at very low temperatures — atomic, molecular systems and, a fortiori, macroscopic seem naturally excluded from non-local quantum entanglement. This could be explained by the fact that structured matter participates in the very definition of space as we perceive it, while the elementary particle, as a fundamental quantum entity, does not reveal any accessible internal structure. Its properties — mass, charge, spin, flavor or color—could then be interpreted as resulting from extrinsic relationships rather than observable internal mechanisms.

In this perspective, only photons, devoid of mass and charge and endowed with a full spin, would be capable of conserving quantum correlations durably after their production, in the absence of interactions with massive particles. If we postulate that the set of photons in the Universe share a common origin and remain linked by a global ground state, then the observed correlations would not violate the Bell inequalities but would fit into a non-local context where space loses its classical meaning. Relativity would appear there as an emergent description, limited to the phenomena of decoherence.

178

Difficult to talk about quantum entanglement for atoms. Only light particles (photons) that have the distinction of being massless, uncharged and full spin would produce, when they divide, photons correlated over time. These should remain so, especially since they will not have interacted with mass particles. If we consider that all the photons of our Universe have the same origin and remain linked since due to their ground state (they are representative of non-gravitational primordial energy), the Bell inequalities are not violated. For the light particles thus entangled, space is somehow erased and relativity is left out. In quantum mechanics, space becomes an uncertain datum that no longer has the meaning we classically give it in terms of displacement and location.

This prompts us to draw a parallel with antimatter, which is governed by the same quantum mechanics. The antimatter would also possess an 'apparent' temporality but in an unrecognized space. Inaccessible to observation, antimatter would therefore be in application of a broader principle of non-locality, discretely correlated and unfinished with its symmetry. It is also a way of defining, related to quantum entanglement, what chirality following an original symmetry break is.

To summarize, quantum symmetry (matter/antimatter) would be discretely at the base of the dynamics of our Universe. It would fall under a virtual time for us and which is not the relative, spatialized time that we know. The problem is that it takes less than that to upset our understanding!

The foregoing considerations would be likely to shed light on certain gravitational anomalies (see chapter XIV). The exclusion of antimatter from usual cosmological models led, by default, to postulate the existence of unknown components such as dark matter and dark energy. However, these hypotheses, although descriptively operational, remain devoid of direct experimental basis. They respond more to a need for model closure than to an observational validation.

In this context, the idea of a Cosmos conceived as a reservoir of latent energy, not manifested and without directly observable physical reality, possibly marked by a fundamental symmetry break, is not intended to constitute a definitive explanation. Rather, it aims to serve as a support for the development of a coherent paradigm, capable of continuously articulating microscopic and cosmological phenomena. An excessively simplified formalism risks obscuring essential mechanisms, while an excess of complexity can lead to a loss of readability and interpretative dead ends.

Regarding wave-corpuscle duality, the commonly accepted approach is to consider that the wave or corpuscular character of a quantum system depends on the experimental setup and the observer's point of view. This apparent contradiction led to the introduction of the wave packet reduction postulate during measurement. In order to mitigate the counterintuitive nature of this duality, it may be useful to resort to an analogy derived from classical physics, while explicitly recognizing its limitations.

Thus, the analogy with oceanic wave trains allows to illustrate some formal aspects of undulatory behavior. Waves carry energy without global displacement of matter, which imperfectly recalls the propagation of electromagnetic field excitations. At great distances or on a large scale, the sea surface may appear flat, while at the local scale each water molecule describes a closed oscillatory motion. Conversely, at the molecular scale, the global phenomenon of swell becomes imperceptible. This analogy emphasizes the dependence of observed phenomena on the observation scale but must be strictly limited to this illustrative function: quantum waves do not correspond to material deformations localizable in classical space and time.

More generally, quantum symmetries can be considered as forming a network of deep correlations, linking together different components of physical reality. Photons and neutrinos, as weakly or uncharged particles with zero or near-zero mass, would play a privileged role in the transmission of these correlations, especially in a context of broken symmetry. Electron neutrinos, in particular, could be interpreted as excitations resulting from interactions pertaining to the electroweak sector, exhibiting certain functional analogies with photons, without however being reduced to them.

The relative stability of material structures is based on the establishment of a load balance ensured by chemical bonds. Kinematic and dynamic quantities—spin, orbital moment, translational and rotational speeds — participate in the fine adjustment of this equilibrium, in a space-time subject to permanent fluctuations of gravitational origin. In this perspective, each particle could be associated, by symmetry, with a corresponding particle of opposite lepton or baryonic number. This symmetry relation — comparable to a left/right or high/low type inversion — would imply, for antiparticles, a

180

magnetic moment and a global angular momentum of opposite direction, a direct consequence of the charge inversion.

These properties make it possible to characterize certain interactions in quantum mechanics, independently of the categories of classical physics. By analogy, the electronic environment of an atomic nucleus can be seen as a filtering area that conditions the absorption, scattering or reflection of electromagnetic quanta. Some photons are absorbed, others are reemitted after partial energy transfer, according to well-established mechanisms. The majority of photons, devoid of mass and charge, do not have sufficient inertia to cross this electronic barrier and interact directly with the nucleus. In the rare cases where a nuclear interaction occurs, it is highly dependent on energy and incidence geometry, and its effects on the nuclear mass are generally limited.

This situation contrasts with that of black holes, which continuously absorb energy and matter. Although directly unobservable, black holes are signaled by non-quantum macroscopic effects. Their behavior nevertheless presents analogies with certain quantum entities, particularly regarding the retention of information and the absence of accessible internal spatial structure. When a particle is absorbed by a black hole, the non-local quantum correlations it maintained with other systems seem to disappear from the observable field. The black hole accumulates energy without spatial occupation in the classical sense, which suggests that the space-time region associated with it could be interpreted as an effective construction, modifying the global topology of the Universe.

In these conditions, the notions of pressure, compactness or even entropy become problematic when applied to entities for which unoccupied space seems excluded. The elementary particle, described as a wave packet, formally possesses properties analogous to those of light. Now, at the speed of light, proper time tends towards a null limit, which makes inseparable the notions of time and occupied space. The elementary particle would thus be located at the space-time boundary and could be considered as a limit object between our observable Universe and a broader cosmic framework.

In a similar way, the black hole returns any form of absorbed energy to a ground state, which raises the question of the very meaning of entropy within it. In this extreme environment, time and space lose their operational significance in terms of location and movement.

181

It follows that space-time could be interpreted as transitional, intermediate between two limit regimes: that of the elementary particle and that of the black hole. This situation makes it particularly difficult to integrate them into a unique cosmological model, since space-time is an essential condition for any observation. Space/time remains nevertheless essential to describe the interactions of structured matter and to report, from the observer's point of view, on the evolution of the Universe.

Although devoid of significant spatial and temporal dimensions, elementary particles and black holes register, for the observer, in interstellar space. This lack of dimension is not fundamentally more problematic than that of the geometric point, commonly used without major difficulty. It is therefore not these entities themselves that manifest directly, but the effects of their interactions with matter at charge equilibrium, from the atomic scale to macroscopic structures. Inside a black hole, the distinctions between particles, bosons and fundamental forces lose their relevance, even though these interactions accounted for most of the energy associated with the mass of absorbed bodies. The energy thus captured can be considered temporarily unavailable, placed in a latent state.

This reflection leads, in the more general hypothesis of a cosmic system in quantum symmetry, to consider an extension of the principle of energy conservation.

XII <u>A standard model that does not explain everything</u>
(And is still looking for new particles)

The following development is based on the table of elementary particles presented in the annex, which remains likely to be discussed as part of a possible revision of the standard model. This chapter recalls, as a synthesis, the foundations of contemporary astrophysics, considered here in continuity with quantum mechanics.

The first-generation elementary particles, the lightest and most stable ones, constitute the essential part of ordinary matter. It is about the up quarks, the down quarks and the electron. The up and down quarks assemble to form the nucleons — protons and neutrons—which constitute the atomic nucleus. This structuring is based on the strong interaction, characterized by an intense attractiveness and very short range. The electronic neutrino is distinguished by the absence of an electric charge and by an extremely low mass (see Chapter XIII). These properties make it a singular particle, capable, according to certain hypotheses, of adopting an effective behavior comparable sometimes to that of a particle, sometimes to that of an antiparticle.

Third generation elementary particles are characterized by high masses and energies. They include the top and bottom quarks, the tau lepton and the tauic neutrino. These particles are unstable and decay quickly, after their creation, into lighter particles belonging to the lower generations. Their existence thus seems mainly related to the extreme energy conditions of the primordial Universe.

Between these two extremes is the second generation, composed of intermediate mass particles: the charm and strange quarks, the muon and the muonic neutrino. Hadrons containing charm or strange quarks (mesons, kaons, pions, etc.) are unstable and decay via weak interaction, notably by producing neutrinos and antineutrinos. This hierarchization in generations translates an organization of matter according to increasing levels of energy equivalent to mass.

An atom generally consists of a nucleus grouping protons and neutrons, around which an electron cloud is distributed. Each proton and neutron is a hadron composed of three fundamental constituents: quarks. These are not

183

observable in isolation and must be considered as elementary degrees of freedom, carriers of fractional charges and quantized energy. Up quarks have an electric charge of +2/3, while down quarks carry a charge of -1/3. The other varieties of quarks, more massive (c, s, t, b), may have been present at certain stages of cosmic evolution, but their instability has made them transient on the scale of the current Universe.

The atomic mass corresponds to the sum of the masses of protons and neutrons contained in the nucleus. In order to make the description of physical phenomena intelligible, it is customary to introduce a conventional distinction between:

- the quantum regime, corresponding to elementary particles, whose character is partially virtual and not directly observable.
- the observable regime, that of constructed baryonic matter, formed of atoms and molecules.
- an intermediate regime, including composite particles (protons, neutrons, mesons) and electron clouds, which ensure the transition between quantum and macroscopic. This separation makes it possible to analytically distinguish weak and strong electromagnetic, nuclear interactions, without claiming a strict ontological boundary.

But why 3 quarks to make a nucleon? *One could say oneself – considering that it is only an image - that each of the 3 quarks represents one of the 3 dimensions necessary to define space: height, width, depth (or why not, a dimension of time: past, present, future). Thus 2 quarks would be insufficient because they would then achieve a flat surface and 4 quarks would not correspond to the idea of a 3D volume. A triplet of quarks proves to be necessary and sufficient. Chance dictates that one arbitrarily assigns a color to each of the 3 quarks, in this case blue, green, and red. Curiously, it turns out that these 3 colors combined give white. Moreover, this white color which is not really one, realizes the synthesis of all the colors of the prism associated with the different intensities of energy. Obviously, it's primary (just as much as these 3 colors), but we find a certain sometimes disjointed logic that is our own!*

Quarks thus appear as entities devoid of accessible spatial dimensions and do not lend themselves easily to a description in terms of trajectory or lifespan in the classical sense. This characteristic partly explains why the usual relativistic concepts are difficult to apply to their internal dynamics.

A proton consists of two up quarks and one down quark, which gives it a positive overall electric charge. Its mass is about 938 MeV.

A neutron consists of one up quark and two down quarks, which gives it a zero electric charge, for a mass of about 939 MeV.

The very small difference in mass between proton and neutron — of the order of 1 MeV —, although minimal, plays a decisive role in the stability of matter and in nucleosynthesis processes, conditioning the formation of atoms more complex than hydrogen.

A stable atom is electrically neutral: it contains as many protons as electrons, the latter carrying an opposite charge. In a speculative interpretation, the electron can be considered as an elementary particle functionally related to quarks, but freed from the confinement constraint, which allows it to detach from the nucleus and participate in the formation of chemical bonds. This ability is at the origin of the assembly of atoms into molecules and, more broadly, of the structuring of matter.

Muons and tau leptons, although of the same charge and nature as the electron, have significantly higher masses. Their instability and rapid disintegration inscribe them in an evolutionary logic comparable to that of heavy quarks (c, s, t, b), now absent from ordinary matter. Their existence nevertheless attests to the richness of the spectrum of possible states of matter under extreme energy conditions.

The photon, as a particle devoid of rest mass, propagates along null trajectories and does not possess inertia specific to the classical sense. It therefore only interacts with what is on its actual trajectory. However, this trajectory can be deviated by the gravitational deformations of space-time, in accordance with the predictions of general relativity.

This property leads to several types of physically remarkable interactions.

Photon – electron interactions

When a photon interacts with an electron bound to an atom, it can be absorbed by it. The electron then acquires additional energy, which translates either into a transition to a higher level of energy, or through its ejection from the atom: this is the photoelectric effect. Conversely, when an electron loses energy, it can emit a photon and return to a lower energy state.

In the case of very high-energy gamma photons, part of the kinetic energy associated with radiation can interact indirectly with the atomic nucleus. When this interaction does not result in the complete absorption of the photon, excess energy can be converted to mass energy by creating particle-antiparticle pairs, primarily electron-positron pairs. The positron, antiparticle of the electron, quickly annihilates upon contact with a surrounding electron. The energy thus transitorily released is classically described, in the weak interactions scheme, by the intervention of intermediate bosons, notably the Z boson, whose effective mass reflects the energy involved in the process.

Photon interactions – atomic nucleus

Several cases can be distinguished:

- When a gamma-ray photon of energy at least equal to 1,022 MeV interacts with the field of a nucleus without being absorbed, it can transform into an electron-positron pair. After annihilation of this pair, two 511 keV photons are emitted in opposite directions, in accordance with the conservation of the pulse. These photons remain correlated, forming an entangled system. This type of process contributes to the gradual degradation of electromagnetic radiation energy upon contact with matter.
- When the energy of the photon is high enough to be absorbed by the nucleus, it can change its internal configuration. It can notably occur a photo-nuclear process in which a proton is converted into a neutron, analogous to an electronic capture induced by an external energy input.

- In some extreme conditions, a proton can meet an antiproton. Their annihilation does not lead to a disappearance of energy, but to its redistribution in the form of intermediate particles, classically described by the W bosons, vectors of weak interactions. These bosons, like the Z boson, make observable processes otherwise inaccessible to direct observation.
- During a beta decay, a neutron transforms into a proton by converting a down quark into an up quark. To preserve global charge neutrality, the atom can capture a free electron, while an electronic antineutrino is emitted. This antineutrino carries a minute fraction of the process energy without changing the charge parity.

These mechanisms contribute to a continuous redistribution of energy in the Universe: high-frequency electromagnetic energy tends to transform into mass energy or lower-frequency radiation. In the long term, this dynamic suggests a Universe dominated by residual electromagnetic waves of very long wavelength, weakly interactive, gradually absorbed under the effect of gravitational fields.

Particle, wave and probability field

When a particle is described as a wave packet, it can no longer be assimilated to a locatable point in space. The notion of precise position becomes incompatible with an undulatory description, which can only be expressed in terms of probabilities of presence. The particle must then be understood as a spatially extended field of energy, whose boundaries are neither clear nor fixed.

In order to make this abstraction more accessible, it is possible to resort to a pictorial representation, explicitly recognized as non-physical.

Also, a way of conceiving in recognized terms, the energy field of a particle would be to compare it with a bubble of localized influence, without delimited dimension and all the more remarkable as we consider its most central part in mass data. This "wave-packet bubble" would be dressed in all the colors of the rainbow, in proportions supposed to decline the information of charge, intensity, flavor, spin... characterizing any particle.

187

The "bubbles" of influence mentioned here do not correspond to physical entities in the usual sense and are not part of the space that constitutes our observation environment. They must be understood as a heuristic representation of abstract quantum structures, devoid of spatial localization and independent of classical notions of distance. These quantum bubbles that do not have the perception of time, would symbolize the necessary passage or mode of access allowing to exit a binary system of universes in symmetry, through 'the bottom'. The quantum world would thus lead to the multiverse Cosmos. What we could call quantum teleportation is nothing other than a process of discrete exchanges between an unrecognized right symmetry (arbitrary choice) and a left symmetry that would be ours.

This left/right symbolism in a universe of bubbles is just an allegory. And talking here about laterality is not really appropriate.

Assimilated to knots or energy bubbles (chapter IX) or replaced by strings, particles remain elusive in all cases.

We can approach them in another way, considering only the binding forces. The particle could, in this way, be compared to the attachment points of links of various calibers, in a tangle of mismatched chains, particularly elastic. These links, which do not have the oblong shape for all that, can open, assemble, stretch and close together in as many interactions. The Universe is then no longer an expanding soap foam.. Without physical reality, such links could, for lack of better, be defined as related "D-branes", to use a term already used, or to mark the difference, as "links/branes".

This idea of energy in the form of assembled links, intertwined at all levels and present in symmetry, distances us somewhat from the standard cosmological model we have adopted. This is of course just another image, but all these metaphors allow us to dress up in an acceptable way phenomena that have no equivalent in our daily reality. Thus, we avoid resorting to the concept of wave packet or beam of wave packets so difficult to integrate and developed here.

In the current state of knowledge, any rethought cosmological model necessarily remains partially speculative and cannot claim a complete experimental validation. The standard model itself relies on efficient, but incomplete, assumptions. The present work is explicitly part of this exploratory perspective: to propose coherent avenues of reflection, capable of articulating quantum mechanics, cosmology and fundamental symmetries, without claiming a definitive status.

XIII <u>Stealthy and exemplary of discretion</u>
(Insignificant particle that are released from borders)

The inert mass is here interpreted as the expression of the total kinetic energy associated with a set of entangled quantum waves, confined in the form of a wave packet constituting a particle moving at constant speed. Any change in the state of motion—whether it be acceleration or deceleration — necessarily implies an energy input or output. The acquisition of additional kinetic energy then increases the inertial capacity of the particle and modifies its initial trajectory. Moreover, accelerating amounts to increasing the overall kinetic energy of the system, which translates into an effective increase in inertial mass. It is well established that considerable amounts of energy are required to produce even a marginal variation in this inertial mass.

This relationship raises a fundamental difficulty: how could a material body acquire enough kinetic energy to approach the speed of light, as the continuous increase in its inertial mass more and more effectively counters any further acceleration? To this relativistic constraint is added a gravitational limit: beyond a certain threshold of critical mass, a body ceases to be able to maintain its structural integrity and collapses under its own gravitation. Such a collapse most often leads to the formation of a neutron star.

Now, the free neutron has an average lifetime of around fifteen minutes. It is therefore necessary to explain why neutrons are stable when integrated into the atomic nucleus, and why a neutron star can constitute a macroscopically stable structure on cosmological time scales. One approach consists in bringing the strong nuclear force and gravitation closer together, by postulating that they could be two manifestations of the same fundamental mechanism linked to charge effects, understood in a broader sense.

Within the atomic nucleus, oppositely charged quarks interact through electromagnetic interaction, while being confined by the strong force. Similarly, in a neutron star, electrons are captured by protons during electron capture processes (often likened to a form of reverse beta decay), leading to the transformation of protons into neutrons with antineutrino emission. The atoms in it are destroyed as distinct structures, and the overall charge neutrality is maintained. This generalized conversion of protons into neutrons puts an end to classical nuclear reactions and marks an advanced stage in the concentration-cycle evolution of matter in the Universe.

The neutron star thus formed has an energy density stratification: pressure increases towards the inner layers, while intense magnetic fields persist at the surface. In many cases, this configuration is only a transient state. When the mass of the object exceeds a critical threshold—estimated at a value greater than about ten solar masses — gravitational collapse continues and leads to the formation of a black hole.

In this context, only extremely low mass particles, such as neutrinos, can move at speeds close to that of light. Their low mass and absence of electrical charge make them little sensitive to ordinary gravitational fields and considerably reduce the probability of interactions with the crossed matter.

Although located at the extremes of the mass spectrum—neutrinos on one side, black holes on the other — these two objects have enough common features to justify the hypothesis of their central role in cosmic evolution. The primordial neutrino could represent an embryonic form of matter, anterior to current baryonic particles, while supermassive black holes would constitute the ultimate outcome of the collapse of baryonic matter in a cooled Universe at the end of evolution.

Like the photon, the neutrino can be considered as an exchange vector between different quantum symmetries. Both are accessible to observation only indirectly, by the effects they induce on matter: scattering, diffraction, refraction or photosynthesis for electromagnetic waves, and weak nuclear interactions for neutrinos. These can nevertheless be detected indirectly by the emission of Cherenkov radiation when they interact with water molecules.

Without electrical charge, neutrinos contribute to the energy adjustment of nuclear systems without altering their charge neutrality, in a manner analogous to the role of photons in energy exchanges. During beta decays, they carry the part of energy, momentum and spin that cannot be attributed to either the electron or the final nucleon, thus ensuring the overall conservation of physical quantities. A heuristic representation consists in considering them either as electrons having lost their electric charge during

191

a nuclear interaction, either like gamma photons having acquired an infinitesimal mass while giving up part of their kinetic energy during an electronic capture process.

Due to their extremely low mass, most of the energy of neutrinos is carried by their motion, their speed being close to that of light. They seem to be among the most abundant massive particles in the Universe. Although they may change effective mass by oscillating between different generations, or "flavors", they do not seem to disappear or completely disintegrate. Their weak interaction with matter and with the strong nuclear force makes them sensitive only to the most intense gravitational fields, capable of modifying their energy equivalent to mass.

These interactions can lead to oscillation phenomena, by which an electronic neutrino transforms into a muonic neutrino, then a tauic one, and vice versa. This property allows the hypothesis of the existence of less stable intermediate states, as well as more massive neutrinos, possible vestiges of the first phases of cosmic nucleosynthesis.

This multiplicity of states partly explains the historical difficulty in detecting and characterizing neutrinos. They seem to evolve in a superposition of energy levels, making it difficult to distinguish clearly between neutrino and antineutrino. It is conceivable that the neutrino could manifest sometimes as a particle, sometimes as an antiparticle, without however being able to adopt these two statuses simultaneously. All the neutrinos observed have a left helicity, that is to say a negative projection of their spin on the direction of their movement. If one admits that a massive particle can, in principle, present both helicities, the absence of observed right helicity could indicate that it belongs to a complementary quantum symmetry, associated with antineutrinos.

The decay of a free neutron illustrates this global conservation: a neutron transforms into a proton, an electron, and an antineutrino, without matter or energy disappearance, but by simple reorganization of the interaction terms.

If particles and antiparticles are generally described as possessing the same mass, spin, opposite charge and inverted helicity, it is likely that the underlying symmetry is more complex than a simple inversion of these parameters. The neutrino, due to its electrical neutrality, could

adapt to the most appropriate quantum symmetry to accompany
processes involving electrons in nuclear interactions.

Very high energy neutrinos, although fewer in number, are produced during
the most violent astrophysical events, such as supernovae, hypernovae and
processes taking place near the event horizon of black holes. These
phenomena are accompanied by intense gamma radiation. Conversely,
elementary nuclear reactions, such as the formation of deuterium from two
hydrogen nuclei, only produce very low-energy neutrinos.

Mergers of neutron stars or neutron stars with black holes are manifested by
gamma-ray bursts, but no neutrino emission was detected there, as they
cannot be extracted from the extreme gravitational fields generated during
these events.

In summary, neutrinos, like photons, are supposed to discretely cross the
two fundamental symmetries of matter and antimatter, playing a mediating
role in their interactions. As such, they would participate in the overall
process of transformation and progressive deconstruction of the Universe.

> Can we consider that the current particles of matter, in particular
> fermions, have emerged from the first moments of the Universe in their
> definitive form? In cosmology, such a hypothesis seems unlikely:
> fundamental processes generally unfold through gradual transitions,
> involving successive intermediate states. It is therefore legitimate to
> consider an evolutionary scenario in which primitive particles would
> have preceded the appearance of observable fermions today. Neutrinos
> could be considered as plausible candidates for the primo-particle state.

In this hypothesis, the primordial neutrino would be comparable to a
minimal structure consisting of entangled quantum waves, devoid of
electrical charge and not yet exhibiting marked internal symmetry. It could
thus have constituted the first form of massive particle resulting from the
Big Bang. A fraction of these primo-neutrinos would have continued to
interact with the extremely energetic radiation dominating the early phases
of the Universe. Subjected to an environment saturated with energy, some
of them could have participated in transformation processes gradually
leading to the emergence of charged elementary particles and their
antiparticles constituting current matter.

Before the formation of the first hydrogen and helium nuclei, these primitive neutrinos can be interpreted as precursor states for quarks and electrons, as well as their antimatter counterparts. They would thus constitute a common origin for matter and antimatter. Unlike the photon, which does not have a mass or charge structure and whose symmetry is essentially linked to electromagnetic interaction, the neutrino, due to its non-zero mass and electrical neutrality, could conceal a specific quantum symmetry still imperfectly understood. This singularity clearly distinguishes it from other known elementary particles.

Indirect observations associated with extreme astrophysical phenomena, such as certain hypernovae, suggest the existence of neutrinos reaching exceptionally high energies, on the order of several hundred teraelectronvolts. It is then reasonable to assume that, in the primordial Universe, neutrinos could have reached even much higher energy levels, potentially several billion teraelectronvolts. These ultra-energetic neutrinos, now extinct or undetectable, could have played a decisive role in the emergence of electromagnetism and the first elementary charged particles.

During this initial phase, relatively brief on the cosmic scale, the newly formed neutrinos would have interacted with each other and with ambient diffuse radiation, giving rise to the first free charged particles, identifiable as a primary generation of quarks and electrons. These primitive elementary particles, initially dispersed and highly energetic, would have gradually assembled as the Universe cooled, leading to the formation of protons, neutrons, and electron clouds. The association of nucleons with their peripheral electrons would then have allowed the formation of atomic nuclei, fundamental structuring elements of matter.

Primordial nucleosynthesis, controlled by the gradual decrease in temperature, could only produce light nuclei. It led mainly to the formation of hydrogen atoms, helium and, in much smaller quantities, lithium, each accompanied by its most stable isotopes. The heavier elements appeared later within stars, by stellar nucleosynthesis. The condensation of hydrogen clouds enriched with helium and lithium was able to take place thanks to the presence of binding electrons, which allowed compensating for electrostatic repulsion between positive nuclei.

During cosmic evolution, primitive neutrinos necessarily lost a large part of their initial energy, due to the innumerable nuclear interactions between

194

protons and neutrons, as well as electroweak interactions that have marked the history of the Universe. The weak force, closely involving neutrons and neutrinos, favored the durability of protons in large quantities. A free neutron, if it is not integrated into an atomic nucleus or a neutron star, is indeed unstable and spontaneously transforms into a proton. In this context, the ultra-energetic primo-neutrinos no longer occupy an identifiable place in the current model of elementary particles, which is essentially limited to three generations of fermions.

In the current era, nuclear reactions mainly produce low-energy electron neutrinos. Certain more energetic interactions may, however, incidentally generate more massive muon and tauic neutrinos, as well as the corresponding unstable charged leptons — muons and taus—characterized by a very short lifetime. These heavy leptons, belonging to the second and third generations, decay by giving up part of their energy and give rise to light electrons of first generation. It is these first-generation particles—electrons, up quarks and down quarks — that ensure the relative stability and structuration of ordinary matter observed today.

This organization does not exclude the transient or localized appearance of unstable structures necessary for global equilibrium, such as composite particles, atoms or so-called "exotic" molecules, involving particles of heavier generations.

Finally, the evolution of electromagnetic radiation, characterized by a continuous elongation of wavelengths and a decrease in frequencies, implies a gradual reduction of the ability of this radiation to ensure sufficient energy transfers to maintain the charge balance between massive particles in the primordial Universe. This evolution contributes to explain the current atomic structure and the charge equilibrium observed between first generation fermions in the contemporary baryonic Universe.

XIV <u>Dark Matter and Dark Energy</u>
(Everything would be clear if it turns out they have no purpose)

Let us recall some figures which are cornerstones and which, faced with the most recent observations, block.
The Universe would consist of:

- ✔ 68-69% dark energy of unknown nature
- ✔ 26-27% dark matter of unknown shape
- ✔ 5% identified as baryonic matter

These estimates, accepted by a large part of the scientific community, currently lead to a deadlock because the first two presumed components, which are not the least, are lacking in the direct observation.

Wouldn't our appreciation of the energies in presence indicate an approach that is both too simplistic and too restrictive, based on the belief that our Universe would be expanding on the one hand and that antimatter would have mostly disappeared on the other?

<u>First and foremost, let us return these figures to their fair value</u>
In this distribution, which would represent the totality of the content forming our Universe, we add energy to matter. Even if there is a form of equivalence between energy and matter (E= mc 2), how can we understand that we can put as a percentage in the same equation, x% of black energy with x% of dark matter? Energy and matter are not substitutable terms as such, in this type of approach.

On the other hand, regarding the baryonic material (5%), the mass represents the inertia capacity. But most of this mass energy is not inherent in assembled elementary particles. The intrinsic residual energy of the particles assembled to make the constructed matter, would be only 1%. The other 99% would reside in the kinetic energy that contributes to the building of matter and in the binding forces. It is mainly about the strong force considered here as the result of close electromagnetic interactions in a confined medium represented by the atomic nucleus. Also partly contribute to it, the weak electromagnetic and nuclear forces that participate in the cohesion of atoms and correct imbalances.

What we think we know about our Universe is based on what is open to our observations or deductible from them. Given certain observational anomalies, it would seem that this part that is accessible more or less directly

196

to us represents only 5% of the content in matter and energy of it. A supposed so-called dark matter estimated at 27% and a more than hypothetical so-called dark energy estimated at 68% are supposed to constitute the rest of the energy content of our Universe.

A particle can be described as a physical state that does not lend itself to characterization in terms of density or space occupancy volume. In this perspective, the elementary particle remains a formal object, whose existence is mediated by theories and measuring devices. The quantities attributed to it — mass, electric charge, spin, energy—do not describe an extended material object, but constitute operating parameters, introduced in order to account for experimental results. They thus reflect a certain degree of 'virtuality' in the sense that they do not necessarily imply a classical spatial reality.

The spin, as a discrete quantity, cannot be interpreted as a mechanical rotation of the particle on itself. Indeed, an entity without a defined spatial extension cannot be described using kinematic properties from macroscopic physics, such as the proper rotation around an axis or the precise location in a given volume. In the same way, spin cannot be reduced to a simple geometric property associated with classical spatial symmetry.

The spin orientation specific to each type of elementary particle can therefore be considered as resulting from an internal polarization state, associated with a loop dynamic. Spin must then be understood not as an intrinsic rotation, but as the expression of an intrinsic magnetic moment. It would represent the effective angular momentum associated with the coherent set of entangled waves that make up the matter particle. Combined with the orbital movements and relative displacements of closely interacting particles, spin participates in determining their observable electromagnetic properties.

This apparent rendering of rotation, which is spin, would thus result from internal interactions within closed wave systems. However, the formal

197

description of a wave packet remains more complex than that of a strictly corpuscular entity, due to the fundamentally evanescent and contextual nature of quantum measurements. This difficulty explains that in practice, and for the sake of descriptive efficiency, quantum physics frequently calls upon the notion of a corpuscle, although it constitutes an ideological approximation.

In a more speculative approach, the elementary particle of matter presents several analogies with the black hole, considered as an ultimate stage in the evolution of matter:

- neither of them can be assimilated to bodies in the sense of complex and extensive objects.
- they cannot be characterized by usual thermodynamic concepts such as hot or cold.
- both can be considered as energy concentrations or limit configurations of wave packets, the elementary particle corresponding to an initial state, and the black hole to a final state of cosmic evolution.
- although we speak of fields as potential interaction spaces concerning them, neither the elementary particle nor the black hole constitutes in themselves representations of space, even if, by changing scale, they are integrated into it.
- finally, both of them can be considered devoid of proper temporality, although they are part of a global process of transformation that we interpret as time.

A major difference, however, lies in the electric charge, which, at the atomic scale, tends to neutralize itself for the particle of matter, unlike the black hole. This property, remarkable at the quantum scale, can be interpreted as a symmetry substitute, preserving the memory of an earlier broken symmetry. The electric charge confers to the particle its "volatility" in quantum mechanics and makes possible the interaction with electromagnetic waves, an interaction which is manifested in particular by their capture by massive bodies. In this perspective, it is envisaged that at a certain scale, electromagnetic force can constitute the effective source of gravitational phenomena.

Therefore, if the particle, considered as a packet of waves folded back on itself, does not contain free space, and if, at the other end of the

198

deconstruction process, the black hole is located outside classical space-time, the very notion of space becomes problematic. One can then be led to consider space not as a fundamental entity, but as an observation framework, necessary for our representation and understanding of the phenomena that, from elementary particles to black holes, structure the evolution of the Universe.

<u>Dark matter</u> constitutes, in the current state of knowledge, an explanatory hypothesis intended to account for the abnormally high orbital velocities observed for stars within galaxies, as well as for galaxies themselves inside galactic clusters. In the absence of direct proof of its existence, this mass component postulated, but remaining undetected, can also be interpreted as the symptom of an incomplete evaluation of the masses actually involved or of a still imperfect understanding of the large-scale gravitational mechanisms, whose origin could exceed the limits of the standard cosmological model.

Estimates of the average mass density of the Universe are based on the compilation of masses of galaxies, their clusters and a diffuse background of particles. It is however necessary to distinguish the luminous mass, deduced from electromagnetic emissions, from the dynamic mass, inferred from the observed gravitational effects. The recurring observation is that the sum of the masses evaluated by classical methods often remains significantly lower than the total mass required to explain the gravitational dynamics of the systems under consideration.

There is undoubtedly a correlation between the electromagnetic emissions of a body and its mass, density and composition. However, this link becomes inoperative in the case of black holes, whose observable emissions are not intrinsic, but depend on the activity of their accretion disk. The intensity of these emissions cannot therefore constitute a reliable indicator of the mass-equivalent of the black hole.

Moreover, astrophysical measurements aggregate signals from very ancient events, observed at great distances, and phenomena closer in time and space. This superposition raises the question of our ability to correctly correct, through the application of relativistic laws, the deformations induced by the joint effects of gravitation and electromagnetic fields traversed by signals

199

during their propagation. It therefore appears that our ability to determine with precision the mass of large cosmic structures, and a fortiori that of the Universe as a whole, remains limited, all the more so since the fraction actually accessible to observation remains undetermined.

The apparent insufficiency of matter could be explained, at least in part, by the existence of populations not or poorly illuminated and insufficiently inventoried, such as neutron stars, isolated black holes or binary systems, as well as brown dwarfs. To this could be added low-density hydrogen clouds, neutral or ionized, distributed not homogeneously in space. Isolated extragalactic black holes in gas-poor regions would be highly likely to lack an accretion disk, making them particularly difficult to detect. Their gravitational effects, relatively confined, would then only manifest themselves through tenuous spatial distortions, mainly observable by the effect of gravitational lensing on background objects. Yet the systematic exploitation of this effect proves to be complex due to the multiplicity of stellar bodies located along the lines of sight, which limits the detection of black holes other than the most massive.

One can also suppose that the intergalactic space, although appearing largely empty, is occupied by baryons and elementary particles very dispersed, difficult to detect individually. This diffuse component would contribute to the global mass effect and intervene in maintaining a minimum temperature, preventing it from reaching absolute thermodynamic zero. To this are added vast structures of diffuse gas and heavy molecules, associated with galactic central regions, likely to modify the local gravitational potential. Finally, cosmic residues commonly referred to as star dust, mostly composed of carbon and silicon, come from the ejection of outer layers of stars at the end of evolution. Although faint, these dusts contribute significantly to the mass content of galaxies. Fermi bubbles also represent a significant number of diffuse gases and heavy molecules, centered on either side of the axis of rotation of the galaxy. These bubbles likely modify the gravitational weight of the central region of it. Add to this the stray cosmic residues called stardust and made up largely of carbon and silicon. These 'heavy' dust is what remains of stars like our sun and which have reached the end of their life, cooling down, losing brightness to finally eject their outer layer into space.

It is also not established that the apparent mass of a black hole is strictly proportional to its effective size, which could grow less rapidly than the mass attributed to it. This hypothesis finds support in two types of observations.

On the one hand, gravitational analysis of dwarf or starless galaxies suggests the presence of central black holes with a particularly high mass-to-size ratio. These galaxies, often ancient, could have seen their central black hole gradually accreting a significant fraction of the surrounding mass. The precise determination of the mass and energy content of such an object, based solely on gravitational measurements, remains however uncertain. Unlike stars, a black hole does not lend itself to a classical definition of density, making its physical properties difficult to quantify. It therefore seems inappropriate to consider black holes as mere stellar bodies, since they can be part of a fundamentally quantum physical regime, despite their apparent size.

On the other hand, when merging two black holes, the size of the resulting black hole can give the impression of a mass loss. However, since a black hole does not occupy space in the classical sense, the energy density of the final system should not be lower than that of the original black holes, even if some of the energy is released as gravitational radiation during the event.

In any case, the mass-equivalent of the supermassive black hole located at the center of a galaxy seems insufficient, on its own, to ensure the overall cohesion of the galactic structure. It is then conceivable that the cumulative gravitational forces of all the bodies making up the galaxy produce a collective amplifying effect. This resulting attractiveness would be all the more marked in active galaxies and would contribute to strengthening the attraction exerted on peripheral regions, although the precise mechanisms of this phenomenon remain difficult to isolate.

Another aspect that may affect the interpretation of observations is the spatial distribution of gravitational effects in a spherical or quasi-spherical system. The intensity of these effects is not necessarily uniform at the surface or in the environment of the system and may vary depending on whether one considers polar or equatorial regions. In galaxies, the flattened distribution of bodies orbiting predominantly in a plane perpendicular to the axis of rotation gives the gravitational field a privileged extension into the equatorial plane. This configuration could help explain the relative stability of bodies located at the periphery of active galaxies, despite the high observed velocities, and limit their dispersion.

201

Finally, it should be noted that gases located near the galactic center are generally hotter and denser than those from external regions. Their increased thermal agitation implies a higher energy content. The mass deduced from the radiation of these hot gases may thus appear underestimated, which potentially contributes to the observed differences between luminous mass and dynamic mass.

The more observation is directed towards distant regions of the Universe, the more it corresponds to the exploration of ancient phases of its evolution. It is therefore coherent to note, in these observations, an apparent deficit of matter today not directly observable, this matter having been able to structure and densify during more recent eras. This lag essentially reflects the time required for information from a distant event to reach the observer.

Although electromagnetic waves propagate locally at the speed of light, their effective trajectory, from the observer's point of view, is not rectilinear in a relativistic space, heterogeneous and affected by variations in gravitational potential. It is thus plausible that the light emitted by a galaxy located at a distance of one million light-years required a time greater than one million years, measured in the observer's reference frame, to be detected. Furthermore, waves of different wavelengths do not necessarily follow identical trajectories. The radiation received therefore neither corresponds strictly to that which was emitted in the past, nor to an exact superposition of the gravitational effects associated with the source at the time of the emission.

As a reminder:

- a year belonging to a distant cosmological past cannot be directly assimilated to a year measured in the current conditions of the Universe.
- the fields of energy traversed by the electromagnetic waves emitted several million years ago have interacted continuously with them. In addition to the gravitational effects of the bodies encountered, these waves were subjected to the influence of multiple radiation sources. The amplitudes, emission peaks and frequencies of fossil radiation can thus only provide a partial and necessarily imprecise extrapolation of the initial state of the Universe.

It follows that the recorded image corresponds to a distorted past, which we inevitably relate to a space-time referring to our local present. The Universe observed at the limits of the cosmological horizon has undergone major transformations since the time corresponding to this 'instant image', taken at a time when its properties deeply differed from those of the nearby Universe.

Distant galaxies thus appear deformed, hotter and marked by intense thermonuclear activity. These galaxies, belonging to a phase where time expansion was less, seem to present excessively high rotation and displacement speeds. This impression results from the fact that speeds measured in an ancient past are related to a current time scale, slowed down by the subsequent evolution of the Universe. The corrections applied to these observations are then based more on hypotheses than on firmly constrained parameters, which can lead to an overestimation of the exhaust velocities calculated in a necessarily local reference context.

In the ancient phases of the Universe, space was more densely populated with high-intensity radiation, while matter remained more diffuse, with a smaller population of white dwarfs, neutron stars, of black holes and other high mass density stellar bodies. The relative speeds of rotation were likely to be higher. Now the rapid rotation of a star, by accentuating the local deformation of space-time, modifies its gravitational effects. The speeds we observe today therefore belong to the ancient history of these systems. If the received image could be updated, it is likely that a gradual decrease in dispersion and rotation rates over time would be observed.

Many parameters necessary for a rigorous correction of these effects still escape us. It is legitimate to question our real ability to adequately take into account the global "aging" of the Universe.

If light takes a finite time to reach us, altered by all the interactions encountered during its propagation, gravitation constitutes on the other hand a universal phenomenon that affects globally all regions of the Universe. Of theoretically unlimited scope, it could define, on a large scale, an average space-time referential, independent of the mixture between past and present specific to the field of the observable. Such a reference system, resulting from a smoothing of gravitational effects, would be representative of the global evolution of the Universe and could constitute an indicator of its age, without providing the operational means to determine precisely its lifespan.

A parallel has been proposed with electromagnetic waves, postulating that the space-time deformations associated with gravitation manifest as gravitational waves. These deformations become perceptible when very massive bodies interact violently, producing measurable perturbations of the space-time structure.

This interpretation in terms of waves specific to gravitational effects constitutes a coherent response. However, it seems more appropriate to speak of deformation of space-time than of propagation of waves analogous to electromagnetic waves. Unlike the latter, electromagnetic waves undergo gravitational deformations of interstellar space, while indirectly contributing to it through magnetic fields associated with electrical phenomena. Gravitational disturbances cannot therefore be unreservedly assimilated to waves propagating in a medium in the classical sense.

We could proceed by analogy with a closed body of water (our Universe) subject to fine and regular rain. The impact of each raindrop (any stellar mass) marks the surface of an aureole that spreads (gravitational effects) in concentric circles attenuating with distance. Seen in its entirety, from very high up, the body of water shows a barely trembling relief (comparable to the so-called empty space in depression) presenting the same surface aspect everywhere at the same moment.

The minimum speed required for an object to free itself from the gravitational influence of a body decreases with the distance between them. This relationship, well established for an isolated body, should however be nuanced when it comes to a complex gravitational system whose mass is distributed heterogeneously, as is the case with galaxies. As one moves away from the galactic center, the amount of mass between the point under consideration and this center increases, due to the cumulative presence of stars, gases and other diffuse components. This mass is neither punctual nor uniformly distributed but organized according to a globally concentric distribution.

Such a configuration implies that the gravitational force exerted at the periphery of a galaxy should be greater than that produced by an equivalent mass concentrated at a single point in the center, where the supermassive

204

black hole resides. In this context, the high velocities of stars located at galactic boundaries can give the impression that they are about to escape the gravitational potential of their galaxy. It is nevertheless likely that a fraction of these stars indeed has sufficient speed to leave the galactic halo, without disappearing from the cosmic landscape, since they can be captured later by a neighboring galaxy.

The increase in the speed of a body results in a quadratic growth of its kinetic energy. The question then arises as to whether the gravitational capacity of a rapidly rotating body evolves in a correlated manner. The combination of gravitation and rotation gives most stellar bodies an almost spherical shape, with a flattening at the poles induced by centrifugal effects. This deformation suggests that the intensity of the gravitational field is not strictly isotropic and that it could be strengthened in the equatorial plane. These dynamics, observable at all scales—from the atom to galactic clusters, via pulsars, some neutron stars performing more than a thousand rotations per second — could modify the gravitational effects felt in the plane of rotation of galaxies. These additional gravitational contributions would likely be attenuated in older, less active, less populated and thermally colder galaxies.

Mass estimates of galaxies and their clusters are largely based on the analysis of their spectra, affected both by gravitational lensing effects and by the properties of space traversed by light signals. The image thus received corresponds to a vestige of an ancient local space-time. It is therefore not surprising that the masses deduced from these observations are insufficient to explain the rotation rates of stars located at the periphery of distant galaxies.

The introduction of hypothetical weakly interactive particles, commonly referred to as Wimps, constituted an attempt at resolution suitable for this mass deficit. These particles would be characterized by the absence of detectable electromagnetic interactions, neither emitting nor absorbing radiation, their only manifestation being their gravitational contribution. This hypothesis nevertheless raises the question of whether these exotic entities would not constitute, under another formulation, a roundabout way of evoking an antimatter escaping direct observation.

The ease with which the Wimps hypothesis was adopted is reminiscent of that preceding the discovery of the neutrino. However, unlike the Wimps,

205

the neutrino had been theoretically predicted with well-defined properties and then experimentally confirmed. The case of Wimps differs in that they are introduced largely empirically, in response to observed deviations in general relativity predictions. The existence of a so-called sterile neutrino, devoid of interactions other than gravitational, has also been considered, but this hypothesis remains devoid of solid experimental foundations.

Dark matter thus appears as a default solution intended to account for poorly understood gravitational effects, highlighted in the majority of galaxies. The identifiable and cumulative masses of all observable components of these rotating structures are insufficient, explaining only about 20% of the gravitational effects deduced from peripheral orbital velocities. Dark matter therefore primarily demonstrates our current inability to exhaustively identify all relevant mass contributions.

In the absence of the instrumental means necessary to revise these estimates rigorously, it is nevertheless possible to examine the consequences of an alternative hypothesis. If, for example, the fraction of material identified were arbitrarily increased from 5% to 16%, while maintaining the assumption of approximately 68% dark energy, the required fraction of dark matter would be reduced to 16%, a value comparable to that of the ordinary material identified. Such a distribution would then suggest that so-called dark matter could correspond to an antimatter present in equivalent proportion, discretely contributing to gravitational effects after the upward reassessment of galaxy mass.

Antimatter would therefore preferentially concentrate in regions of high material density, particularly in the vicinity of galaxies and black holes. Present in the background of interactions ensuring global symmetry, it would manifest indirectly from a certain scale of observation, each quantum symmetry feeling the additional effects of its opposite symmetry.

Some recent approaches even propose the existence of a parallel universe, in interaction with ours, in which dark matter corresponds to a form of antimatter belonging to an "anti-universe" remaining inaccessible to direct observation. Excluding antimatter a priori or denying it any gravitational influence then leads to multiplying the hypotheses around an elusive dark matter.

Evoking dark matter indeed amounts to postulating the existence of an entity that we are not able to represent concretely, nor even to confirm experimentally. Conversely, antimatter constitutes a well-defined physical reality, observed occasionally during particle collisions, and whose presence on a large scale in the Universe remains theoretically expected. The analysis of space-time deformations that the observable matter is not enough to explain should logically make it possible to identify the origin of the responsible phenomena. However, if dark matter actually corresponds to antimatter located in another dimension of space-time, its location would not necessarily imply its observational accessibility.

Dark energy, sometimes assimilated to an energy of the "void", constitutes a hypothesis whose ontological status remains as uncertain as that of dark matter. It is introduced in order to account for the accelerated dispersion of galaxies in a Universe generally described as expanding. With reference to the standard cosmological model, this component would represent about two thirds of the total energy content of the Universe.

On the formal level, the introduction of dark energy leads to the need for a cosmological constant, denoted Λ, appearing as an additional term in the gravitational field equations. This constant does not originally correspond to an identified physical entity, but to a mathematical parameter intended to model an effect interpreted as a global expansion of space. It is associated with the hypothesis of an unknown energy exerting an opposite influence to that of gravitation. Dark energy thus constitutes a logical hypothesis making it possible to satisfy the interpretation of an inflation of space.

Historically, the cosmological constant was introduced by Albert Einstein in order to maintain a static Universe, compensating for gravitational attraction and avoiding global collapse. It is however admitted that Einstein himself doubted the physical necessity of this term, seeing it more as a formal artifice than a fundamental reality, especially after the emergence of dynamic cosmological solutions proposed by Friedmann. The subsequent renunciation of this constant can be interpreted as recognition of its provisional character, masking a deeper inability to explain what will later be interpreted as a galaxy leak.

207

In this perspective, it is proposed to replace the idea of an expansion of the Universe with Such an approach leads a gradual decrease of the energy density of the so-called "empty" space. To focus not on the geometric expansion of space, but on the evolution of its energy content. Talking about density then amounts to considering the transformation of latent primordial energy into baryonic matter at the end of a radiative entanglement phase, as well as the irreversible capture of different forms of energy by black holes.

The Universe would therefore appear as a non-static system, but whose dynamics would not necessarily be related to a real inflation of space. The expansion effect would above all be an observational effect, resulting from the redistribution and gradual concentration of energy and matter. This interpretation leads to prioritize an energy dynamic of a concentrationary nature, marked by a growing gathering of matter and a local increase in density, dynamics described here as "retrograde dispersion".

If the cosmological constant is interpreted as a vacuum energy, it could then refer to a vacuum that is not devoid of content, but which would constitute a framework of discrete interactions involving two chiral dimensions of space-time, revealing a fundamental quantum symmetry. In this perspective and rejecting the idea of quantum symmetry, the cosmological constant would become necessary. Interpreted as the manifestation of an unobservable state of matter particles in opposite symmetry, it would cease to be a universal constant to become a variable quantity, adjusted to the concentration camp evolution of the Universe.

Cosmic evolution could then be described as a gradual depletion of gas in galaxies, the collapse of stars, the fusion of celestial bodies and their incorporation into stellar or galactic black holes. At the same time, the so-called empty space would be stripped of its energy fields, while the population of black holes would continue to grow. Such an evolution would lead to an increasingly smoothed space-time, tending towards a non-multi-referential state, in which the very notions of space and time gradually lose their operational significance.

In this reading, dark energy appears as a default response to a phenomenon interpreted as an accelerated expansion of the Universe, based mainly on the analysis of the radiation of distant supernovae with a strong redshift. These observations necessarily rely on a local reference frame, chosen as the basis for interpretation. From then on, even corrected, it is legitimate to question

208

the relevance of the measures resulting from highly distorted images of a distant cosmological past, fundamentally different from our close environment.

In the hypothesis of a retrograde dispersion not involving real inflation of the Universe, dark energy loses its necessary character and the cosmological constant can be abandoned. The impression of a slowdown in expansion, sometimes evoked in some models, could be interpreted either as a gradual decrease in dark energy, or as the consequence of the gradual disappearance of antimatter, which, like matter, would end its baryonic cycle within black holes. In this ultimate, non-reversible phase, antimatter would lose its mass properties and dissociate from a space-time that would gradually empty itself of the energy it contains, giving the illusion of an expansion.

If one admits that the Universe does not really expand, neither dark matter nor dark energy impose themselves as necessary components. The quantity of baryonic matter in our universe symmetry is then estimated from the observed gravitational effects, supposed corrected for distortions related to gravitational lenses and propagation disturbances, corrections whose reliability remains uncertain. It then becomes conceivable that the measured gravitational effects are attributable in equal parts to two opposite quantum symmetries.

Cosmic expansion, dark matter and dark energy can be simultaneously dismissed, along with the cosmological constant, in favor of a symmetry-based model. The Universe would then be made up in a balanced way of matter and antimatter, in equal proportions, interacting discretely within a shared field of electromagnetic waves. We are doing here a spell on expansion, dark matter as well as dark energy and abandoning the cosmological constant in a context of symmetry. Who could complain about it?

To conclude, the Universe would therefore be made up of 50% of matter and 50% of antimatter in discrete interactions in a shared field of EMW.

If the cosmic expansion model allows for a coherent interpretation of a certain number of observations, it nevertheless raises several difficulties as to its validity and explanatory scope.

Firstly, how to account for the fact that a supposed expanding Universe, whose observation is essentially based on signals from an extremely distant past, has a remarkably homogeneous temperature on a large scale? A proponent of the expansionist model might argue that the observable portion of the Universe is likely to be only an infinitesimal fraction of its totality. In this hypothesis, possible thermal heterogeneities would exist at scales much higher than our observational horizon, making temperature variations locally undetectable.

Secondly, expansion assumes an initial quasi-zero-dimensional singularity, prior to the very existence of space. It therefore seems legitimate, in view of the current state of the Universe and the phenomena observable at different scales, to seek to estimate the rate of expansion and, by extension, the age of the cosmos. However, this approach encounters serious inconsistencies.

The standard expansionist scenario postulates an initial phase of extremely rapid expansion, followed by a gradual slowdown correlated with the decrease in the average temperature of the Universe. Some hypotheses go so far as to envisage an initial expansion greater than the speed of light. Such a proposal poses a problem with regard to the relativistic principles, which closely link space and time from the early phases of radiative entanglement and according to which the propagation speed of photons depends on the energy state of the space they pass through. In this perspective, the speed of light could be interpreted as an indicator of the state of structuration—or of deconstruction — of the Universe, and thus indirectly of its age. It thus acquires the status of an effective constant for events close in time, without necessarily constituting an absolute limit on a cosmological scale.

After the initial phase of radiative entanglement, the early Universe was characterized by a high density of free particles, relatively homogeneously

210

distributed. In this diffuse and opaque medium, the primordial particles gradually began to aggregate. This process led to the formation of the first composite structures, then the first atoms and the first molecules. During this period, the absence of significant gravitational effects implied a relatively uniform, or "smoothed" temporal evolution. Electromagnetism then played a preponderant role in the organization of matter, favoring the constitution of vast molecular sets. By gradually differentiating, these sets have revealed regions that are increasingly impoverished in terms of materials.

The impression of "accelerated" time in these low-density regions can be linked to the local emergence of space curvatures, previously non-existent, which modify temporal properties. These first groupings gave birth to primordial galaxies of low mass, but particularly large dimensions, beyond the measure of the galaxies observed in the current nearby Universe.

The concept of space-time constitutes a theoretical representation in which gravitation, by modifying the geometry of space, directly influences the course of time. The contraction of spatial lengths — interpreted as a tightening of empty space — associated with temporal expansion — the apparent slowing down of the flow of time — leads to conceive a Universe characterized by a plurality of referentials. Each event thus has its own spatiotemporal framework. This approach corresponds to general relativity, which integrates the invariance of the speed of light into all inertial frames of reference, a principle inherited from special relativity.

The speed of light thus acquires a central status. If its local invariance at a given moment does not seem to be called into question, several questions remain: why is it an unbridgeable limit for any material entity? And how to justify its precise numerical value? These questions refer to the status of postulates, understood as unproven principles but necessary for the mathematical construction of theories whose predictive effectiveness is proven, without conferring them an absolute universal character.

The hypothesis explored here is based on the energy level of space described as 'empty'. Relativity, based on the distinction between observer and observed object, would lose its relevance if the Universe were apprehended as a whole. In such a non-relativistic and global context, the depressive evolution of energy associated with interstellar space would become a determining factor in the speed of propagation of electromagnetic waves. It would result in a speed of light variable over time, dependent on a multi-referential space whose energy density gradually decreases. General relativity should then be reconsidered as a theory indexed on the dynamic evolution of the Universe itself.

In this perspective, the observed speed of a light signal could remain constant in a local reference frame, while appearing variable when compared to a more global external scheme. This apparent consistency would mask an underlying evolution related to the energy depletion of the so-called empty space.

In this perspective, the young Universe, from which we now perceive light coming from the most distant galaxies, would have passed through a particularly unstable phase, characterized by a rate of agitation or retrograde dispersion significantly higher than that currently observed. The energetic deepening of space would have been faster, while local gravitational effects remained relatively weak. This combination of factors leads to the impression that the Universe expands all the more rapidly as more remote regions are observed. Indeed, the more cosmological distances are observed, the more objects seem to move away at high speed, sometimes to the point of suggesting apparent speeds higher than that of light.

Such an observation comes into tension with the invariance of the distance/time ratio associated with the propagation of electromagnetic waves, which should not be exceeded by any body. It also seems contradictory with the idea that the continuous depletion of space in energy, consequent to the regrouping of matter, should promote a gradual rapprochement of structures under the effect of gravitation. In a space whose energy density is decreasing, the intensification of gravitational interactions would indeed lead, in the long term, to fusion phenomena rather than an indefinite dispersion.

In reality, it is not the objects of the past themselves that are moving away more and more quickly, but rather the image we receive from them. This image corresponds to a signal conveyed by electromagnetic waves, in particular visible light, emitted or reflected by the observed object. The impression that these signals could exceed the speed of light, set locally at about 300,000 km/s, is explained by the fact that a second measured in the observed past time frame does not necessarily correspond to a second of our present time.

In other words, if the proper time of the reference frame of the observed object were to flow at a faster rate, for example twice as fast as ours, then the distance travelled by light during a local unit of time would be, in this reference frame, proportionately greater. This shift, characteristic of the relativity between space and time, does not imply a real extension of distances when we observe regions from a distant past. However, any measure necessarily remains related to our own temporal frame of reference, which introduces an inevitable distortion in our appreciation of speeds and displacements. The observed effects thus relate more to a biased interpretation than to an effective dynamic of the objects concerned. *We are a bit like the traveler lost in the middle of the desert at the hottest hours and who sees, through the effect of convection, the particularities of the landscape distorted at distances they do not have.*

This apparent and accelerated escape of galaxies may lead to the assumption that after an initial phase of strong expansion, it would have slowed down before experiencing a new increase. Regardless of the inherent confusion between present and past, such a hypothesis raises the question of the origin of the energy needed for this supposed revival of expansion. In the absence of a clearly identified mechanism, the commonly proposed response relies on the existence of an undetectable energy form of unknown nature. This hypothetical entity, introduced to preserve the consistency of the model, was referred to as "dark energy".

Moreover, this inflationary vision of the Universe does not allow to determine if the current expansion will continue indefinitely or if it will eventually slow down or even reverse. In the latter case, the Universe would enter a global contraction phase, leading to a final state symmetrical of the initial state, commonly referred to as the Big Crunch. Although this scenario evokes a global collapse, the recessive mechanism envisaged differs from that proposed in the approach developed here.

In these conditions, it seems legitimate to reconsider the idea of cosmic expansion, all the more so since in astrophysics an apparently obvious interpretation does not necessarily constitute an established truth.

At the moment of the Big Bang, photons and primitive particles were indistinguishable due to a lack of clearly differentiated intrinsic properties. The first electrons would have appeared as a result of state transitions affecting certain primo-particles, themselves supposed to derive from an abundant background of primo-neutrinos. It was only from this

213

differentiation that the first hydrogen atoms — consisting of a nucleus formed by assembled quarks and an electron — then helium, in their most stable configurations, were able to form. The formation of vast clouds of gas, followed by the emergence of more or less dense and hot stellar bodies, was an essential preliminary step for the formation of black holes, considered here as an ultimate phase of this evolutionary process.

It should be emphasized that the expansionist hypothesis hardly agrees with the idea of a Universe devoid of defined perimeter, center and border, and therefore without measurable global volume. No larger set can serve as an external referent to our Universe. Under these conditions, the very notion of a unit of measurement likely to validate an expansion of space poses a problem: how to measure the expansion of a whole that does not fit into any container?

The energy from the Big Bang is not intended to remain uniformly distributed. Each energy concentration associated with an elementary particle can be interpreted as the local onset of a space energy depression. This trend is accentuated when the quarks assemble into hadrons, then when the number of grouped particles increases. As matter concentrates, the energetic depression of the surrounding space intensifies, potentially giving the illusion of a global repulsive force, without a clear frame of reference allowing to define in relation to what this repulsion would exert.

In order to make intelligible the apparent coexistence of two opposing dynamics — gravitational attraction and retrograde dispersion—it is possible to refer, on a strictly analogical basis, to biological systems. Living organisms are hierarchical structures resulting from the assembly of molecules, differentiated cells and specialized organs, organized according to complementary functions. These structures develop, reproduce and maintain themselves from a precise genetic model, the DNA.
If organic chemistry can explain certain recurring mechanisms, many interactions remain difficult to identify. Yet they play a role in this highly structured assembly of particles, which constantly collects, selects and eliminates the constituents necessary for living organisms to persist. The considerable amount of energy required to form and maintain a complex organism also imposes extremely demanding environmental and thermal conditions.

214

This analogy is only valuable because of the comparable difficulty in identifying the root causes of interactions which, at first glance, seem difficult to reconcile. In the same way that life emerges from an unstable balance between organization and dissipation, the gravitation of bodies and retrograde dispersion can be considered as the two inseparable aspects of the same fundamental process. Resulting from the first radiation, life—a simple particular assembly of molecules — remains paradoxically vulnerable to the destructive effects of these same radiations. It is only maintained through reproduction, favoring the transmission of a genetic model rather than the indefinite prolongation of the individual. Each generation thus returns, in a renewed form, to a common starting point. Similarly, the Universe could succeed itself within a continuum of systems in quantum symmetry, without keeping an explicit memory of its previous states. In this extrapolation, each cosmic "generation" would constitute a functional return to an initial state.

What we observe about the Universe necessarily relates to a more or less distant past. The current and effective composition of the Universe is only accessible at the scale of an immediate present. A significant part of matter, formed and aggregated later than the epoch we observe in the distance, therefore escapes our global estimates. Can we really extrapolate the total mass of a universe with unknown dimensions from a sample of local observations? Moreover, observation based on past EMW does not allow direct determination of the present state of remote areas. Interstellar clouds and diffuse materials, often barely visible, blur or obscure the received signals. To this are added the space-time distortions induced by the effects of gravitational lens and shear. The intrinsically obsolete nature of data from the distant contributes to minimizing the apparent insufficiency of matter in our cosmic neighborhood, making less obvious the need for a hypothetical dark matter on a small scale.

There is no way to directly update information from distant cosmological regions today. Noting the transformations that have occurred since the emission of the observed signals would require access to the present proper of these regions, which will only be conceivable, if necessary, in several billion years. And even then, the time lag issues would remain. If one admits that the Universe has neither center, nor edge, nor measurable volume, and that any measure must be corrected for evolutionary relativistic effects, it becomes evident that the notions of inflation or cosmic expansion remain open concepts, far from having delivered all their explanatory scope.

215

XVI <u>To comprehend more precisely the gravitational effects</u>
(A phenomenon that attracts, above all, curiosity)

An approach to quantum mechanics that ignores supra-atomic structures—molecules, stellar bodies, galaxies and macroscopic aggregates of matter — would make it particularly difficult to grasp the notions of the relativity of time and space. Conversely, an understanding of general relativity detached from fundamental quantum processes would be equally incomplete. The situation would be analogous to that of a knowledge of solfège dissociated from any experience with acoustic phenomena.

The fact that a particle of matter can, under certain conditions, annihilate upon its encounter with its symmetric particle constitutes a point of convergence between quantum mechanics and relativistic physics. Gravitation, by promoting the processes of gathering matter, could establish the conditions conducive to such extreme confrontations, leading, from a speculative perspective to a global collapse of the Universe. It therefore appears inseparable from a quantum description which, for its part, frames this evolution at the most elementary scales.

The relativity formulated by Albert Einstein is based on a deterministic framework aimed at predicting the trajectory of a body from space-time distortions induced by the presence of other masses. Conversely, quantum mechanics adopts a probabilistic description: the wave function does not provide a unique position, but a distribution of probability amplitudes associated with possible positions of a particle. These two localization modes appear a priori incompatible, while they pursue a common goal: to account for the fundamental interactions governing the evolution of the Universe.

These theoretical options are based on principles that seem difficult to reconcile. The same goes for the association, here proposed, between an unbounded Universe and the hypothesis of a virtual multiverse Cosmos. The ambition of this essay is precisely to go beyond this apparent border between, on the one hand, established physical laws and interpretations resulting from systems observed and experimentally validated, and, on the other hand, more speculative logical constructions, located on the sidelines of the dominant paradigms. The unification of electromagnetism and gravitation, as well as the coherent relationship between space and time,

216

constitute necessary conditions for the development of a unified theory envisaged here on an exploratory mode.

Gravitation remains one of the main points of tension in contemporary astrophysics. Does it justify the introduction of a hypothetical dedicated particle, the graviton? Gravitational force embodies above all the dynamics of assembling matter at all scales. Its origin could only become intelligible by examining the deepest structures of subatomic space, where force fields would play the role of space-time referentials within a quantum "dimension".

Describing the movement of a body essentially amounts to analyzing the variations in its trajectory induced by the gravitational effects it undergoes and generates, including during violent interactions such as collisions. For illustrative purposes, the experiment attributed to Galileo suggests that a cannon ball and a feather, released simultaneously in a uniform gravitational field and in the absence of an atmosphere, would reach the ground at the same time. This formulation is however idealized: the higher mass of the ball introduces an additional gravitational effect, certainly negligible, but non-existent for the feather. It follows that the inertia of a body is constantly modified, both in speed and direction, and that orbital trajectories at any scale are never perfect circles or ellipses, as illustrated by the problem of systems with three bodies or more.

Gravitation, material density and EMW intensity

The curvature of space induced by gravitation continuously modifies the movement of bodies, affecting their trajectories, their spatial organization and, in a relativistic context, their effective mass and density properties. On Earth, relative to the planetary center of gravity — region where gravitational effects compensate but average density is highest — materials of lower density are at a greater distance: oceans are located above rocks, and the atmosphere, even less dense, extends beyond liquids. And gases (atmosphere), even less dense, are more distant than liquids.

The majority of stellar bodies exhibit a proper rotational motion. This movement is not uniformly distributed in their internal structure: it generates

217

shear zones between layers, accompanied by temperature and density gradients. In terrestrial planets, the surface regions are generally colder than the inner areas. In stars, such as the Sun, thermonuclear reactions, combined with density effects, can lead to temperature maxima located in intermediate layers.

Gravitation can thus be described as a dynamic of contraction of distances, all the more pronounced as one approach the center of inertia of a massive body. Conversely, when one changes scale and the observation focuses on increasingly distant regions, distances seem to lengthen, with the impression that distant objects move away all the more quickly as they are distant. This impression results from the fact that our observations do not directly correct relativistic effects affecting distance and time measurements.

By analogy, when a vehicle moves away, the sound waves it emits stretch. However, the situation can be reformulated by considering not the movement of the vehicle, but the actual lengthening of the route, if it became increasingly winding. In a comparable way, the observation of the distant — therefore of the past — involves an information journey whose relativistic complexity largely escapes us.

In this perspective, it would not necessarily be the observed galaxy that moves away, but the distance separating us from it which seems stretched due to a past temporality flowing faster than ours. In a younger and less gravitationally structured Universe, photons constituted the referents of a faster effective time. The distances observed today therefore appear to us to be increased, not because we share this same temporal reference frame, but because we remain anchored in the proper time of a region of the Universe already strongly "hollowed out" by gravitation.

In this past Universe, the global gravitational effects were weaker, and the curvature of space less pronounced. Although the Universe is, on a large scale, isotropic and homogeneous, it is neither static nor immutable. The fact that past time seems to pass more quickly to us leads to the impression that the speed of light increases when we observe regions more and more distant. Galaxies and stellar bodies thus give space-time its flexibility through the gravitational effects they induce, which results in a gradual elongation of wavelengths from the past.

This phenomenon, commonly interpreted as a Doppler effect, led to the idea that one could go back in time and estimate the age of the Universe. It also inspired the hypothesis of photon aging, which, although partial, is part of a non-expansionist representation of the Universe.

Fossil radiation (RFC), often invoked as an argument in favor of the Big Bang scenario, does not validate the existence of an initial point singularity devoid of volume. If it testifies to a dense and energetic past, it does not necessarily imply an expansion of galaxies in relation to each other. Such an interpretation neglects the evolution of the curvature of the "empty" space and underestimates the central role of general relativity in describing cosmic evolution.

In other words, the radiation that transmits to us the image of distant objects or events are EMW from the past. Their spread across increasingly differentiated regions of space-time affects them all the more as it extends over long periods. The observed elongation of wavelengths is then interpreted as an increase in distances, while it also results from the double variability — local and cumulative — of time and space, as formalized by general relativity.

A black hole, in which energy is confined to an extreme degree, could be distinguished from any other astrophysical object by the absence of differentiated density strata. Everything that crosses the accretion disk would remain only in the form of energy, deprived of classical interactions, in a limit state comparable to a radiative plasma devoid of effective mass, viscosity and internal gradients, due to an almost perfect energetic homogeneity. This state cannot, however, be described as a «liquid» in the usual physical sense, the energy density reaching values incompatible with the states of matter that we know characterize.

The macroscopic aspect of a black hole suggests an inert object, stable and devoid of observable internal activity, although the final destination of the energy it absorbs remains unknown. In the absence of direct observational data, its intrinsic state remains inaccessible. Its position, on the fringes of space-time as we experience it, might no longer allow for distinguishing states analogous to 'super solid' or 'superfluid', and may not correspond to any familiar physical category.

Could the elementary particle—considered as a packet of waves — intended to cross the horizon of a black hole, also fall under a comparable state indetermination. This ambiguity, assimilable to a superposition of possible states, raises the question of its origin. For both the elementary particle and the black hole, the decoherence process may not apply, as these systems escape any direct physical representation. It follows that no wave function, in the usual operative sense, can exhaustively describe either an isolated elementary particle or a black hole. In both cases, these are systems of a quantum nature, closed to experimental introspection and irreducible to a classical phenomenological description.

We can qualify states such as solid, liquid, gaseous or plasmatic because they are observable and characterizable experimentally. On the other hand, we have no solid information allowing us to imagine what a black hole or an elementary particle are, fundamentally and intrinsically, as ultimate quantum objects.

Inside a black hole, atomic structures would be deconstructed. The energy would reach a maximum in a state comparable to a frozen disorder. In a manner analogous to the supposed conditions of the primordial Universe, the mediating photons of electromagnetic interaction would no longer be distinguished from the other energetic components. The black hole could then be understood as a black body in the strict sense, devoid of proper thermal radiation, apart from that coming from its accretion disk and the relativistic jets it emits. It can also be considered as a sink with no return, without a definable spatial dimension or temporality, in which any structure returns to a primordial state.

Nevertheless, from the observer's point of view, the black hole is not dissociable from our space-time, since it manifests itself indirectly through accretion phenomena, jet emission and gravitational lensing effects. In these extreme conditions, the relationship $E=mc^2$ loses its operational scope, due to a lack of measurable parameters allowing to distinguish mass, energy and spatiotemporal referentials.

Matter and antimatter would be expected to merge into a restored virtual state, where Planck's magnitudes like the current physical laws would cease to have any operational meaning. In this ultimate state, kinetic energy, matter, space and time would no longer be differentiated, suggesting the absence of intrinsic physical reality of the multiverse Cosmos.

This final singularity would restore to the multiverse Cosmos the energy it contained. It is then conceivable that a new binary system of the Universe, in quantum symmetry, could emerge elsewhere, independently of any direct causal relationship between collapses and successive Big Bangs. In this perspective, there would not be a single continuum, but an uncountable multiplicity of disjoint space-time couples.

Gravitation, space and time in a context of symmetry

The two symmetrical states envisaged would share a temporality that does not correspond to that of our physical experience. In the absence of direct representation, one can propose an analogy with a reflected and shifted image in a system of non-coplanar mirrors: the image is not accessible to the direct visual field, but it reacts in a strictly correlated manner to the model, with a delay close to corresponding to the light propagation time.

Every particle has an associated antiparticle. In their proper symmetry, both can be considered as emanating from entangled waves, whose fluctuations and oscillations would reflect a fundamental cosmological imbalance. The atom constitutes, in this perspective, the minimum accessible representation of these aggregates of wave packets, although it is only a derived manifestation.

To illustrate this relationship, particles and antiparticles can be compared with two discontinuous lines of complementary properties, parallel and closed on themselves, whose segments compensate each other. In the absence of a slight shift in symmetry — a residual chirality—these two lines would merge into a single continuous trajectory, neutralizing any observable distinction.

Fermions, with the exception of near-zero mass neutrinos, carry electrical charges that allow the relatively stable assembly of matter. Thus, the proton-electron couple can remain permanently, outside of major nuclear events. Such events occur notably during the gravitational collapse of a supernova, leaving behind a neutron star. In this process, protons are converted into neutrons by electron capture, with neutrino and positron emission, ensuring the conservation of charges and energy balances.

221

The neutron star, like any high mass object, attracts surrounding matter, which is gradually neutralized. Nothing indicates, however, that these objects are perfectly homogeneous. Convection zones and load irregularities could remain, allowing the presence of heavy quarks (strange, charm, bottom, top). The interactions between these non-uniform regions could produce X-ray and gamma-ray emissions. Unlike a black hole, the gravitational field of a neutron star is insufficient to confine this radiation, which reaches us in the form of periodic signals aligned on the magnetic axis of the star: these objects are then observed like pulsars.

The neutron star can thus be described as an unfinished black hole. It is however distinguished by the emission of detectable radiation and by the absence of a dominant accretion disc. Its rapid rotation and the shift of its magnetic poles structure these emissions.

For masses between about 1.4 and 3.2 solar masses, electrons and protons assemble into stable neutrons. Beyond this threshold, composite structures disintegrate, merge, and energy returns to a primordial, cold and undifferentiated state, characteristic of black holes. An increased porosity of the transition zone between quantum symmetries could then explain the appearance of particle-antiparticle pairs, observed during the emission of high-energy jets by certain galactic black holes.

The limits of gravitation

Experiments involving two conductive plates closely spaced show that the space between them tends to be depleted of free particles — notably electrons — and electromagnetic waves (EMW), in particular photons. This phenomenon suggests that matter can, by drawing energy from what is commonly called a "void", induce an extremely weak but real rapprochement with the surrounding matter. The result is a local increase in the energy density of bodies, correlated with a depression of space at the intra-atomic, intermolecular and interstellar scales.

In counterpoint, the Coulomb repulsion between atomic nuclei of the same positive charge, mediated by the sharing and exchange of electrons, opposes these approximations and prevents the spontaneous fusion of atomic and molecular structures. This electromagnetic interaction thus constitutes an effective limit to the gravitational collapse of ordinary matter.

Gravitation, by locally increasing the energy density, generates an increase in molecular agitation, accompanied by a rise in temperature. The energy extracted from space then manifests transitorily in thermal form. At the atomic level, and even more so in nuclear fusion processes, the atom can be seen as a strongly depressionary system, which tends to strip the surrounding space of its energy. This dynamic is amplified at the molecular level, then macroscopic.

Gravitation is generally understood as a phenomenon associated with mass. This property of stellar bodies gives space-time its flexible and dynamic character. Conversely, the fact that elementary particles do not seem to occupy space in the classical sense gives the impression that they do not modify its properties in the same way as massive objects. It results in an apparent incompatibility to link, on the one hand, the gravitation that structures space-time in relativity, and, on the other hand, the electromagnetic and nuclear interactions that underpin quantum mechanics and seem, for their part, to disregard space-time.

This tension however does not prevent considering a rapprochement between these fundamental interactions, notably through their integration into the electromagnetic force, considered here as a possible unifying vector.

Gravitation is fully expressed only at the scale of stars. Beyond a certain threshold of density, the electrostatic repulsion between atomic nuclei becomes insufficient to counterbalance the cumulative gravitational pressures. These latter result from the intensification of internal interactions in nuclei during nuclear fusion processes, interactions that can be assimilated, in this approach, to an effective manifestation of the so-called strong force. When this equilibrium is broken, gravitational collapse leads to the formation of a black hole, often at the end of a hypernova. In such an object, atoms and molecules, deprived of the space indispensable for any interaction or movement, cease to be identifiable as distinct structures.

223

In order to resolve the apparent contradiction between an omnipresent gravitation and the idea of an expanding Universe, and to explain the origin of mass, the existence of a new particle has been postulated: the Higgs boson. This is supposed to confer mass on bosons that are initially devoid of it, thus allowing the appearance of massive intermediate bosons. The latter would make it possible, indirectly, to assign a mass to fermions, depending on the nature of their electromagnetic and nuclear interactions.

In this perspective, the mass of matter particles would be explained by the existence of adapted force-vector particles, capable of accounting for variations in mass and density that were hitherto difficult to interpret. The Higgs boson would thus be invoked to justify the coherence and synergy of the world of elementary particles.

However, if a particle of matter can be described as a probability wave function, the very notion of mass raises questions. Mass, associated with gravitation, is generally interpreted as the expression of inertia, that is to say the resistance of a particle to any variation in speed or direction. Was it then necessary to postulate the existence of an additional particle to explain this property?

Such an approach neglects the fact that the so-called "empty" space is not empty. Inseparable from time, space constitutes an emerging context stemming from quantum mechanics, accessible to macroscopic observation. This apparent void hides considerable energy, in the form of fluctuations and potential pairs particle–antiparticle. Space-time can thus be seen as an event macrocosm where the energy of vacuum manifests itself in an attenuated manner.

In this perspective, mass could be understood as an indirect manifestation of discrete interactions between two symmetrical states of matter. The mass of a particle would then be the observable expression of an entangled wave packet, in potential interaction with its symmetry. In the absence of antimatter, matter could not manifest mass, and perhaps even could not exist. The Higgs boson could then be interpreted as a formalization of the effects of antimatter on matter, accounting for the resistance of bodies to changes in state of motion.

The space vacuum also contains free particles, neutrinos, as well as a tangle of electric and magnetic fields. It is therefore not neutral with regard to the observed interactions. General relativity is precisely based on the idea of multiple and evolving frameworks, where no magnitude is absolute. This dynamic occupation of interstellar space could be sufficient to explain the emergence of a mass effect, all the more marked as particles are aggregated into atoms, molecules, then stellar bodies.

On a large scale, this effect manifests itself in the form of gravitation, which locally determines the mathematical structure of space and time. The notion of mass then appears as a measurement parameter intended to account for the effects of vacuum energy on matter, by attributing this property to elementary particles. What is valid at the microscopic scale is also valid for stellar bodies, whose gravitational collapse marks the limits of a process structuring the dynamics of the Universe.

The Higgs boson could thus be reduced to an entity similar to a reconfigured neutrino, of high mass, extremely unstable, conferring by its presence a mass on the other particles. However, it would only explain a minute fraction of the observed mass and would not provide any definitive insight into the nature of gravitation. Its existence is based on delicate indirect observations, susceptible to oriented interpretations.

Devoid of charge, spin and intrinsic kinetic or magnetic moments, this boson, not directly observable, is invoked to explain mass transfers during particle collisions. One may then wonder if it is not above all an elegant hypothesis, developed to respond to persistent difficulties in conceiving otherwise.

This question invites a parallel with other hypothetical particles, such as the neutralino, resulting from supersymmetry. This theory, in search of completion, proposes an extended symmetry that questions the very role of bosons as vectors of force.

It is therefore legitimate to question the multiplication of particles postulated in the extensions of the standard model. While this approach has often been successful, it can also be a default response to poorly understood phenomena. The Higgs field, as described, evokes an ambient environment in which particles would bather, recalling the ancient hypothesis of ether intended to characterize the space vacuum.

225

We can finally consider that mass is not an intrinsic property of matter, but an emergent property resulting from the primordial radiative entanglement between particles and antiparticles. A symmetry chirality still not recognized could then account for not only the mass, but all the distinctive properties of matter particles, which would exist only by reference to antimatter.

It clearly appears that the physics of the infinitely small is today in search of rapprochement. An interpretative approach compatible with the idea of symmetry would be to suppose that the properties of quanta are determined by the nature of the osmotic interactions they maintain with a "mirror" Universe, governed by complementary quantum symmetry. In this approach, the observable characteristics of the particles would not be intrinsic but would result from exchange conditions between two symmetrical states of reality.

In the Standard Model, the Higgs field is assumed to interact with elementary particles in order to give mass to subatomic composite hadrons and their associated electron clouds. However, it is well established that most of the mass of a composite particle comes from the binding energy between its elementary constituents. These binding forces contribute to the mass of hadrons, nuclei, atoms, molecules and, by extension, all astrophysical objects, from planets to galaxy clusters. The Higgs boson cannot therefore, on its own, account for the global origin of mass.

Yet, understanding the mass is a central issue. It is even fundamental, because it conditions the possibility of lifting the apparently contradictory and counterintuitive character of quantum mechanics, by establishing a link with the gravitational effects that structure relativistic space-time. In other words, the question of reconciling the four fundamental interactions in a unified theory remains open, but could lose its relevance, formulated differently.

The hypothesis of a radiative entanglement, considered here as a constitutive principle of the elementary particle conceived as a packet of infrangible waves, leads to postulate the existence of a binding force of very high intensity and extremely short range. This force would ensure the particle's integrity and internal stability. It then becomes legitimate to question the reasons why such a force, intrinsic to the elementary particle and confined to the "dimension" quantum, would have no effect on the non-quantum

226

phenomena that we classically describe in terms of time flow and movement in space.

In this perspective, mass and moments of inertia would only be observable manifestations, for the macroscopic observer, of unrecognized internal forces that structure and maintain the wave packets constituting elementary particles. These intrinsic forces, common to the particle and the antiparticle, escape any spatiotemporal description, since the fundamental properties of elementary particles belong neither to time nor space. Rejecting the hypothesis of radiative entanglement then leads to introducing an omnipresent energy field—in this case the Higgs field — supposed to fill the space and oppose the movement of particles. One can therefore wonder if the Higgs boson does not primarily meet the need to preserve the internal consistency of a standard model in lack of coherence.

Quantum mechanics is partly based on postulates and conventions. This is notably the case of the introduction of bosons unobservable directly but deemed necessary to account for phenomena otherwise difficult to explain. It then became logical to assign a mass to some of these bosons — in particular the W and Z bosons — in connection with the nature of the interactions studied. Some of them even present an electrical charge, which raises the question of whether these properties are not introduced to meet the requirements of the formalism.

The Z boson, a close relative of the photon with an extremely short lifetime, allows modeling the decay of lepton–antilepton and quark–antiquark pairs. The W boson, equally ephemeral, is considered as the mediating agent of electroweak interactions between quarks and leptons and intervenes in nuclear fusion processes. The existence of these bosons, which have never been observed directly, retains a speculative character, although it is firmly rooted in mathematical coherence aimed at introducing intermediate force vectors between elementary particles.

These bosons are essential insofar as they allow the modeling of fleeting transitions, particularly those associated with the change of quark flavor, a process attributed to weak interaction, which can be unified to electromagnetism at certain energies.

The photon and the gluon are the two bosons of the standard model that are distinguished by the absence of mass and charge. They are associated
227

respectively with electromagnetic interaction and strong interaction, the latter itself being likely to be related to electromagnetism. The Higgs field, transparent to these two bosons, could then be interpreted not as a universal inertial medium, but as a specific exchange zone between particles and antiparticles, located at the border of two quantum symmetries of the Universe.

Bosons can thus be considered as quantum entities, agreed upon, intended to represent the exchanges of energy or information between particles. Even though the Higgs boson may have existed as a high mass particle, it only appears today in a fleeting and indirect way during high energy collisions, such as those carried out at CERN in 2012. Modern accelerators indeed allow the artificial production of extremely unstable exotic particles, which only exist temporarily and reflect local imbalances in forces.

It then remains to understand why the Higgs boson would possess a proper mass, even though its supposed function implies that mass is not an intrinsic property of isolated particles, but a contextual effect. Unless we give this enigmatic particle a different interpretation and not matched by who holds the pen, it is often by using such shortcuts that at the limits of the abstract, things fall within the framework that we want to give them.

The cohesion of quarks is ensured by strong interaction. It might be tempting to see in it the ultimate origin of gravitation. It is indeed established that this interaction increases in intensity when the quarks, generally assembled in groups of three, seem to move away from each other. This restoring force could be explained by the fact that quarks are entities outside time, and that the nucleons they compose do not occupy a space where an effective separation would be possible.

For photons, considered as corpuscular representations without mass of electromagnetic waves, time does not exist as a quantity of duration, and space does not have the usual meaning we assign to it. This would be enough to explain that the particles endowed with mass, conceived here as packets of synthesized entangled waves, are themselves on the fringes of space-time. What is out of time then becomes spatially virtual, which reflects the experimental impossibility of 'breaking' an elementary particle.

This situation does not exclude that, by changing scale, quantum mechanics eventually fits into the space-time of relativity, starting from the atomic

228

level. In this hypothesis, gravitation would become a phenomenon of emergent quantum origin. The atomic nucleus could then constitute the threshold at which gravitation begins to become significant, stripping space of vacuum energy. Vacuum energy, carried mainly by free electromagnetic waves, indeed represents a considerable mass potential according to the relationship $E/c^2=m$. On a large cosmic scale, this energy, just like matter, could be gradually absorbed by a growing population of black holes, contributing to the ultimate dynamics of the Universe.

Attempt to relate electromagnetic interaction and gravitation

In the approach proposed here, the electromagnetic interaction—considered together with the weak interaction in the context of electroweak unification — is envisaged as being able, at a fundamental level, to also encompass the effects attributed to the strong interaction. This hypothesis is part of a speculative unifying perspective, which does not correspond to the current state of standard models, but seeks to propose an alternative reading of it.

The energy transported by photons can, in accordance with the principle of mass equivalence–energy, be formally associated with an equivalent mass. In this perspective, mass is not interpreted as an intrinsic and primitive property, but as an emergent relativistic quantity, reflecting the degree of entanglement and radiative coherence of wave packets constituting so-called massive particles. Macroscopic material systems would then be understood as complex sets of such interacting wave packets.

On this basis, it is postulated that the gravitation associated with material bodies would find its origin in the collective state of entanglement of the waves that constitute matter. These waves, originally, would be of the same nature as those described here under the term "EMW", considered as vectors of electromagnetic interaction. The distinction between radiation and matter would then result from a different regime of organization of waves with the same fundamental component.

229

When these electromagnetic waves are absorbed or mediated by massive, charged particles, in particular electrons, electric fields are formed. These reflect energy transfers between charges and are described by field lines whose geometry depends on the configuration of the charges: open lines in the case of charges of the same sign, closed lines connecting charges of opposite signs. The intensity of the electric currents thus generated depends on the amount of energy extracted from atoms by interaction between incident radiation and electrons, as in the photoelectric effect.

A central point of the reflection concerns the systematic association between electric fields and magnetic fields. Unlike electric fields, magnetic fields do not directly perform energetic work on the charges. Their local intensity nevertheless depends on the value of the electric current and the distance to it. Magnetism is then interpreted not as an autonomous source of energy, but as a geometric manifestation resulting from the collective movement and intrinsic angular momentum (spin) of electrons circulating between atoms and molecules in the form of a current.

In this hypothesis, the magnetic fields would correspond to a local modification of space-time properties induced by these currents. Above a certain threshold of scale and collective coherence, these distortions could produce effects similar to those described by gravitation. Gravitation would therefore not be an independent fundamental interaction, but the macroscopic expression of quantum electromagnetic effects integrated into a geometric description of space-time.

This interpretation leads to consider gravitation as an emerging consequence of omnipresent magnetism associated with charge interactions, assuming that global quantum symmetry would play a structuring role. The hypothesis of chirality matter-antimatter, involving particles and antiparticles of opposite charges, reinforces the idea that gravitational effects in each sector of symmetry would fundamentally rely on corresponding magnetic fields.

In this type of reflection, a magnetic field can be described as the result of a local deformation of the space under the effect of an electric current passing through a medium of finite conductivity. This deformation would modify the trajectories of bodies without it being necessary to invoke a force in the Newtonian sense. Electromagnetic fields and gravitational effects would thus be two manifestations, at different scales, of the same ability of matter and radiation to alter the geometric properties of space-time.

230

General relativity having established that space and time are not absolute entities, this reading suggests that gravitation could, in the final analysis, be related to phenomena pertaining to quantum mechanics and charge interactions. She thus proposes an interpretative quantum extension of the classical geometric description of gravitation.

Concerning antimatter, it is consistent to assume that it can generate electric currents from oppositely charged particles, such as positrons. Although direct interactions between antiparticles remain largely inaccessible to experimental observation, the magnetic fields associated with antimatter should, in this hypothesis, contribute to modifying the properties of space in a manner analogous to those produced by ordinary matter. This could result in extremely weak polarization effects of electromagnetic waves, manifesting as spatial disturbances that are difficult to detect.

On a very large scale, such disturbances could affect the filamentous structure of cosmic matter, inducing slight field variations resulting in minute oscillations of these filaments. Such an observation would however suppose access to a currently unobservable sector of the Universe, where antimatter would be spatially or dimensionally separated.

Finally, a formal analogy reinforces this attempt at rapprochement: just as the gravitational mass of electrically neutral bodies determines their gravitational interaction, electric charges and currents determine magnetic fields in electromagnetic interaction. The two types of interaction share the property of modifying the characteristics of space. Coulomb's law, like the law of Newtonian gravitation, involves an inverse dependence on the square of the distance, suggesting a common mathematical structure. Although electromagnetic effects tend to neutralize at great distances, they remain, like gravitation, of theoretically infinite range.

In a more conventional interpretation, gravitational effects can be understood as resulting from the collective organization of matter. The particles constituting celestial objects, by binding through electromagnetic and nuclear interactions, would locally reduce the degrees of freedom of the so-called empty space. Each particle could then be interpreted as a local space-time singularity, analogous, on an extreme scale, to a gravitational depression. This deformation, described by general relativity, manifests itself as a contraction of lengths and an expansion of time near masses, effects perceived by an external observer as gravitational attraction.
231

Similarly, a body subjected to prolonged acceleration can acquire increased effective inertia, producing effects comparable to an increase in gravitational mass.

Synthesis on the electromagnetic origin of gravitation and the limits of current theoretical frameworks.

In the perspective developed here, gravitational interaction is considered not as an independent fundamental force, but as an emergent phenomenon resulting, at a deep level, from electromagnetic interactions that ensure, in quantum mechanics, the cohesion of the atomic nucleus and the overall balance of charges within the atom.

Since matter is made up of particles bearing positive and negative charges, electrical neutrality generally appears to be achieved at atomic, molecular, and macroscopic scales. However, this global neutrality does not imply the strict and instantaneous cancellation of the effects associated with load distributions. It is then postulated that the set of positive charges and the set of negative charges constituting a body could exert, separately and simultaneously, an extremely weak collective electromagnetic influence on the charged constituents of other material systems. Although this interaction is theoretically infinite in scope, its intensity decreases rapidly with distance, making it undetectable on a large scale. Nevertheless, cumulated and integrated over macroscopic volumes, it could contribute in a highly diluted way to the deformations of space-time attributed to gravitation.

In this same interpretative mode, the strong nuclear interaction is considered as a specific intranuclear manifestation of electromagnetism, adapted to extreme conditions of confinement and energy. This interaction, which guarantees the relative stability of atomic nuclei, could, by extrapolation, be associated with the attractive effects observed at the macroscopic scale, described in general relativity as a curvature of space-time. This hypothesis thus establishes a continuous link between electromagnetic interaction, strong interaction and gravitational phenomena.

232

Electromagnetism, an intrinsically quantum interaction, is then interpreted as a fundamental structuring principle at the origin of the very notions of space and time. In this perspective, quantum mechanics and gravitation would not be fundamentally incompatible. Their apparent disjunction would mainly result from a change of scale and context. Descriptions from general relativity would then become inadequate to account for quantum phenomena, without this invalidating their phenomenological effectiveness on a large scale.

This coexistence of successful but not necessarily related theories, led to the introduction of various hypothetical elements intended to preserve the formal coherence of the models. Among these are notably:

- the hypothesis of a dark matter not directly observable,
- that of a dark energy of unknown nature,
- the introduction of an initial singularity of infinite density as the origin of the Universe,
- the idea of an indefinite cosmic expansion,
- the unexplained disappearance of antimatter,
- the use of still incomplete theoretical models such as string theory or loop quantum gravity,
- partially substantiated conjectures about black holes,
- and the notion of indivisible elementary particles associated with interacting mediating particles.

The accumulation of these hypotheses suggests the possibility that our current understanding of the Universe is, in significant part, incomplete or biased. The very position of the observer imposes a description of reality limited by the conditions of measurement, perception and representation specific to our scale of existence, as the history of science has shown repeatedly.

Despite the abundance of theoretical avenues concerning cosmic evolution, attempts to unify all fundamental interactions into a coherent theory remain limited. It is then legitimate to question the adequacy of our exploration methods and tools. Although gravitation, the electroweak interaction and the strong interaction are described by distinct mathematical processes, it seems unlikely that they do not share, at a fundamental level, common mechanisms.

233

In the standard cosmological scenario, the initial phase assimilated to the Big Bang, described as a singularity without defined temperature, is followed by a nucleosynthesis phase characterized by extremely high temperatures. Under these conditions, the strong interaction is distinguished from a more global unified interaction, referred to as electroweak. The gradual decrease in temperature then leads to a differentiation of this electroweak interaction into weak interaction and electromagnetic interaction. Conversely, in environments of extreme pressure and temperature, the distinctions between electromagnetic and weak interactions tend to fade, while the strong interaction remains confined to the atomic nucleus. This feature suggests that the strong force results from intrinsic and nuclear-specific charge interactions. It is then conceivable that beyond a certain threshold of energy and scale, the three interactions merge into a single interaction.

Although the classical analogies between atomic structure and planetary systems are today recognized as inadequate, gravitation can nevertheless be considered as a phenomenon of quantum origin, in principle falling under a quantum field theory. The major difficulty lies in reconciling this idea with the description of particles as entities without spatial extent, and with the existence of singularities such as the Big Bang or the final gravitational collapse. On the other hand, any observable interaction necessarily involves spatiotemporal involvement.

This paradox highlights the intrinsic limits of our representation of reality, based on measures of space and time and on observable phenomena, while ignoring non-accessible sectors, such as antimatter. Entities or processes without observable spatiotemporal referents remain in the state of hypotheses. Considering objects existing only in their potential state, such as elementary particles before measurement, runs against the very foundations of our mode of description. This difficulty manifests itself in the impossibility of simultaneously determining, at a given moment, the exact position and the speed of a particle, thus revealing the fundamental limits of any classical representation of quantum reality.

XVII <u>A gravitation that takes effect at the root of time</u>
(Of quantum origin, it draws the topology of the Universe)

If one assumes that the energy revealed during the Big Bang manifested from a multiplicity of simultaneous points of release, it is not necessary to introduce the classical notion of metric expansion of space, nor that of an associated escape velocity. The evolution of the Universe is no longer interpreted primarily in terms of kinematic distance from objects, but as a continuous set of interferences, exchanges and energy interactions. The Universe can then be considered as globally finite, without its effective dimensionality being strictly defined, and governed by non-local correlations linking all of its constituents.

In this perspective, a binary system of universes in opposite quantum symmetry can be interpreted as an energy substitution process: the progressive conversion of a latent primordial energy into a potential structured energy, carrier of quantum symmetry. The energy initially released during the Big Bang, comparable to a founding kinetic energy and prefiguring the EMW, is uniformly distributed in a space-time continuum that it itself helps to define. This dynamic is perceived as an apparent expansion, or more precisely as a retrograde dispersion: a diffusion process that nevertheless remains globally contained by the very structure of the system.

Conversely, potential energy acts as a cohesion factor. It promotes the aggregation of matter and the capture of radiation, a phenomenon that we interpret on our descriptive scale as gravitation. Apparent expansion and gravitation thus appear as two complementary manifestations of the same underlying mechanism, difficult to conceptualize directly: an energy depression associated with space commonly referred to as "void".

If this scheme is valid for a given symmetry, it must also be so for its opposite symmetry. The joint effects of dispersion and gathering, grouped under the term retrograde dispersion, could then result from discrete interactions between two inseparable and complementary quantum symmetries. Gravitation ceases to be an isolated fundamental phenomenon to become the macroscopic expression of these inter-symmetrical exchanges.

235

Certain particles play a particular role in this device. Massless and charge-free bosons, such as photons, as well as gluons envisioned as their transposition into the field of strong interactions, are generally considered to lack proper quantum symmetry. Other particles, such as neutrinos, of extremely low mass and electrically neutral, are sometimes supposed to carry both symmetries simultaneously. These characteristics give them the status of privileged vectors for possible "osmotic" exchanges between quantum symmetries.

Thus, at a certain energy threshold, the absorption of a photon by an electron can lead to the appearance of an antielectron (positron). The meeting of these two particles of opposite symmetry then leads to their almost immediate annihilation, accompanied by the emission of gamma radiation of equivalent energy. Similarly, neutrinos reveal their associated symmetry, in the form of antineutrinos, during certain specific nuclear interactions.

High-energy collisions between nucleons can also lead to the transient appearance of antiprotons and antineutrons, giving rise to annihilation processes via mesons with extremely short lifetimes, locally modifying the composition of the atomic nucleus. In this context, photons and neutrinos can be interpreted as major mediators, capable of crossing the boundaries between symmetries.

Dynamic evolution of dispersion

At the first moments of the Universe, the retrograde dispersion of particles can be represented as the set of more or less inflected trajectories within a multitude of nested energy fields, without fixed dimensions. Gradually, under the cumulative effect of gravitational interactions, the displacement directions tend to become more and more tangential with respect to dominant energy centers. This evolution results in increasingly curved trajectories in a so-called empty space whose energy density decreases continuously.

Such dynamics could suggest, by extrapolation, a global regression of the aging Universe towards a Big Crunch-like state, conceived as the temporal inverse of the Big Bang. However, this interpretation should not be retained in its classical sense, because it does not consider the complexity of the energy and metric mechanisms at work.

Paradox of apparent expansion

The apparent expansion of the Universe manifests itself through an observable paradox. On the one hand, we perceive vast apparently empty regions, dotted with clumps of matter seeming to move away from each other. On the other hand, we find that matter tends to gather in an increasing way, reaching locally extreme densities which lead to changes of physical regime. We also know that the initial dispersion cannot be maintained after the appearance of massive particles, and that the energy intensity of the EMW decreases over time.

The impression of increasing distances could then result not from an effective expansion of space, but from the increase in the equivalent mass of black holes and the gradual densification of stellar bodies destined to collapse gravitationally. This evolution would induce a form of negative pressure on interstellar space, leading the observer to interpret the phenomena without properly adjusting the analysis scale. Like galaxies, black holes, increasingly numerous and massive, seem to move away from each other. This apparent expansion could thus only be an illusion related to the progressive energy depletion of space-time, energy — in all its forms— being gradually captured by a growing population of gravitationally dominant structures.

If the Universe cannot be strictly assimilated to a black hole, its evolution could nevertheless converge towards an analogous global state, resulting from the ultimate fusion of all black holes, forming a singularity on the scale of the Universe itself.

Observational limits and metric of space

Formulated differently, apparent expansion could result from properties of the space metric that we only partially master. This scenario remains

237

compatible with general relativity, which predicts that gravitation distorts space-time and forces movements to follow geodesic trajectories. These trajectories, however, do not necessarily correspond to the intuition of a flight movement in a Universe presenting no material escape.

The determination of an object's intrinsic properties, and in particular its mass, depends in practice on its distance from the observer. The parallax method, based on the position of the Earth at six-month intervals, is the most direct tool for measuring these distances, but its accuracy decreases rapidly with the distance from objects. Moreover, the line of sight cannot be assimilated to a line: the analyzed radiation was continuously deflected by the curvature of space and interfered with other emissions, in a constructive or destructive way.

The apparent homogeneity of the Universe on a large scale can thus give the illusion that the images observed belong to our current reality and have traveled without significant alteration. Now, the light spectrum of a star essentially comes from its outer layers. While spectral analysis allows the inference of the composition and global evolution of the planet, it does not guarantee that information from the deepest regions is fully accessible. The question therefore remains open as to the real capacity of these observations to reveal the nature of the physical processes that take place at the heart of astrophysical objects.

Spectrometry is a central tool for estimating the displacement and distance of astrophysical objects. It is based in particular on the use of distance indicators serving as references, called standard candles. The principle consists in comparing the apparent magnitude observed of an object with its supposed known absolute magnitude, after correction for instrumental and environmental effects. This method is particularly applied to supergiant stars with regular variations in brightness, such as the Cepheids, whose cycles are considered sufficiently stable to serve as cosmological standards.

However, it is not certain that the theoretical absolute magnitude, deduced from local and contemporary data, actually corresponds to the intrinsic luminosity that a distant star possessed at the time when the radiation we observe was emitted. The measured magnitude remains, by definition, an apparent magnitude, affected by the environments crossed, the gravitational
238

fields encountered and multiple phenomena of interaction with the cosmic environment. Then arises the question of our real ability to reliably correct all these effects, especially for the most distant objects.

This uncertainty is further reinforced in the case of supernovae, whose brightness peaks also serve as distance markers. The interpretation of these light signatures presupposes a precise knowledge of the physical mechanisms, characteristic durations and initial conditions specific to each event. However, these parameters remain partially undetermined, which limits the robustness of reconstructions based on these observations. It is also not demonstrated that the relationships established for the evaluation of the absolute brightness of nearby cepheids can be extrapolated without bias to the most distant standard candles.

We also know that gravitation distorts space-time and, consequently, the distances themselves. The question arises as to whether this deformation is correctly considered in current cosmological estimation methods. There is reason to fear that, in this context, general relativity will be applied in a simplified manner or even reduced to an insufficient approximation for very large-scale evaluations.

The astronomical line of sight, often amplified by gravitational lensing effects, corresponds in reality to a time travel. It cannot be assimilated to a rectilinear trajectory, even if the Universe gives the impression, on a very large scale, of evolving slowly and in a homogeneous manner, without immediately discernible local phenomena. This apparent regularity however makes the interpretation of its evolution all the more delicate.

This difficulty joins a widely accepted hypothesis according to which a Universe devoid of privileged center and definable borders should present, globally, no major disparity. Conversely, if one posits that our Universe was born from a singular point and is in continuous expansion, it may seem paradoxical that it appears similar everywhere, even after correction of local anomalies. This objection leads to consider, as retained here, that the dynamics of the Universe does not necessarily imply an expansion in the strict sense.

When speaking of expansion, we implicitly assume a change in the volume occupied by the Universe. Such a notion implies that one can infer the global
239

properties of the Universe from the only observable part. However, this extrapolation becomes problematic if one admits that the Universe initially represented, during the Big Bang, only a non-localizable point, devoid of spatial reference. The evolution of such an initial state towards a geometry with closed curvature, without clearly defined limits, is more a postulate than a conclusion supported by direct observational data and does not in itself constitute a demonstration.

Then a central question arises: how can the Universe, within the limits of the observable, give the impression of being globally homogeneous and isotropic, even though we know that it is continuously evolving, and that what we observe results from a mixture between a present of proximity and a distant past? The Universe necessarily had to present different states in this past that we try to reconstruct by observing its boundaries. Before answering it, it is necessary to question the meaning of the observable part of the Universe with regard to a total whole of which we have no direct representation.

This reflection inevitably leads to the imagination that the relativistic Universe would possess dimensions giving it volume, and therefore limits. However, it is not possible to determine whether these properties evolve, nor in relation to which reference frame they could be defined. Therefore, considering the Universe as an object with a precise geometric shape becomes problematic, since it is supposed to have neither identifiable center nor measurable edge.

The space-time dynamics of the Universe is based on relativity. When it is claimed that the Universe is homogeneous, this claim really only concerns its observable part, and associate events from very different times. How, then, could the complex image received from the confines of a younger Universe be identical to that we have of a closer, and therefore more recent, Universe? The reconstruction of cosmic history from these temporally shifted, altered and distorted images remains inherently uncertain.

The hypothesis of homogeneity, according to which the Universe would evolve in the same way at all points, then raises two fundamental questions.

On the one hand, how could the most distant events — remnants of a past accessible only to the confines of the observable Universe, or even beyond —be related to the current local dynamics without reconstructing the entire past evolution of the Universe? On the other hand, since time is not an absolute quantity, how could events located in distant frames of reference be considered as connected, and even more so as simultaneous?

These questions do not call into question the general principle of causality, which remains compatible with the relativistic theory of gravitation, but emphasize its interpretative limits on a large scale.

A peculiarity of gravitation is that it can be considered as being exerted everywhere according to the same fundamental laws, in globally comparable conditions, despite its inverse dependence on the square of the distance. At scales much higher than macroscopic, it becomes conceivable to consider that gravitation animates the Universe in a relatively uniform way, when considering all of its cumulative effects.

Gravitation can then be interpreted as the manifestation of the energetic relief of the Universe. Any concentration of mass locally modifies the structure of spacetime and acts as an energy sink, likely to influence any form of energy present in a wide neighborhood. This influence can lead to the capture or deflection of incident energy, depending on local conditions. However, this representation in terms of gravitational wells scattered throughout the Universe remains partial. It can be replaced by a more global vision, in which the entire Universe could be assimilated to a single energy structure, oriented towards a more fundamental level, possibly associated with an underlying multiverse.

In the Universe, the dominant dynamics seem to favor systems with the highest concentration of energy in the smallest accessible volume. This predominance can be interpreted not as a «law of the strongest» in the metaphorical sense, but as the physical consequence of the role played by energy density. A body or system with highly concentrated energy exerts a decisive influence on its environment, notably through gravitational interactions.

A fundamental question then remains open: does matter intrinsically attract matter, or does this apparent attraction result from a gradual evolution of the dynamic properties of quantum vacuum? In this second hypothesis, the interstitial space, continuously deprived of a part of its energy, would see its characteristics evolve in such a way that material structures seem to get closer, while the wavelengths that propagate there lengthen. The apparent
241

expansion of space could thus be correlated with an energy depletion of vacuum, rather than an effective increase in its geometric extent.

Cosmic evolution, marked by gravitational effects, can then be interpreted as a global quantum process aimed at correcting a chirality associated with an initially broken symmetry. This dynamic would play the role of an intrinsic adjustment mechanism, through which the Universe would tend towards a symmetrical equilibrium state. To support this hypothesis, several considerations can be put forward.

- First, fermions can be described as wave packets in the stable vibration regime, constituting closed and durable systems.
- Secondly, these wave packets would result from radiative entanglements that appeared at the first moments of the Universe, when the energy densities of the primordial plasma made possible such quantum correlations.
- Thirdly, as the intensity of the initial radiation is no longer sufficient today, these radiative entanglements can no longer reproduce significantly.
- Fourthly, the residual radiation, essentially of a kinetic nature and corresponding to currently free OEMs, is gradually captured by the material already formed.
- Fifthly, this process leads to a continuous conversion of kinetic energy—associated with the OEM and characteristic of so-called empty space — into potential energy, embodied by the material structures that make up the stars. The observed thermal, mechanical, chemical, electrical and nuclear phenomena can be interpreted as manifestations of this energy conversion.
- Sixth, atoms can be considered as open microsystems, capable of federating packets of elementary waves to form more complex molecular structures.
- Finally, these aggregation processes lead to an increase in the mass density of bodies and, correspondingly, an accentuation of the energy depression of the surrounding space, commonly referred to as emptiness.

Gravitation therefore appears as a fundamentally quantum phenomenon, rather than as a strictly geometric or classical interaction. This idea meets, in some aspects, the theoretical attempts developed within the framework of

242

loop quantum gravity. In this approach, space is described as a field made up of elementary quanta of space, comparable to pseudo-particles, giving the space a discrete and energetically modular structure. Gravitation then becomes the quantified expression of this structure, at the cost however of explicitly setting aside the three fundamental interactions described by the standard model.

The quantification of space in the form of indivisible elementary units, not directly observed but supposed to be able to interact with matter, raises many questions. This hypothesis nevertheless leads, as developed here, to considering a globally finite Universe, and does not exclude the existence of regions or components of space-time associated with antimatter.

Loop quantum gravity is mainly based on a mathematical construction and statistical considerations. Its strongly counterintuitive character leads to question its status: does it constitute a fully supported physical theory or an exploratory formal framework? This approach is notably derived from a particular field equation, the Wheeler–DeWitt equation, which seeks to unify quantum mechanics and general relativity by explicitly eliminating the notion of time while redefining that of space. The complexity of this formulation makes its delicate interpretation and its experimental validation particularly difficult.

From the perspective developed here, the integration, to classical properties of particles—such as mass or charge — of discrete interactions governed by an opposite quantum symmetry could allow to account for gravitational phenomena without resorting to this type of global field equation. Such an approach would aim to reconcile quantum and relativistic descriptions from local interactional mechanisms, while maintaining continuity with observable phenomena.

243

XVIII <u>Unite gravitation with the 3 fundamental forces</u>
(Bet or challenge?)

This chapter addresses concepts likely to cause reading difficulties due to their level of abstraction and the complexity of the ideas mobilized. It is therefore necessary to specify explicitly the hypotheses and the logical sequences.

Stable atoms, whether light or heavy, exhibit overall electrical neutrality. This neutrality results from the balance between a globally positively charged atomic nucleus and a procession of negatively charged electrons. In the hypothesis developed here, the atomic nucleus is considered as the effective seat of gravitation. In the absence of electrons ensuring a screening and mediation role, the atomic nuclei would repel each other due to their positive charges. Electrons, linked to nuclei by electromagnetic interaction, thus act as regulatory elements in interatomic interactions.

Electrons are attracted to nuclei due to their opposite charge and can absorb the energy provided by photons, which determines their energy state. Their orbital movement is not to be interpreted as a classical trajectory, but as a manifestation of their quantified inertia, which keeps them at an average distance from the nucleus compatible with the atom's equilibrium. This configuration is not fixed: electrons can, under certain conditions, be transferred from one nucleus to another, which founds the phenomena of chemical bonding. In the absence of adequate equilibrium conditions, an electron cannot remain permanently within a molecular system.

In interactions between nuclei and electrons, the dynamic quantities of particles (velocity, angular momentum orientation, orbitals, spin) are continuously adjusted. This permanent tuning, analogous to a self-stabilized dynamic system, ensures the cohesion of atoms and their ability to assemble into stable molecular structures. It is not, however, an intentional or reflective behaviour of matter, but a play of reciprocal influences governed by the laws of fundamental interactions. Any imbalance tends to be compensated by the available interaction mechanisms, notably by the weak interaction in certain readjustment processes.

244

The currently accepted cosmological model is based on the existence of four fundamental types of interactions, distinct in their scope and intensity, but possibly linked by a common structure.

<u>the strong nuclear interaction</u>, the most intense known to date, presents attractive characteristics that evoke, by analogy, certain effects of gravitation, although it is exerted exclusively at the scale of the atomic nucleus. Of extremely short range, it acts mainly between the elementary constituents of the nucleus. This interaction, although distinct from electromagnetic interaction, presents sufficient formal similarities to suggest the existence of shared properties, a hypothesis that will be discussed later.

The strong interaction ensures the cohesion of quarks within nucleons, where they are stably grouped by three. It is described, in the current theoretical model, by the exchange of gluons, entities without mass or electric charge, introduced as indispensable mediators for the modeling of this phenomenon. Quarks assembled into baryons thus form an inseparable structure: any attempt to isolate a quark leads to an immediate reorganization of the system under the effect of strong interaction, preventing its free existence.

At the very high energy scale, beyond the limits defined by Planck's conditions, the primordial processes that marked the evolution of the Universe led to the gradual differentiation of four fundamental interactions: electromagnetic interaction, responsible for phenomena related to electrical charges; the strong interaction, ensuring the relative stability of nuclei; the weak interaction, involved in nuclear transformation processes; and the gravitational interaction, governing the attraction of masses on a large scale. It remains however legitimate to question the existence of a deep link between these interactions, considered as differentiated manifestations of a common underlying dynamic.

In this speculative perspective, the gluon could be interpreted as an expression of electromagnetic interaction devoid of any classical spatio-temporal referential, confined within hadrons and to the interactions between their constituents. Like the photon, the gluon does not possess an antiparticle, which would be coherent if one considers that it represents an internal and neutralized form of electromagnetic interaction, ensuring the

245

cohesion of the nucleus while indirectly participating in the overall neutrality of the atom.

A fundamental question remains: why do quarks acquire a stable existence only when they are assembled by three within a nucleon? Any attempt at separation leads to a constraint imposed by the strong interaction, which immediately restores a bound configuration. One can then formulate the hypothesis that the effective or virtual presence of antiquarks, and a form of osmotic permeability to electromagnetic waves between matter and antimatter, would participate in the emergence of strong interaction.

In quantum chromodynamics, quarks are endowed with additional properties, called «colors», introduced to formalize their complementarity and the stability of their assemblies. The combination of these colors and electrical charges makes quarks practically inseparable within a nucleon. Nothing prohibits considering an extended multiplicity of these internal properties, enriching the spectrum of known particles. If one applies a symmetrical reasoning to antiparticles, endowed with anti-colors, the superposition of the two symmetries would lead to a globally neutral state, or 'colorless', reflecting a form of fundamental balance between matter and antimatter (to stay on this "rich in color" image).

The weak interaction intervenes at different levels of matter organization by modifying certain intrinsic properties of quarks. It manifests itself in the form of localized processes, which can be interpreted as mechanisms for readjusting instabilities occurring in atomic equilibrium, an equilibrium that is usually ensured by electromagnetic interaction. At certain energy levels, the weak interaction and the electromagnetic interaction cease to be distinguishable and are then described within a unified force, that of the electroweak interaction. This convergence suggests a continuity between the fundamental forces, analogous to that postulated here between electromagnetism and strong interaction.

The weak interaction is mainly responsible for nuclear transformations during which a neutron is converted into a proton, or vice versa. Its scope and intensity depend on the nature of the nuclei concerned, and therefore indirectly on their mass and internal structure. The phenomena of fission and nuclear fusion thus lead to the disintegration or recompositing of atoms, whose cohesion is conventionally attributed to strong interaction. In the hypothesis developed here, this strong interaction is interpreted as the result of internal electromagnetic interactions, giving the atomic nucleus its stability.

Electrons, by temporarily leaving the atoms that have captured them, ensure neighborhood bonds between atoms, allowing the formation of molecules. However, when electrons are stripped from an atom as a result of collisions with free particles, particularly those derived from alpha radioactivity, the electric neutrality of the atom is broken. The loss of an electron, not compensated by the loss of a proton, then makes the atom unstable. A comparable state appears when, within the nucleus, a proton transforms into a neutron following the capture of an electron, a phenomenon associated with beta radioactivity. In these two situations, the atom is said to be ionized. The weak interaction then acts as a corrective mechanism, operating through radioactive processes in order to restore certain charge instabilities that electromagnetism does not directly compensate.

The cohesion between protons and neutrons within the nucleus can be disrupted and restored according to several mechanisms.

In the case of fission or nuclear scission, several processes are observed:

- When a neutron is released from the nucleus, it decays into a proton, an electron, and an antineutrino in a beta decay process. The β radiation then corresponds to a flow of negative charges made up of electrons, reflecting a conversion between opposite symmetries.
- When a heavy nucleus becomes unstable, it can split into lighter and more stable nuclei, such as the helium nucleus composed of two protons and two neutrons. This process corresponds to alpha decay, and the α radiation is then a flow of positive charges consisting of heavy ions.
- Finally, the excess energy released during these decays is frequently evacuated in the form of very high-energy photons, corresponding to gamma radiation, of purely electromagnetic nature, devoid of mass and charge.

Conversely, nuclear fusion processes consist of the assembly of light nuclei. In stars, hydrogen nuclei fuse to form deuterium, which then participates in

247

the formation of helium nuclei and heavier elements. The control of nuclear fusion would offer an abundant and low-polluting source of energy, outside the constraints related to neutron capture by surrounding structures. The major difficulty lies in initiating and maintaining the fusion of deuterium and tritium nuclei within a plasma confined by powerful magnetic fields, under extreme conditions of temperature and pressure, on the order of 150 million degrees. To date, no other process than magnetic confinement allows for such control. The technological challenge is then to design materials and cooling systems capable of sustainably preserving the integrity of infrastructures. The ITER project is part of this ambitious perspective and could constitute, if successful, a decisive step in the energy and technological history of humanity. However, it would still be necessary to rationalize costs, control global thermal effects and manage the replacement and decontamination of irradiated chambers.

Humanity today knows how to exploit the energy resulting from nuclear fission and seriously considers that resulting from fusion. One can therefore question the theoretical possibility of exploiting the energy resulting from coalescence or disintegration matter–antimatter. While such a perspective is not prohibited by the known fundamental principles, its practical implementation modalities remain, at present, beyond technological reach.

The mediators of the weak interaction are massive bosons, noted W and Z, whose existence is linked to the very nature of this interaction. Their role is mainly to allow a quantified and coherent description of the processes involved. These vector particles would adopt the most favorable quantum states for their interaction with all the concerned particles.

It appears that the weak interaction acts preferentially on so-called left particles and, in a correlated manner, on so-called right antiparticles. CP symmetry, associated with invariance by mirror inversion and charge conjugation, is therefore not strictly conserved in this type of interaction, especially when antineutrinos are involved. This violation of symmetry can be interpreted as the consequence of a fundamental chirality governing the weak interaction.

The electromagnetic interaction manifests particularly clearly at the atomic level. It includes all the interactions necessary for the balance of

248

electric charges between particles. Although of relatively short range, its effects confer to the atom and matter a lasting stability. Gravitation, on the other hand, would extend these effects without limitation of scope, by exerting itself continuously on a large scale. In the hypothesis developed here, the atom would constitute the point of origin of gravitation, even if electromagnetic waves do not have a gravitational mass. By arbitrating the charge interactions at the presumed origin of gravitation, electromagnetism would thus make it possible to consider a unifying model linking quantum physics to classical relativistic physics.

The mediator of electromagnetic interaction is the photon, a light quantum without mass or charge, which allows it to ensure neutral exchanges between particles and antiparticles. The symmetry between these would play a determining role in the origin of the electric charges carried by massive particles.

The photons resulting from the annihilation of an electron–positron pair can, under appropriate conditions, be converted back into that same pair of particles. In the early Universe, these processes corresponded to energy levels considerably higher than those associated with the depleted photons of the current Universe. Although electromagnetic interaction is of a quantum nature, its relativistic implications suggest that it may be associated with gravitational effects. Covalent bonds, resulting from electron transfers between atoms, allow the assembly of matter in molecular form. The resulting interactions between electrons and magnetic fields could thus contribute indirectly to gravitational effects, which result in an expansion of time and a contraction of lengths, leading to the apparent rapprochement of distant objects.

Finally, one can formulate the hypothesis that a potential antimatter, not directly observable, interacts with ordinary matter. This interaction would suppose the existence of a discrete chiral symmetry, related to a superposition of states of matter in a parallel spatiotemporal dimension, shifted but inseparable from our observable reality. Any interaction involving energy and charge exchanges between particles, antiparticles and quantum symmetries, these processes could modify the topology of the so-called empty space. The quantum charge interactions between these symmetries, although inaccessible to direct observation, could thus participate in the mechanisms responsible for gravitational approximation of bodies.

<u>The gravitational interaction</u>, as previously stated, manifests itself by a deformation of space and acts without limitation of scope, giving the Universe a dynamic and evolutionary topology. The distribution of matter in space, as a function of the masses present and the distances that separate them, determines the local intensity of gravitational effects. At the level of atoms and molecules, these effects are negligible and remain practically undetectable. Gravitation, on the other hand, becomes manifest at the scale of astrophysical objects. However, it cannot be considered as an exclusively macroscopic phenomenon: processes of rapprochement and structuring of matter occur at all scales.

Thus, in the primordial Universe, clouds of hydrogen atoms and ions, corresponding to the HI and HII regions, have evolved towards more complex molecular structures. This evolution is explained by the sharing of electrons between several nuclei; a phenomenon whereby electronic fields ensure interatomic bonds and the progressive assembly of molecules. These transfer and charge balancing interactions belong to the electromagnetic interaction, which joins the low to very high energy interaction thus achieving electroweak unification.

Nuclear fusion, by increasing the number of nucleons in atomic nuclei, increases their mass and, consequently, the number of electrons associated with the atom. This process contributes to the progressive densification of matter and the grouping of increasingly massive structures. It then becomes possible to account for the formation of stellar objects of increasing mass, such as stars and planets, whose ultimate evolution can lead, for the most massive ones, to the formation of black holes. In this perspective, gravitation could be interpreted as an emerging mechanism, of quantum origin, fundamentally arising from electromagnetic interactions.

Quantum mechanics profoundly challenges classical intuitions. Before any measurement, a particle is described by a wave function representing a superposition of possible states. Any modification of the quantum state of a particle can instantly affect the properties of other particles with which it has shared a common history, these particles remaining entangled within the same quantum system. Two particles from the same splitting process, whether recent or old, can thus maintain correlated states regardless of the distance that separates them. Moreover, the quantum state of a particle is not

250

intrinsically defined before the act of measurement, and can correspond to several potential values, with the measure selecting a state compatible with the experimental context and the observer's interpretation.

Based on these principles, consider two celestial objects far from each other. Like the atoms that constitute them, these bodies are generally electrically neutral, although they are composed of elementary particles carrying positive and negative charges in equal proportions. Suppose now that these two massive bodies, denoted A and B, share particles which remain entangled despite their spatial separation. Such a situation is not exceptional in quantum dynamics. A particle X belonging to body A, whose state is correlated with that of a particle Y of body B, could then be engaged in electromagnetic interactions with particles of body B exhibiting opposite charges. Reciprocally, the particle Y would share the properties and interactions of the particle X within the body A.

Although these non-local interactions are of extremely low intensity, the fact that they are free themselves from distance could generate a cumulative effect interpreted as an apparent rapprochement between distant celestial objects sharing a large number of entangled particles. The more massive the bodies, the more likely they are to contain such quantum correlations, and the more these interactions could contribute to a collective effect akin to a form of quantum gravity. This underlying quantum electrodynamics would thus participate in a local decrease of the vacuum energy, by inducing space-time deformations associated with field extension at the subatomic scale.

Gravitation does not necessarily appear as a force propagated by an exchange of particles in the classical sense, even if one is tempted to introduce a mediator boson to account for it. The gravitational effects of massive bodies modify the geometry of space, which influences the propagation of electromagnetic waves and generates deformation fronts perceived as undulations of space-time. This phenomenon can give the illusion of the existence of gravitational waves analogous to electromagnetic waves, as well as that of an associated particle, on the photon model.

It is in this logic that the idea of graviton was proposed, designed as the vector boson of gravitational interaction. However, its introduction relies more on a formal analogy than on an experimental validation, and its existence remains highly speculative.

251

The global architecture of the Universe seems to rely mainly on the presence of electromagnetic waves and on the role of this kinetic energy transported by massless particles, photons. By interacting with matter, these waves participate in a gradual evolution towards states of increased concentration of matter. Electromagnetic waves regulate, at equilibrium, the charge interactions that ensure the relative stability of atoms, molecules, and stellar structures. However, by interactively associating with the constituted material, they locally reduce the available energy of the vacuum. The result is more or less depressionary regions of space-time, giving the impression of a curvature of the trajectories followed by waves and material bodies. This interpretation can be understood as an emergent geometric effect, difficult to distinguish from a force effect in the classical sense.

In this perspective, the strong interaction, often related to electromagnetic interaction by some of its properties, could itself involve charge interactions between quarks, as well as between protons and neutrons. Gluons, introduced as cohesion agents of atomic nuclei, would then become superfluous theoretical constructions. Gravitational effects and strong interaction would therefore appear as secondary manifestations of electromagnetism, in the same way that weak interaction is already integrated into the idea of electroweak force. Gravitation would thus not be a fundamental force in the strict sense, but the collective result of electromagnetic interactions modifying locally the energy of vacuum and the geometry of space-time.

The energy density of space can be described as exhibiting local variations, perceptible by a distant observer in the form of deformations. However, this description depends on the point of view adopted. In general relativity, any contraction of space is accompanied by a correlative dilation of time. For an observer located locally in the region concerned, these effects compensate each other so that local physical laws and measures remain unchanged. Therefore, the question arises as to whether it is rigorously relevant to qualify these phenomena as gravitational waves.

What is commonly referred to as gravitational waves can be described more precisely as dynamic deformations of space-time, perceived in the form of geometric undulations. The gravitational effects associated with a massive body lead, from the point of view of a distant observer, to an apparent contraction of distances. However, this contraction is inseparable from a

252

slowing down of time, so that the ratio between distance covered and measured duration remains invariant.

The emission of high-energy electromagnetic waves produced during extreme astrophysical events, such as supernovae, neutron star mergers or black holes, have contributed to the imagining of gravitational waves. In this interpretation, it would actually be marked space-time undulations, comparable to isobaric disturbances, which accompany cataclysmic events and influence the trajectories of the associated electromagnetic waves.

By propagating through the Universe, these space-time perturbations interfere with each other and locally modify the propagation of electromagnetic waves. The gamma-ray bursts observed during violent phenomena could thus suggest the existence of a specific gravitational radiation. However, the gravitational deformations of space-time, which constitute the heart of general relativity, are neither reflected nor absorbed by matter in the same way as electromagnetic waves. They therefore do not necessarily exhibit the classical wave characteristics defined by a frequency or a wavelength, which makes their direct assimilation to electromagnetic waves questionable.

A universal constant, denoted G, is introduced in order to quantify the intensity of gravitational interaction as a function of the product of masses and the inverse of the square of distances. This formulation remains, however, an approximation, insofar as it cannot integrate all the incident gravitational contributions due to the presence of all bodies, near or far, which act simultaneously. These competing influences cannot be rigorously taken into account, so that the results obtained, although significant, necessarily remain approximate, particularly for very large systems.

From hydrogen atoms, mainly produced during primordial nucleosynthesis, and stable helium, thermonuclear reactions gradually generated heavier elements. This chemical and nuclear evolution created the conditions for a global cosmic dynamic, of which gravitation constitutes the macroscopic expression.

When an object is dropped from a certain height, it falls towards the ground. This fall can be interpreted as a gradual decrease in the distance separating the object from the earth's surface. In the absence of an atmosphere, this movement would not be slowed down. An observer immobile on the ground

253

surface does not fall, but instead feels a downward force, equivalent to a constant acceleration. This situation is analogous to that of an observer located in a reference frame under uniform acceleration, such as inside a spacecraft with constant propulsion. One can then consider that the ground and the released object tend to approach each other, this approximation being interpreted as a dynamic of concentration or densification of matter on a planetary scale. This phenomenon, valid at any point on the Earth's surface, can be understood as a local manifestation of global processes of matter structuring.

The Standard Model of physics is based on identifying four fundamental interactions. This classification, although coherent and operational, can nevertheless be questioned as to its fundamental or arbitrary character. The main difficulty lies in the apparent absence of a direct link between gravitation and the three other interactions. This situation changes, however, if one adopts the hypothesis that a deeper quantum symmetry would allow gravity to be requalified, and if one considers that the strong interaction, interpreted here as an emanation of electromagnetic interaction, initiates the first gravitational effects associated with material bodies.

Seeking to integrate gravitation as a fourth force within a unified theory can lead to some confusion. Gravitation does not correspond to a force in the usual sense, but to a dynamic of space-time itself. From this angle, descriptions from general relativity and those from quantum physics find a point of convergence.

The apparent diversity of fundamental interactions could result from a broken symmetry associated with a fundamental chirality. This symmetry break would be reflected by the diversity of observed properties — mass, charge, spin, color, dynamics—and by the complexity of interactions, as well as by the multiplicity of particulate entities identified. In an ultimate perspective, a final state of collapse could erase these disparities.

The electromagnetic interaction and gravitation share the property that their effects decrease with the inverse of the square of the distance. Their essential difference lies in the scale to which they are exerted and in the fact that gravitation modifies the geometry of space-time, which requires redefining distances and the flow of time by changing coordinates.

The distinction between electromagnetic field and gravitational field could then be partly formal. The electromagnetic field describes a space where kinetic energy can exchange with matter and is manifested by interfering waves in a complex dispersive regime. The gravitational field, on the other hand, describes a space where energy tends to be absorbed by matter, giving the Universe a dynamic topography that shapes the fields of energy associated with the so-called empty space.

At temperatures above 10^{31} K, such as those that characterized the early Universe, electromagnetic, strong and weak interactions did not differentiate significantly. Nuclear forces and gravitational effects would have gradually emerged with the formation of the first quantum excitations constituting matter. Nuclear interactions would have manifested during primordial nucleosynthesis, while gravitational effects would become observable with the assembly of the first molecular structures.

In this perspective, gravitation appears as an incident phenomenon, emerging from the first localized collapse processes of ionized hydrogen, hydrogen isotopes and free electrons. This slow and continuous process, guided by electromagnetic interaction, is at the origin of structured matter. It leads to the formation of molecules, then stellar objects, planets and stars. Although clearly perceptible at the macroscopic scale, gravitation would thus be the transposed expression of quantum charge interactions, without an immediately observable direct link with them.

A gravitational field can be compared, by analogy, to certain dynamic phenomena observed in the atmosphere, in which the formation of high-pressure zones induces, by compensation, regions of lower pressure. This analogy suggests that the distribution of mass energy in the Universe evolves according to a dynamic of progressive concentration. Matter would thus tend to gather and densify, at the expense of regions of space characterized by an increasingly low vacuum energy. In this perspective, antimatter, if it exists in a form that is difficult to observe, would participate in this global dynamic and could manifest indirectly through additional gravitational effects, now attributed to dark matter.

Another interpretation adopted here is explicitly based on the potential role of antimatter. Its presence «in mirror» of ordinary matter would lead to distinguish gravitational mass, deduced from the measured attraction effects, from inertial mass, defined as the resistance of a body to

255

acceleration. From a certain threshold of matter density, the effective gravitational mass could no longer grow proportionally to the inert mass. Such dissociation would help explain the anomaly observed in the rotation curves of rotating astrophysical systems, particularly galaxies and galactic clusters, where the peripheral orbital velocities appear higher than the predictions from classical Newtonian dynamics.

The theory proposed by Milgrom, known as modified Newtonian dynamics, also constitutes an alternative to the dark matter hypothesis. It takes up, in another form, the idea of modifying low-acceleration dynamical laws, without however involving antimatter. The absence of this reference nevertheless raises a difficulty: without introducing an additional component of the energy content of the Universe, it becomes difficult to justify a fundamental modification of the relationship between mass and acceleration, as it is formulated in Newton's second law.

Three fundamental interactions and their mediators

In the Standard Model, three interactions are described using mediating bosons. Photons ensure the description of electromagnetic interactions. Gluons, interpreted here as a particular form of electromagnetic interaction confined to the atomic nucleus, account for so-called strong interactions. The W and Z bosons are involved in weak interactions, their high mass reflecting the inertia necessary for managing nuclear transformation processes.

These bosons, also referred to as interaction vectors, are supposed to be emitted and absorbed by fermions, that is, the constituent particles of matter. They are not directly observed but constitute a suitable tool for describing interactions other than by an instantaneous action difficult to formalize. The fermions themselves are only accessible to observation through measurable effects that give them notably a mass. By interacting through the emission or absorption of bosons, fermions acquire the variations in motion and transformations observed experimentally.

Bosons are thus conceived as markers of exchanges between fermions, classified into quarks and leptons. However, a question remains open: are bosons fundamental physical entities at the origin of fermion movements, or are they theoretical constructs introduced to account for pre-existing interactions between these fermions? At the currently accessible scale, the analysis stops at fermions, described as possibly entangled wave packets, without it being possible to explore their internal structure. The hypothesis of more fundamental substructures, such as the preons, remains purely speculative and does not provide, at this stage, any additional insight into the ultimate nature of matter.

Our description of the real is thus based on a set of particles allowing to represent otherwise imperceptible interactions. These entities constitute a formal language intended to make intelligible underlying energy and dynamic phenomena.

Towards a unification by changing scale

Electromagnetic and weak interactions can be correlated within a unified framework. Although their respective mediators, photons and W and Z bosons, have distinct properties, this distinction disappears when the energy reaches values greater than about 100 GeV. In this regime, the bosons W and Z lose their effective mass, converted into kinetic energy, and behave like electromagnetic excitations. We then speak of electroweak interaction.

By analogy, it is proposed that gravitation and strong interaction can be part of a common mechanism. Gluons and gravitons would then constitute theoretical skins of the same phenomenon of gathering matter, resulting from discrete exchanges between two complementary quantum symmetries of the Universe. This interaction could be described as "nucleo-gravitational" dynamics. However, if one admits that the strong interaction, like gravitation, is an emerging manifestation of electromagnetic interaction, the distinction between these forces becomes essentially a question of scale.

The hypothesis according to which electromagnetism, as it is exercised within the atomic nucleus, would be at the origin of gravitation, would allow to reduce certain inconsistencies of the standard model. Matter can only persist and structure itself in a context of global load balance.

Electromagnetic waves play a central role in this equilibrium, by regulating and neutralizing charge interactions at all scales.

By generating electric currents and magnetic fields, electromagnetic waves suggest that the strong interaction could be interpreted as a form of nuclear electrodynamics essential for the stability of atoms. By changing scale, gravitational effects could then be understood as cosmic electrodynamics, operating over considerably larger distances and masses.

The relative stability of atomic nuclei could thus be explained without necessarily resorting to the hypothesis of specific binding bosons, such as gluons. Although the latter are postulated in eight distinct forms, they have never been directly observed. In an alternative approach where the strong interaction is assimilated to electromagnetic interaction, the cohesion of the nucleus would rely on interactions between quarks with opposite but generally positive charges, confined in a volume stabilized by an envelope of negatively charged electrons.

Neutrons, whose number determines the isotopy and relative stability of chemical elements, would also participate in this internal cohesion of the nucleus. In this perspective, the nuclear structure would result from a dynamic equilibrium of internal electromagnetic interactions, rather than from the action of a distinct fundamental force.

----

1. Hypothesis of quark cohesion within the proton

One can formulate the hypothesis that a proton would be described as a composite system consisting of two up quarks of positive electric charge and one down quark of negative charge. These constituents would each possess an intrinsic spin and participate in a collective baryonic state resulting from an internal movement not directly observable.

In this speculative representation, the quarks would not be localizable according to defined trajectories but engaged in an internal dynamic comparable to a global virtual rotation of the system, without explicit reference to a measurable time flow at this scale. It would therefore not be a classical motion, but a stationary quantum state characterized by conserved magnitudes (total spin, charge, energy), without the detail of internal motion being observed or described in classical orbital terms.

2. Hypothesis of quark cohesion within the neutron

On an analogous model, the neutron can be considered as a system composed of two down quarks of negative charge and one up quark of positive charge. Again, the cohesion of the system would be ensured by a non-localizable internal quantum dynamic, described by a global baryonic state, without it being relevant to talk about individual trajectories or classical rotation.

This description remains voluntarily qualitative and speculative and essentially aims to provide a coherent picture of the confinement of quarks, without claiming to substitute for the formal description provided by quantum chromodynamics.

3. Hypothesis of cohesion between protons and neutrons within the nucleus

The protons, carrying a positive electric charge, are subjected to mutual electrostatic repulsion. The presence of neutrons within the atomic nucleus nevertheless allows for the stability of the nuclear assembly. Neutrons, electrically neutral, contribute to the cohesion of the nucleus by acting as mediators of strong nuclear interactions, without introducing additional Coulomb repulsion.

One can propose a complementary qualitative interpretation, according to which each electron of an atom would be associated with a specific proton, thus participating in the global charge equilibrium. In this speculative perspective, the proton associated with a given electron would be statistically positioned, on average, opposite to the nucleus center with respect to the region of maximum probability occupied by the corresponding electron. The presence of neutrons would then shield the direct electrostatic attraction between this proton and the electron, indirectly contributing to nuclear stability.

This description would allow to interpret the difference in behavior observed between the hydrogen atom and the more complex atoms. The nucleus of hydrogen, consisting of a single proton and devoid of neutrons, does not benefit from this mediation mechanism. This results in a greater ease of transition between the ground state and the ionized state, which is manifested astro-physically by the formation of ionized hydrogen regions (H II regions). It can also be considered that the hydrogen atom has an increased propensity for excitation and bond establishment, due to this minimal nuclear structure.

4. Wave function, localization and stability of the hydrogen atom

In quantum mechanics, the electron is described by a wave function that prohibits any point location. This wave function extends over the entire orbital associated with the quantum state under consideration. In practice, exact equations are analytically solvable only for single-electron systems, which limits rigorous mathematical descriptions to the hydrogen atom.

Given the electron's relativistic velocity and the low probability of direct interaction with the nucleus, electron-nucleus collisions are extremely rare. When such an event occurs, it may correspond to a conversion process involving the weak interaction, temporarily modifying the nature of the particles in presence. In the case of hydrogen, these processes statistically lead to a return to the initial state, which contributes to the remarkable stability of this atom.

It should be remembered that annihilation is only possible between two particles of opposite properties belonging to conjugated symmetries (matter/antimatter). Two particles of the same symmetry cannot annihilate each other and only diffuse or exchange energy.

260

5. Primacy of the wave-like aspect and wave-particle duality

The Universe can be considered as fundamentally undulatory, although our description is based on wave-corpuscle duality. Long-wavelength waves manifest clearly as wave-like phenomena, while high frequency and high amplitude waves are perceived, observationally, as corpuscular entities.

A massive particle can thus be interpreted as a strongly entangled wave packet, of large amplitude and high frequency. This entanglement makes the wave aspect difficult to access for indirect observation, which explains the predominance of corpuscular description in the macroscopic domain. It is precisely this underlying wave nature that allows particles such as the electron to exhibit diffraction phenomena, analogous to those observed for electromagnetic waves.

This interpretation meets the associated wave hypothesis formulated by Louis de Broglie and mathematically formalized by the wave function introduced by Erwin Schrödinger.

6. Quantum information, observation and emergence of reality

The wave-corpuscle duality translates above all a limit of our access to quantum information. Confronted with a relativistic space-time and gravitational effects, the observer constructs a reality based exclusively on measurable information, expressed in terms of position, trajectory and duration. This construction is in tension with the concepts of superposition, entanglement, and non-locality.

Quantum information describing the possible states of an elementary particle is both local and non-local. This property becomes less paradoxical if one considers the particle as a bundle of entangled waves, for which the classical notions of space and time are not relevant. On the other hand, when a large number of particles assemble into complex systems (atoms, molecules, macroscopic bodies), the information gradually becomes localizable and compatible with a classical causal description.

7. Particle, measure and observer role

We can propose, as a heuristic, to assimilate a particle to a closed quantum cavity containing a finite amount of energy. The entangled waves that constitute this energy are no longer individually distinguishable by their frequency or amplitude. During certain interactions, particularly weak ones, this packet can be recomposed into several lower-energy particles, without violating the principle of energy conservation.

The particle only shows a particular state to observation, determined by a given experimental context. The other potential states remain inaccessible. Any measure imposes a spatiotemporal framework that is foreign to the intrinsic description of the particle. Thus, the superposition of states has no direct equivalent in the macroscopic world, as illustrated by Schrödinger's famous cat thought experiment.

Measurement procedures implicitly select certain properties (position, speed, spin), and the observer, through his interaction with the system itself, constructs a specific quantum environment. The result is an observed reality that is not independent of the experimental device.

8. Cognitive limits and scientific speculation

Finally, time, space, and relativity only reveal themselves to us through observable and traceable exchanges. Devoid of intuitive landmarks, the human mind is poorly adapted to a fundamentally non-intuitive quantum mechanics. All speculation necessarily fits into a set of reflections inherited from our biological condition, our limited cognitive abilities and our imperfect observation tools.

These constraints do not invalidate the theoretical effort but remind us that any representation of quantum mechanics remains a partial construction, dependent on our position as situated observers, and subject to structural limits still largely ununderstood.

262

1. Probabilistic description of the particles and range of the wave function

The precise description of the behaviour of an elementary particle is a fundamental difficulty. In the absence of a definable trajectory in the classical sense, quantum mechanics resorts to the wave function, which provides a mathematical representation of a quantum state comparable to a packet of waves. This representation does not describe a deterministic reality, but a set of possible states, each being affected by a probability.

The wave function from the Erwin Schrödinger equation applies to non-relativistic particles of matter and accounts for their temporal evolution in probabilistic terms. For convenience, a particle can be associated with an average characteristic wavelength, although it does not correspond to a single monochromatic wave.

However, this approach encounters a fundamental limit: when the mass and speed of a body increase—in other words when its total energy becomes important — the associated wavelength strongly decreases. In this regime, the wave aspect becomes unobservable. For a macroscopic body, the multiplicity of internal movements and the superposition of extremely short wavelengths make any undulatory description impractical. It follows that, at the macroscopic scale, state superposition and probabilistic interpretation can be neglected, which explains the difficulty when trying to relate quantum mechanics to our daily experience.

2. Hypothesis of a fundamentally undulatory universe

The mass effect giving matter a corpuscular aspect would mask a universe fundamentally made up of waves in different forms: 'free' electromagnetic waves, packets of entangled waves corresponding to particles, assemblies of these packets in stable charge configurations (atoms), then more complex molecular structures.

263

This perspective sheds light on the mass-energy relation formulated by Albert Einstein, in which the speed of light intervenes as a fundamental constant, characteristic of the kinetic energy carried by electromagnetic waves in free space.

The total energy of a moving body can be interpreted as the sum of several contributions:

- the confined kinetic energy in the form of entangled wave packets, corresponding to the constituent particles of the body at rest.
- the kinetic energy associated with the speed of movement of the body and its internal movements.
- the binding energy ensuring the cohesion of the system, in which electromagnetic waves would participate, interpreted here as an active component of vacuum energy.

3. Mass, radiation and energy

Kinetic energy grows like the square of speed: doubling the speed of a body is equivalent to multiplying its kinetic energy, and thus its relativistic mass, by four. In this perspective, mass and radiation appear as two transposable expressions of the same energy content.

Energy remains, however, an abstract concept, protean and not directly observable. It only manifests itself through the particular state under which it is revealed by a given experimental device or interaction context.

The energy variations observed during electronic transitions—often described as orbital jumps — can be interpreted as resulting from variations in the charge distribution and internal structure of the nucleus. These phenomena reflect dynamic equilibria rather than orbital movements in the classical sense.

4. Role of neutrons and limits of the classical description

Neutrons possess a non-zero magnetic moment, associated with a complex internal structure, as well as a mass slightly higher than that of protons. Their electrical neutrality and aggregation capacity make them essential constituents of nuclear stability. By participating in nuclear interactions, they ensure the cohesion of the nucleus, particularly in complex nuclei such as lithium.

At this scale, classical physics seems inoperative. The concepts of space and time cannot be used intuitively to describe the internal processes of nucleons. It is however remarkable that the energy quantities used to characterize these systems (joule, electron Volt) remain defined in relation to a unit of time. This recourse to time constitutes an essential mathematical artifice for formalization, but it introduces a merely apparent feature of relativity in quantum mechanics.

5. Electrons, electromagnetic fields and stability of matter

The electron can be described as the quantized expression of closed electrical fluxes associated with the atomic nucleus. These fluxes generate magnetic fields whose intensity depends on both nuclear energy and the one involved in chemical bonds. These fields contribute to the stability of atoms and the cohesion of molecules.

Depending on the robustness of the charge balance thus achieved, some atoms exhibit high stability (such as hydrogen or iron), while others, such as uranium, are inherently unstable.

Electrons excited by interaction with photons play a central role in the construction of atomic and molecular matter. Chemical bonds result from the pooling or exchange of electrons between atoms, as illustrated by the formation of simple molecules such as methyl chloride. Homogeneous associations of compatible molecules are generally stable, while heterogeneous associations rely on non-covalent interactions of low intensity (electrostatic interactions, hydrogen or halogen bonds).

6. Towards an electromagnetic unification of interactions

265

If we admit that the strong nuclear force can be interpreted, in a speculative approach, as a complex manifestation of electromagnetic charge interactions, it becomes conceivable to extend this reasoning to gravitation, without formally contravening general relativity. Gravitation and electromagnetism indeed share a dependence inversely proportional to the square of the distance.

In this perspective, gravitation, dominant on a large scale, would take over from an electromagnetic force that manifests mainly at the subatomic level and is quickly neutralized by load equilibria. Although considerably weaker, gravitation could then be interpreted as a consequence with unlimited range of electromagnetism.

7. Role of electromagnetic waves and status of bosons

Electromagnetic waves, which would constitute the essential part of what is called the void around matter, could interact with particles in several ways:

- by contributing to the strong nuclear force, represented in the standard model by the gluon.
- by intervening in the weak nuclear force, mediated by the W and Z bosons, which merge with high-energy electromagnetism by achieving electroweak interaction.

One can also consider that these bosons, which we associate with nuclear interactions, have no reality other than to dress up phenomena that we are unable to interpret otherwise.

Electromagnetic waves would interact symmetrically with matter and antimatter. In the absence of photons, no interaction between quantum symmetries would be possible. Electromagnetic radiation would thus appear as the fundamental, omnipresent and timeless energy vector that structures the evolution of the Universe.

Electromagnetism would therefore present itself as the unique fundamental force, with other interactions constituting only derived manifestations at different scales of energy and organization.

266

Undulatory approach to gravitation based on wave-particle duality

In quantum mechanics, any material entity—an elementary particle, atom or molecule — can be described using wave functions. Organized matter then results from the superposition and interference of these waves at different scales. In linked systems, these interferences are mostly constructive, which promotes stabilization and energy concentration.

The more massive an object is, the higher the number of waves associated with its constituents, and the greater the overall interference capacity of the system. In this perspective, massive objects should be associated with very short wavelengths and high amplitudes. However, such objects do not exhibit electromagnetic radiation proportional to the energy they concentrate. This observation leads to the hypothesis that a significant part of this energy would be confined in the form of non-radiating internal structures, likely to modify the spatial environment of the system.

The gravitational effects of stellar bodies could thus be interpreted as the macroscopic manifestation of this internal undulatory organization, without it being necessary to introduce new specific particles. Gravitation would then emerge as a collective effect related to the distribution and retention of wave energy in massive systems.

Characteristic radiation and concealment of the wave nature of matter

Every material object emits its own radiation. This radiation is not monochromatic: it corresponds to a superposition of multiple, partially coherent frequencies, which meets the principle of the associated wave formulated by Louis de Broglie. The wavelength associated with an object decreases as its mass increases or when its relative velocity to the observer becomes significant.

Particles of matter, interpreted as packets of entangled waves forming closed systems, can be considered as structures retaining part of the information

267

and energy they carry. This retention makes the global radiation of the object difficult to analyze and contributes to masking, on a macroscopic scale, the fundamentally undulatory character of matter.

Thus, the dominant corpuscular description would not be the reflection of an intrinsic reality, but the result of an observational filtering related to the complexity and internal coherence of material systems.

Global energy flows and cases of extreme objects

One can formulate the hypothesis that every material body receives, on average, more energy than it emits, in various forms: electromagnetic waves, free particles and constituted matter. This general trend raises the question of the behavior of the densest objects, such as black holes.

A black hole does not radiate directly in the classical way; the observed radiation comes essentially from its accretion disk. The energy it accumulates does not fit into space-time as accessible to our instruments and our perception of reality. In this context, the notions of frequency, wavelength or information transmission lose their operational relevance. Black holes can then be interpreted as limit states, revealing the shortcomings of our current descriptions and marking a breaking point in the observable structuring of the Universe.

Gravitation as a global phenomenon of space-time deformation

The deformation of space associated with gravitational effects can be considered as a global phenomenon, simultaneously affecting the entire Universe at all scales, as long as local particularities are ignored. Gravitational attraction results from a contraction of space correlated with time dilation, in accordance with the relativistic description.

Gravitation thus appears as a fundamental topological dynamic, structuring the Universe on a large scale. This interpretation is in line with the general relativity formulated by Albert Einstein, while opening the possibility of a complementary reading.

Hypothesis of electromagnetic emergence of gravity

A further step would be to consider that gravitation represents, at the macroscopic scale, the cumulative result of all effects related to electromagnetic radiation. These radiations, in multiple forms, ensure the energy transfers, interaction processes and aggregation mechanisms described by quantum mechanics.

In this speculative perspective, gravitation would not constitute an independent fundamental interaction, but the large-scale expression of underlying electromagnetic phenomena, averaged and integrated over all material structures of the Universe. Such an approach would aim to reconcile the quantum description of microscopic interactions with macroscopic gravitational dynamics, without calling into question the essential achievements of contemporary astrophysics, but by placing them in a unified framework of reflection.

XIX <u>Universe suspected of confusing Time and Space</u>
(At the risk of appearing somewhat confused on this point)

The photon contributes significantly to the ambiguities surrounding the notions of time and space. Our physical thought is indeed organized simultaneously around the duration of phenomena and their spatial inscription. In this context, it appeared necessary to model this quantum of energy, devoid of rest mass, according to a double description: undulatory, when considered as propagating radiation, and corpuscular, when treated as a localized entity capable of interacting punctually with matter.

The energy transported by photons is thus sometimes described as a delocalized radiation, associated with a field effect not strictly circumscribed in space and evolving over time, sometimes as an individualized quantum of energy, likely to be located during an interaction process. This descriptive duality reflects less an intrinsic contradiction than a limitation in our thinking abilities.

A fundamental point must then be emphasized: time and space do not constitute independent quantities in the strict sense, although we frequently treat them as such in mathematical formulations and theoretical discourses. When one moves away from the Galilean idea of a universal time and a homogeneous and isotropic space, the space becomes a dynamic structure able to account, on a large scale, for the gravitational effects observed. This description nevertheless imposes a division of events, essential for their analysis, but which introduces an explicit dependence on the reference frame and the observation duration. The resulting units of measurement then incorporate variations in spatial coordinates in a time that no longer has a universal character.

In this perspective, space can be understood as a gravitational field with variable geometry. Once corrected for local gravitational effects, this fluctuating space would allow the representation of observed phenomena, including position variations, using three orthogonal spatial dimensions and two opposite temporal orientations. However, it remains uncertain whether all the implications of space-time relativity are fully considered, as the phenomena it involves are complex. Time then appears as a relativistic parameter reflecting a more fundamental dynamic, ultimately referring to the structure of space.

270

Conversely, by introducing the notions of past, present—necessarily fugitive — and future, space can be envisaged as the three-dimensional expression of the flow of time. This approach leads to question the possibility that space-time reflects a progressive degradation of a common primordial state, initially timeless, shared by all elementary particles. This state, which can be compared to generalized quantum entanglement, would have connected the first elementary entities without considering distance, before the emergence of nuclear and electromagnetic interactions. Space-time would then represent a phase of transformation of this initial entanglement, likely to lead, in the long term, to a global collapse of the Universe and a return to a ground state.

In this hypothesis, space-time as described by general relativity would constitute an emerging macroscopic manifestation of an underlying quantum dynamics, inaccessible in its foundations to an observer necessarily inscribed in a spatialized time context. Space would thus provide a "medium" of observation derived from a temporality essential to our understanding of increasing entropy, leading to the collapse of material structures, particularly in extreme forms such as compact gravitational objects.

An alternative interpretation consists in considering space as a three-dimensional representation of the arrow of time. The present could then be assimilated to a punctual state, without dynamic extension, while the future would correspond to a directed succession of such states, and the past to an analogous succession in the opposite direction. This spatialization of time would offer an attempt to reconcile the gravitational effects described by relativity with a more fundamental quantum mechanics but largely hidden in the deep structures of the Universe.

The question of time then arises differently for the elementary particles constituting matter. Although a massive particle can be considered as immobile in a given frame of reference, it is never really at rest: it is constantly subject to accelerations resulting from local interactions, and its speed varies continuously according to the chosen frame of reference. The kinetic energy associated with these variations is added to an intrinsic energy acquired during an initial entanglement phase. This intrinsic energy is interpreted as the mass energy of the particle considered at rest in an invariant reference frame and is classically expressed by the relation $E=mc^2$.

271

However, the presence of the term c 2implicitly introduces a reference to time, which calls into question the very notion of absolute rest. This paradox can be interpreted as the consideration of internal dynamics, of radiative nature, associated with the intrinsic angular momentum acquired during the entanglement phase. These dynamics would give particles their fundamental properties, such as mass, electric charge and spin, at least within the framework of an extended interpretation inscribed in space-time. In the absence of any external reference, the expression $E=mc^2$ could then be compared to the relation $E=h\nu$, although the latter also involves a temporal magnitude.

The notions of past, present and future, corresponding respectively to information perceived as irreversible, accessible or indeterminate, are above all a cognitive chronology. They could nevertheless be rooted in the very structure of the Universe. The notion of time indeed becomes problematic when applied to a Universe whose origin and end escape any external temporal reference. Time thus appears too relative to be considered as a fundamental property; it is closely linked to an observer's capacities for perception, memory and anticipation. Without these capabilities, allowing the establishment of causal relationships between distinct observations, the very notion of space would lose its operational significance.

In this perspective, quantum mechanics could constitute the point of emergence of a time parameterized by the limitations specific to observation, while the cosmology of black holes would represent the point of erasure, where any temporality related to an observer ceases to be relevant. Space-time would then appear as an emergent construction, associated with an awareness and a mental modeling of phenomena fundamentally devoid of measurable dimensions and intrinsic temporality.

Time would thus be an extrinsic variable to the elementary particle, derived from its inscription in a space-time context. Its emergence would be linked to speed variations and, more generally, to changes in kinetic energy resulting from local interactions. Time would then become an indicator of these dynamic variations.

Finally, the question of time arises for massless particles, in particular photons. Although they undergo apparent trajectory deviations due to the curvature of space-time, photons do not exhibit acceleration in the proper sense, which justifies their absence of mass. For these quanta of the

272

electromagnetic field, eigentime has no meaning. Their energy is described by the relationship E=hv, which, although it involves a frequency and therefore a temporal magnitude, allows for accounting for the interactions they maintain with massive particles and fields.

The time of events can be considered as an emergent quantity of a three-dimensional space within which no state can be strictly immutable. In this perspective, time does not pre-exist phenomena but appears together with the opening of space and interactions, as suggested by the hypothesis of an initial phase of radiative entanglement preceding the individualization of spatiotemporal quantities. On the other hand, nothing indicates that this temporal variable is applicable to the intrinsic, fundamental and perennial properties of the fermions constituting matter, nor to those of their antiparticles or photons associated with electromagnetism. Time, just like space, would thus result from the first quantum interactions rather than being a primitive property of elementary entities.

Time imposes itself on the local observer as a necessary parameter for any measurement, analysis and causal relation of phenomena, in that it gives depth to a space of events. However, when considering scales tending towards the infinitely small, this notion seems to gradually lose its operational relevance. Subatomic physics thus seems to be circumventing the intuitive conception of time as an ordered and continuous succession of states, which implies that the order of events at this scale is not necessarily predetermined.

Space-time could then be interpreted as a construction inherent in the presence of an observer interacting with a system. It is this spatiotemporal framework that, at the macroscopic scale, leads to attribute a corpuscular appearance, associated with the idea of mass, to phenomena whose deep nature remains quantum. This representation process, although reductive, makes possible an intelligible description of the architecture and evolution of the Universe. Thus, asserting that a proton consists of three quarks is a useful but extremely simplified approximation of a much more complex system. Similarly, the representation of the atomic nucleus surrounded by a procession of electrons is part of a schematic model that masks the profoundly non-classical nature of the interactions involved.

Subatomic interactions, such as those that govern the formation of molecules, rely primarily on quantum correlations. These correlations, not directly perceived as such, produce the illusion of spatial movements and suggest causal relationships, which are associated with temporal continuity. The role of the observer then appears decisive: it is his intervention, and the state of consciousness that accompanies it, which imposes recourse to the idea of space/time, which nevertheless remains fully relevant in a macroscopic context.

The question of the dimensionality of space therefore deserves to be reconsidered. To represent it as a three-dimensional space defined by three orthogonal vectors amounts to neglecting both the gravitational deformations highlighted by relativity and the fundamental indeterminacy that affects any localization and any movement in quantum mechanics. In a Universe devoid of center and definable boundary, where initially parallel geodesics can converge, it becomes difficult to limit space to three arbitrary coordinates. The geometric definition of a space region must then simultaneously integrate:

- the possibility of space-time superposition associated with matter and antimatter,
- the chirality related to fundamental symmetries,
- the relativistic effects of the spatiotemporal structure.

Admitting that space and time cannot be related to any intrinsic property of elementary particles, nor to the macrocosm here envisaged as a multiverse Cosmos without directly accessible physical reality, considerably complicates the cosmological interpretation. This leads to question the relevance of a model linking an initial supposed singularity to a final equally singular deadline, considered as a global collapse.

In quantum physics, two particles that have been correlated at a given moment remain potentially linked by this past correlation. The superposition of states and quantum entanglement thus lead to a questioning of the classical causality principle and the notion of linear time. Matter, by structuring itself and dissociating from its initial symmetry, configures space and defines a temporal orientation. Some formulations of quantum field theory suggest that antiparticles could be described as propagating in an opposite temporal orientation to particles. An alternative interpretation

274

consists in considering that they evolve in a permanent quantum present, apart from the temporal irreversibility specific to the macroscopic world made up of matter.

Talking about a distinct dimension for antimatter, in chiral symmetry with matter, is of course part of an approach of thought. However, this image reflects the fact that antiparticles, when they manifest, seem to emerge from a quantum vacuum within which also evolve electromagnetic waves, with which they interact in a discreet way.

The human condition corresponds to a particular state of matter, which requires the joint integration of the notions of time and space in any reflection concerning quantum symmetries. If it is not relevant to speak of the direction of time for a particle or an antiparticle taken in isolation, space and time nevertheless acquire a specific meaning at the scale of structured matter.

Space and time can then be considered as inseparable vector quantities, allowing, through logical reasoning and mathematical formalisms, to describe phenomena resulting from the interaction of a multitude of energy fields within a Universe without edge or center. The idea of quantum symmetry associated with spatiotemporal chirality, sometimes formulated as the existence of mirrored universes, offers an expanded theoretical framework. By postulating a radiative entanglement at the origin of matter particles, and by considering them as sets of entangled waves devoid of intrinsic physical dimension, it becomes conceivable to establish a link between the infinitely small and the infinitely large.

In this approach, the particle considered as a closed stationary system conceals indissociable and non-discernible variables, precisely because they do not relate to space or time. It can only be understood through its observable interactions, to which are associated physical quantities such as charge, mass, speed or spin, defined by their ability to intervene in interaction processes.

Electromagnetic waves, and in particular light, seem to propagate in a regime where time is practically frozen. In the hypothesis of a space totally devoid of massive particles, the celerity of light would become infinite, making even notions of space and time inoperative. Without a spatiotemporal referential and therefore without a possible observer, the

275

Universe as we conceive it would not exist. This extrapolation makes it possible to outline the concept of a multiverse Cosmos, fundamentally inaccessible to any mathematical description based on measurable quantities.

If electromagnetic waves propagate at a constant speed, why wouldn't it be the same for matter particles? These must be understood as wave packets whose internal dynamics are not directly measurable. These intrinsic movements notably determine the spin and give the particle properties similar to those of a three-dimensional gyroscope. The conservation of angular moments that results in dynamic stability reflected, at the macroscopic scale, by the inertial mass.

Finally, the subjective perception of time in humans can be interpreted as a consequence of the global energy activity involved. An individual in a low-activity situation perceives the flow of time differently from an individual subjected to intense activity or high stress. This variation is explained by the differentiated processing of information and events by neural circuits, illustrating once again the non-fundamental character of time, closely dependent on the conditions of observation and information processing.

At the strictly quantum scale, time does not seem to be a fully relevant variable. It begins, however, to emerge with the first interactions which, from the elementary atomic structures, gradually lead to the construction of organized matter. The notion of reduction — or collapse — of the wave function corresponds to a mathematical formalization by which a particular state is assigned to a particle. This state should not be interpreted as a revealed intrinsic property, but rather as a selection compatible with the observation context, that is to say with the space-time reference system in which the observer necessarily belongs.

The reduction of the wave function, generally postulated as random, is more easily understood in systems with a large number of particles, where quantum correlations become statistically indiscernible. It cannot, however, be formulated in a fully relativistic manner, which leads to the notion of locality being dismissed or at least weakened. The wave function thus appears as an incomplete description since it fails to coherently integrate non-local quantum interactions into a relativistic temporality. The result is a dissociation between the macroscopic Universe, structured by space-time of relativity, and the non-relativistic universe of quantum mechanics, in

276

which the particle only manifests through states implicitly selected by the act of observation. This discontinuity calls into question the idea of a unified model in the classical sense, often seductive but probably misleading.

Then arises the question of the very possibility of observing entities devoid of temporality, dimension and spatial location. It is precisely this difficulty that leads to considering counterintuitive phenomena such as the superposition of states and quantum non-locality. It should be recalled that before the nucleosynthesis phase, when matter was not yet structured in atomic form, neither time nor space could have the operational meaning we ascribe to them today.

In this perspective, time appears above all as a comparator, making it possible to relativize the interactions within the constructed matter. In the absence of stable material structures, the notion of time becomes difficult to define. Compared to the concept of eternity, a fraction of a second like a billion years loses their descriptive relevance. Similarly, any spatial measure becomes unrepresentative in the face of the extremes constituted by the infinitely small and the infinitely large. The notion of space-time thus proves to be ill-suited to notions of infinity, eternal, or timeless.

Paradox of the displacement in space

It is established that no information or energy can travel faster than the speed of light. However, depending on the referential of the observer and that of the observed phenomenon, the duration associated with an event can be perceived as allowing apparent speeds exceeding this limit. This situation stems from the fact that time cannot be calibrated identically at any point in space-time.

This observation leads to question the fundamental character of the speed of light considered as constant in an Universe where no magnitude seems truly absolute. The density or degree of occupation of space could thus play a role analogous to that of a speed regulator. The gravitational effects of massive bodies, by jointly constraining space and time, would impose a limit on the propagation of all forms of energy, including that of electromagnetic waves.

Paradox of the occupation of space

For a composite particle considered out of interaction, as for a black hole beyond its accretion disk, time loses all operational significance, which induces an absence of definable spatial location. If time ceases to be relevant inside black holes and at the heart of elementary particles, it nevertheless remains essential for understanding the majority of observable phenomena.

Between these two extremes — elementary particles and compact gravitational objects—, the spatial positions and the timeline of events remain marred by imprecision, due to the elastic nature of time as measured by an observer facing considerable scale changes. The present, conceived as an interface between a bygone past and an undetermined future, remains inherently elusive. **Time could then be interpreted as a perceptive manifestation of an unrecognized chirality, revealing a binary system of universes in quantum symmetry.**

The perception of time and distances depends closely on the inertial frame considered. Time tends to vanish when the locally accessible space seems to contract, i.e. when the actual distances to be covered decrease. Conversely, space seems to fold in on itself when time expands, as is the case near massive objects such as neutron stars or black holes, or even when time ceases to be a relevant variable, as for electromagnetic waves propagating at the limit speed.

When a reference is changed, the usual units of measurement — second and kilometer — vary in concert. The time dilation associated with the contraction of distances under the effect of gravitation illustrates the fundamentally relativistic character of time, which testifies to the variable topology of space. It follows that two events measured as simultaneous in distinct frames of reference cannot be considered as such in an absolute manner. Without invalidating the principle of causality, this situation introduces a paradox difficult to solve, even by means of mathematical formalism.

It follows that there is only one present effective: the one associated with our immediate environment. The observable Universe is therefore presented as a mosaic of degraded images of more or less distant pasts, whose analysis has made it possible to reconstruct part of its history. On the other hand, extrapolating its future from these remains a highly limited undertaking, comparable to the uncertainties inherent in weather forecasts.

278

For the sake of simplification, we tend to overlook the fact that the curvature of space and the absence of a universal reference make any attempt at absolute positioning in a fluctuating space illusory. Similarly, the absence of universal time invalidates any measure aimed at establishing simultaneity or absolute duration between distant events. The introduction of additional spatial dimensions or the localization of time, as in some contemporary theoretical approaches, constitute coherent mathematical constructs but remain devoid of verifiable physical significance.

In the cosmological paradigm developed here, the following hypotheses are put forward:

- Space-time would emerge from a multiverse Cosmos devoid of temporality—therefore without history — and of spatial dimension—therefore without observable reality for an observer inscribed in a relativistic framework.
- This multiverse Cosmos could nevertheless manifest indirectly at the very heart of space-time.
- The annihilation of a binomial of universes in quantum symmetry would lead to the restitution of the energy corresponding to the multiverse Cosmos, beyond any spatiotemporal description.

A binary system of universes in quantum symmetry can be considered as a closed epiphenomenon, resulting from internal processes in a multiverse Cosmos devoid of definable physical properties. This multiverse Cosmos, deprived of spatial dimensions, temporality and all measurability, would not belong to a reality accessible to observation or experimentation. It is not surprising that this hypothesis evokes, by analogy, ancient representations of an immaterial founding entity, posed as the origin of all things without necessarily possessing a direct physical manifestation. This analogy, although culturally suggestive, nevertheless constitutes here only a formal rapprochement, without explanatory or causal significance. *Our situation is comparable to that of the fish in its jar, unable to imagine what happens outside a restricted ambient environment from which it cannot be physiologically detached.*

In the standard cosmological model, quantum mechanics does not present a formally established link with gravitational phenomena, while general relativity remains difficult to reconcile with the fundamental principles of quantum theory. Below a certain minimum distance (Planck length: about

279

1,6 × 10^{-35} meter) space seems to have no meaning and our physical laws cease to work.

The same is true for the Planck temperature (about 1.417 × 1032 K). General relativity, which describes space and time, and quantum mechanics, which describes the world of particles, are thus incompatible.

One of the major obstacles lies in the unavoidable recourse to the notions of time and space, which structure each of these theories in an incompatible way. The observer then finds himself in a situation similar to that of a system immersed in its own reference environment, unable to conceive an exterior to which he cannot physically access. Any attempt at description is thus conditioned by constraints intrinsic to the type of observation.

If a direct observation of the Universe had been possible in the moments immediately following the Big Bang, it would probably have yielded no exploitable information about its subsequent evolution. Symmetrically, a hypothetical immersion in a cooled and diluted Universe, reduced to the sole presence of extreme compact objects such as black holes or massive dark matter residues, would not allow to reconstruct the initial conditions of its emergence. In such a state, any informational trace of the past would have disappeared, making it impossible to access an exploitable cosmological memory.

The current situation nevertheless presents a remarkable singularity: it brings together physical conditions favourable to the appearance of complex biological systems, capable of self-regulation and abstract representation. Humanity thus has the ability to observe degraded remnants of a partial cosmological past and to formulate limited projections into the near future. This ability remains constrained by necessarily finite means of investigation, not neutral with respect to the observed systems, and subject to instrumental and conceptual biases. Access to the initial conditions of the Universe as well as its ultimate future seems thus to remain, for an indefinite period, in the domain of theoretical modeling and intellectual exercise. It is also likely that the physical conditions conducive to life on Earth will disappear before definitive answers can be formulated and experimentally validated.

In this perspective and with a hint of cynicism, we could say that with his capacity for analysis formatted to his measure, a pragmatism formatted by restricted cognitive abilities and his intuitive logic proceeding by deduction, man remains the abused and credulous witness, of phenomena 'en trompe l'œil'.

XX Is the Time a problem for the relativity?
(Should we ignore space as we invalidated ether?)

Time probably does not possess the autonomous physical reality that is spontaneously attributed to it. It constitutes an inseparable component of the space-time structure. However, if time is not an independent physical entity, it is appropriate to extend this conclusion to space itself within the framework of the relativistic theory of gravitation, where any description of phenomena depends on the observer's reference frame. Considered separately, space and time then appear as observation conditions, elaborated in order to make intelligible discrete and interactive phenomena essentially belonging to quantum physics, in particular those associated with the energy described by superposition states. These constructions are especially adapted to the observation of phenomena manifesting at scales accessible to our experience.

In a Universe where every entity ends up interacting with others, the measurement of time is locally determined by the masses in interaction. Mass can be interpreted as the global expression of the intrinsic and combined movements of the elementary constituents of physical systems. In this perspective, the slowing down of time observed in gravitational fields or at high speed is interpreted as the consequence of the increase or composition of internal and relative velocities. The fact that the degree of occupation of space by energy and matter is correlated with a form of "plasticity" of time confers a relativistic context to gravitational interactions responsible for the progressive aggregation of massive particles.

The measuring devices that allow to establish a chronology of events and describe movements in space are not spontaneously perceived as relative. For cognitive convenience, the observer tends to assign them a universal scope. Now, the observer is an integral, albeit marginal, part of a system of extreme complexity: the Universe. His judgment is necessarily conditioned by a restricted local environment, which does not allow a direct understanding of the system as a whole. Moreover, macroscopic reality seems out of step with quantum description, which seems to ignore the classic distinction between past, present and future, and question the very notion of space through phenomena such as non-locality and the superposition of states.

282

The superposition of states and quantum entanglement thus constitute two fundamental but destabilizing, properties of quantum physics. They reveal the foundations of a particle physics whose development remains recent on the scale of the history of science. The persistent difficulty in articulating this physics with classical and relativistic physics leads to consider the study of the Universe essentially as a question of scale and point of view. This interpretative approach inevitably relies on choices and assumptions of convenience, even when supported by rigorous mathematical formalizations. It is therefore hardly surprising that the standard cosmological model, faced with numerous explanatory limits, is increasingly being questioned.

In this context, it may be relevant to go beyond the already complex mysteries of quantum physics and classical relativity to question a more fundamental dimension of space-time: that describing possible discrete interactions between two quantum symmetries located at the border of a multiverse Cosmos. Such a hypothesis remains speculative but aims to account for emerging properties that are difficult to reconcile with current theoretical models.

The questioning of the nature of time and space leads to consider that these notions could largely result from the activity of the observer himself. Time and space would thus bear the imprint of cognitive structures and referentials specific to the thinking observer. The durations that we judge to be long or short are systematically related to familiar cycles, such as the earth's rotation, the earth's revolution around the sun, or the average duration of a human life. In the same way, extremely short scales, such as microsecond or Planck time, are interpreted from frames of reference that remain anthropocentric. Compared to the global evolution of the Universe, these extreme durations appear just as negligible each other. This observation deeply relativizes our perception of the flow of time and recalls the marginal position of the observer in the Universe, despite his claim to determine its origin and ultimate fate.

The scientific approach consists of locating phenomena, establishing causal relationships and relativizing descriptions in order to integrate them into coherent and, ideally, exhaustive models. The spacetime of general relativity takes shape from quantum phenomena related to fundamental interactions and fully asserts itself with the emergence of gravitational effects of massive

283

structures. Time is a quantity specific to each observer: although the succession of events may seem locally invariant, it remains relative as soon as one changes the frame of reference.

It follows that local time cannot be considered absolute. It is reasonable to assume that it evolves in correlation with the global dynamics of the Universe, notably with the redistribution and concentration of energy and matter. Space, on the other hand, constitutes the framework for locating observed events and describing movements. Similarly, time can be defined as a parameter for estimating speeds and establishing cause-effect relationships. However, when the spatial location becomes indeterminate, time seems to be able to be considered independently of space, which avoids directly confronting the question of the smallest measurable distance, associated with the length of Planck.

In this perspective, the length of Planck can be interpreted as a mathematical artifact intended to discretize space, in order to make possible the localization of idealized particles as dimensionless points. Moreover, quantum entanglement, by ignoring classical notions of displacement, calls into question our fundamental landmarks that are time and space. Time, as used in physics, must be spatialized to be formalized mathematically. One may then wonder if another way of perceiving it, independent of its spatial translation, could be sufficient to describe events correlated in a non-local manner.

Finally, if time is considered irreversible, so are events, and any consequence becomes the cause of a later event. This view must, however, be qualified if one assumes that the global evolution of the Universe is entirely determined from its initial state. In this case, an inevitable final state could be seen as closely correlated to the initial state, to the point of blurring the strict distinction between causes and effects. **Such a correlation could even be speculatively interpreted as the emergence of a secondary system of universes in quantum symmetry. This hypothesis leads to question the fundamentally unidirectional character of time, despite the image provided by entropy in the macroscopic world.**

It is generally accepted that time constitutes the fourth dimension of a space perceived as three-dimensional. It is however possible to formulate the hypothesis that the truly fundamental dimensions would be those associated with the existence of two distinct quantum symmetries. The fifth space-time

284

dimension introduced by Theodor Kaluza could be interpreted not as an additional spatial dimension in the classical sense, but as a formal representation of a hidden dimension, reflecting the existence of an opposite quantum symmetry, inaccessible to the observer and devoid of direct phenomenological reality.

The interactions between particles and antiparticles, if they occur within such a discrete dimension immanent to exchanges between quantum symmetries, largely escape any possibility of direct measurement. Nothing then seems to be able to be immediately correlated with our macroscopic reality. The human observer, by nature pragmatic, is situated in a frame of reference dominated by gravitational interactions of matter, as described by Albert Einstein's relativistic space-time. This reference framework essentially hides the interactions related to possible complementary quantum symmetries.

In an exercise of thought integrating this hidden dimension, the description of the Universe would be profoundly modified. Cosmic evolution could then be interpreted as a process of progressive deconstruction or return to a cosmological state of equilibrium, likely to be understood as a form of time reversibility. In this perspective, time would appear as an emergent quantity, resulting from a spatial illusion specific to an observer confined to a single quantum symmetry. This hypothesis nevertheless clashes with our natural inclination to reject any quantum symmetry that would call into question the familiar space in which our sensitive experience is inscribed, space where antimatter seems absent and where the interpretation of phenomena is based on a logic of material occupation in which we participate as physical entities.

It has been established that gravitation, by redistributing energy and locally rarefying what is referred to as the "empty" space, modifies the topology of spacetime. The concepts of distance and duration must therefore be understood differently depending on the reference frame considered. This relativity manifests itself in a particularly marked way near a black hole: for a distant observer, a clock approaching the horizon of events seems to see its rhythm slow down indefinitely. The clock only stops working when it crosses this horizon, that is to say when it is irreversibly deconstructed.

The relativistic character of time leads to consider that the depressive structure of space-time influences the propagation of electromagnetic waves. The units of measure for distance and duration used to express the speed of propagation of these waves cannot therefore be assumed to be invariant on a cosmological scale. Relativity thus requires taking into account the successive transformations of the Universe that alter the image of the past observed at a great distance.

As prescribed by general relativity, an observer located at the surface of the Earth measures a time that flows more slowly than that of a body located at a higher altitude, for example at the top of a mountain. To assert that space and time are relative is to recognize that the units of measure depend on each observer's frame of reference. In theory, this relativity can lead to an apparent inversion of the order of events, depending on the mass, acceleration and gravitational field conditions affecting the observer and the observed system.

Two typical situations illustrate these effects, in contradiction with the intuition derived from Newtonian physics.

In a first case, a system B separates from a system A, like a rocket leaving Earth or the outer envelope of a star detaching from its core during an explosion. Initially, A and B constitute a single system. Separation involves a transfer of energy from A to B, part of which is converted into heat, thus increasing the effective mass of B. This process breaks the initial symmetry between A and B. The acceleration of B must exceed the gravitational effects exerted by A, which decrease with the distance. Although their initial proper times coincide, the observer associated with A perceives the time of B as slowed down, while B perceives the time of A as flowing faster.

in a second case, two astrophysical bodies or structures A and B, unable to resist the gravitational effects they exert on each other, collide. The rapprochement leads to a partial fusion of their energies, accompanied by a conversion of mass into heat and an addition of their gravitational fields. This process, characteristic of cosmic evolution, restores a form of symmetry between A and B. Their initial eigentimes are then confused, while the time of the merged system A+B appears slowed down in a space whose curvature is accentuated. This phenomenon explains in particular the apparent extension of the lifespan of certain particles interacting with the

Earth's gravitational field, as well as the gravitational redshift of the observed light spectrum.

The twin paradox illustrates these relativistic effects in an emblematic way. Two twins who have evolved in distinct frames of reference can, in theory, find themselves in the same frame of reference with different ages. The traveling twin, to return to its starting point, must provide an energy comparable to that spent on the way, converting part of its inertial mass into kinetic energy and possibly exploiting space curvatures induced by gravitational fields encountered. During this return phase, the twin left on Earth then appears, from the traveler's point of view, as subject to acceleration, which again modifies the relative perception of time. This situation can be assimilated to a form of prolonged temporal slowdown, making it conceivable that interstellar voyages would take place in limited periods, subject to severe physical constraints related to radiation and gravity variations.

The superposition of notions of time and space implies that distance cannot be defined independently of the curvature of space-time. An increase in curvature in a depressionary region of space corresponds, for a distant observer, to an equivalent decrease in measured time. It is therefore necessary to correct the observed durations and distances considering the gravitational and inertial effects affecting the studied systems, as is done for the operation of satellite positioning systems.

The Universe can thus be described as a set of regions presenting energy depressions of variable depth, in permanent interaction. Each of these regions constitutes a gravitational center with its own spatiotemporal reference, participating in the global dynamics of space-time.

In a Universe where no system can be considered strictly isolated, any observed phenomenon should, in principle, be described relatively to an extended gravitational environment. Such a requirement implies, however, the consideration of a considerable number of parameters, many of which are neither measurable nor controllable. This results in structural uncertainty as to the scope and precision of the measurements carried out, an uncertainty that necessarily limits the universal validity of the results obtained.

The difficulty increases when one considers that two distinct regions of space cannot share the same global spatio-temporal context. In these

287

conditions, the notion of simultaneity loses all absolute meaning: events associated with these regions can only be said to be simultaneous within a common reference system, strictly local. Time, thus differentiated according to places and gravitational conditions, cannot be considered as a universal greatness. This lack of temporal uniformity can be interpreted as the manifestation of a chirality associated with distinct quantum symmetries.

The matter/antimatter chirality places quantum mechanics in tension with the classical space-time that structures our macroscopic experience. For its part, relativity accounts for the gravitational effects of matter on a large scale while excluding any global simultaneity for events that do not belong to the same referential. Quantum chirality and relativity thus converge towards the same observation: time does not possess any absolute character. The time to which we have access reflects only phenomena compatible with our capacity for perception and interaction. This situation can be interpreted as the selection of a particular state from among a superposition of possible states for matter observed in interaction, while antimatter is not assigned any recognized phenomenological state in our reference scheme.

Describing the Universe in an immediate present poses a major difficulty. The Universe manifests as a continuous succession of events linked by logical, causal or collateral relationships. A more adequate description would probably require an in-depth revision of the usual vocabulary: the concepts of object, matter, position or structure would benefit from being replaced by those of state, radiation, field of influence or process, respectively, while mass could be understood as a measure of energy level.

Locating a system in space means placing it in its own time, distinct from that of the observer. Conversely, circumscribing an event in time amounts to locating it in a space that is specific to it and does not necessarily coincide with the reference space-time of the observer. Relativity is precisely based on this interdependence of space and time, considered as two correlated variables within a multidimensional context dependent on the local reference system.

In order to illustrate this interdependence, it is possible to use a simplified geometric representation, based on elementary notions of planes and angles. This approach, although schematic, allows to identify certain structural properties of space-time.

First, let us consider an idealized particle as a dimensionless point evolving in a supposedly empty space.

– A single point, taken in isolation and outside any interaction context, cannot be described by spatial coordinates. In the absence of any possible relationship, the notion of space loses its meaning, and that of time also becomes devoid of foundation, for lack of measurable evolution or change.

– Two points considered as the only elements of a system can only be defined in relation to each other. Their geometric relationship is reduced to a line segment, a one-dimensional figure that does not allow any absolute measurement in the absence of a length standard. As long as these two points do not merge, space is limited to this linear relationship, without any observable change being attributable to a flow of time. In this case, it is not relevant to talk about space-time.

– Three points united in a system define a triangle inscribed in a plane, thus realizing a two-dimensional space. The position of each point can only be determined with reference to the other two, by the ratios of distance and the angles formed. In such a flat space, any variation in angles or lengths can be interpreted as an evolution of the system. However, this space remains incomplete: it lacks depth and does not allow the introduction of an observer. The time that manifests there therefore does not yet have the relativistic character associated with the geometry of physical space-time.

– Four points forming an isolated system and arranged in such a way as to constitute a tetrahedron define four plane faces intersecting each other. This geometric configuration allows the emergence of three spatial dimensions. Each point, taken as a vertex, can only be located with reference to the other three, and any modification affecting one of the points leads to a reconfiguration of the entire system. Each summit is thus associated with its own temporality, linked to the evolution of its relative coordinates. Such a minimal system of four particles is then enough to bring forth a space-time structure with variable geometry, analogous to that described by general relativity.

289

– When the number of points increases and several systems share common vertices, space becomes increasingly complex and multifaceted. The fluctuating energy density of what is called empty space, dependent on the effects of mass and the superposition of elementary structures, then determines the rate of time, more or less dilated or compressed according to the regions. Time is distinguished from space only from the point of view of the observer, and this distinction is only possible in a non-flat space, defined by a sufficient number of relations to allow the very existence of an observer.

This perspective naturally leads to the idea that space-time, and by extension the Universe as we describe it, only manifests itself because it includes an observer capable of formulating a representation of it. Such a conclusion explicitly refers to an interpretation of type anthropic.

The phase of radiative entanglement which would have led to the appearance of the first particles of matter cannot reasonably be envisaged as involving only a very limited number of particles. In the hypothesis considered here, this episode, associated with the initial phase of the Big-bang, would have occurred over an extremely brief duration, possibly negligible at the scale of subsequent physical processes. It would then have generated a considerable amount of primordial elementary particles. The massive production of these entities would simultaneously constitute the opening of space-time in which our Universe unfolds.

This representation remains deliberately imaged and does not strictly correspond to the usual modes of formalization of theoretical physics. She nevertheless proposes to link the appearance of time with that of space. In ordinary experience, which is at scales where quantum effects are largely averaged, time manifests essentially from the supra-atomic scale.

In this perspective, and by extrapolation to the fundamental scales, one can formulate the hypothesis that below the characteristic scales close to Planck units, the notion of time loses its operational relevance. The fundamental quantum regime could then be assimilated to a form of «permanent present», in which the distinction between past and future would have no physical meaning for elementary particles. The temporal variable would, in practice, be inoperative. The discrete interactions likely to occur between quantum symmetries would thus remain largely beyond the scope of direct observation and difficult to integrate into a single model.

290

Conversely, when considering interactions involving particles of matter belonging to the symmetry of our Universe, the introduction of a chronology makes it possible to describe these processes. Time then appears as a parameter allowing the ordering of physical interactions. In this hypothesis, however, this relative time would not quite apply, in the same way for interactions involving particles belonging to an opposite quantum symmetry.

From the thermodynamic point of view, the physical properties of material systems can be described in terms of momentum, energy and temperature. The Big-bang could thus be interpreted not as a punctual event in the strict sense, but as a thermodynamic threshold marking the actual appearance of physical time. Symmetrically, the final state of a Universe having reached an extreme level of cooling could correspond to a situation where the temporal dynamics gradually extinguish. The disappearance of time would then be associated with the erasure of a chirality resulting from an initial symmetry break.

According to this interpretation, a global process dominated by gravitation would lead the Universe towards a state where the relativity of proper durations tends to diminish. In a space characterized by a progressive dilution of energy, the eigentimes associated with the different reference systems would be increasingly dilated. This evolution could tend towards an asymptotic standardization of measurable durations. Within the limits of a strongly cooled Universe, all the referentials would converge towards a form of global synchronization. At this limit moment, characterized by the absence of a distinct physical reference frame, to assert that time disappears would hardly make more sense than talking about an infinite speed. Quantum chirality would thus indirectly participate in the relativity of the space-time structure.

In the envisaged process, the primordial energy would have been partially converted, by radiative entanglement, into particles of matter. This transformation would have introduced a characteristic property of matter: mass, at the origin of gravitational effects. The non-uniform distribution of these gravitational effects would then lead to local differences in the rate of time as measured by an observer.

291

Thus, in a region of space where gravitational effects would be extremely weak, the eigentime could theoretically flow at an extremely fast rate relative to that measured in regions of strong gravitational field. Conversely, in the vicinity of a black hole, the gravitational dilation of time can reach values such that its flow seems practically suspended for an external observer. In these two borderline situations, the notion of space-time becomes difficult to interpret in the usual terms. It would nevertheless be abusive to see in it a simple illusion: we undeniably occupy a portion of space and exist for a certain duration.

In this speculative model, the photon, mediator of electromagnetic interactions, could be interpreted as the vestige of primordial energy that would not have been converted into fermions during the initial entanglement phase. Due to its absence of mass at rest, the photon does not have a proper time and cannot be associated with an ordinary spatiotemporal trajectory. However, its interactions with matter contribute indirectly to the energy dynamics of material systems and their effective mass properties.

At the macroscopic scale described by relativity, the speed of light nevertheless constitutes a fundamental constant allowing to link distance and time measurements. The question then arises whether the temporal variable has an intrinsic existence or if it only emerges at the scale of the observer and the material structures to which he belongs.

This question leads to consider that mass can be understood, in some descriptions, as a mathematical parameter characterizing the energy state of a system rather than as an intrinsic irreducible property of particles, especially when they are described as wave packets.

In this context, the notion of Cosmos multiverse can be introduced as an extended theoretical framework allowing to think physical realities where classical variables of space and time would no longer be fundamental.

The fact that space and time do not possess an absolute character is not incompatible with the observation that the Universe appears globally isotropic and homogeneous on a large scale. However, this finding raises questions about the real scope of our observations. Our perception of cosmic reality remains inseparable from our status as conscious observers. The models we develop are therefore inevitably based on interpretations and hypotheses.

292

When these hypotheses cannot be tested experimentally, the result is frequently a lack of scientific consensus and the emergence of competing theories. Despite the significant progress in contemporary physics, our understanding of the Universe remains based on increasingly elaborate mathematical representations, whose limits gradually appear.

This situation invites us to critically re-examine certain aspects of the standard cosmological model, which today shows certain theoretical and observational tensions. The difficulty, however, lies in the fact that our description of the world is fundamentally anchored in a four-dimensional box of reflection —three space dimensions and one time dimension — which corresponds closely to our physical experience.

Therefore, the question arises whether the human mind is really able to conceive the deep structure of the Universe, its origin and its eventual purpose.

In our representation of the world, three notions play a central role:

- space, which allows the location of objects.
- time, which orders events.
- the mass, which determines gravitational effects.

They structure our understanding of body movement and energy transfers. Even if the current cosmological model is based on solid observational foundations, it remains possible that it only constitutes an interpretation compatible with our cognitive abilities and mathematical tools.

In this perspective, space and time could be emerging concepts whose relevance depends on the scale considered. Their physical significance would truly become operational from the atomic scale, then fully assert itself with the organization of matter into macroscopic structures.

Our perception of reality is above all based on the intellectual interpretation of an immediate and tangible environment. However, scientific reflection also mobilizes the imagination in the form of thought experiments intended to explore theoretical hypotheses.

At the scale of macroscopic objects, every material entity is assigned a physical property called mass, which allows quantifying gravitational effects and describing the local geometry of spacetime.

On the other hand, at the subatomic scale described by quantum mechanics, the notion of object becomes much less relevant. Time as we conceive it seems to lose its intuitive meaning and space ceases to appear as a continuous support. Physical phenomena are then described as interactions between quantum fields organized in the form of wave systems.

The notion of a particle no longer necessarily corresponds to that of a corpuscle occupying a specific point in space. This interpretation has notably inspired certain theoretical developments, such as string theory, in which point particles are replaced by one-dimensional vibrating objects.

Experimentation in particle physics has led, due to the inability to definitively decide on the ultimate nature of matter, to associate the notion of corpuscle with that of a particle of matter. However, the recognition of wave-corpuscle duality has gradually led to introduce into theoretical models the existence of additional particles playing the role of interaction vectors. These entities, called bosons, make it possible to describe the mechanisms by which matter particles endowed with mass—fermions — interact and exchange information remotely.

Thus, within the Standard Particle Model, several mediating bosons have been introduced in order to account for the fundamental interactions observed.

– The gauge bosons Z^0, W-, and W+ were postulated to explain the nuclear decay processes associated with weak interaction. At sufficiently high energies, these particles fit into the theoretical framework of electroweak force, which unifies electromagnetic interaction and weak interaction.

– For these bosons to fulfill their role in field theory, their description assumes the existence of an additional coupling mechanism, represented by the Higgs boson. This boson corresponds to the quantum excitation of a scalar field, called the Higgs field. In the current model, this field interacts with fermions (with the probable exception of neutrinos in certain formulations) as well as with bosons of weak interaction. This interaction is

294

interpreted as the mechanism by which particles acquire an effective mass, this being generally understood as a manifestation of an inertial resistance.

– The photon, a mediator boson of electromagnetic interaction, can be interpreted as the quantized manifestation of the electromagnetic field. In quantum field theory, it also represents a possible excitation of vacuum energy, that is to say the fundamental quantum state of a space devoid of massive particles. The photon plays a central role in charged particle interactions and can, under certain energy conditions, give rise to particle-antiparticle pairs. This phenomenon contributed to the introduction of the concept of zero-point energy of quantum vacuum. Due to its involvement in a very large number of well-established physical phenomena, the photon occupies a fundamental place in the description of energy transfers and electromagnetic interactions.

– Finally, the strong interaction responsible for the cohesion of atomic nuclei is mediated by gluons. In quantum chromodynamics, these bosons ensure the cohesion of quarks within hadrons, notably protons and neutrons. This interaction can be interpreted as the result of a complex dynamics of color charges confined in strongly coupled systems.

From these elements, one can question the real nature of what we call mass, traditionally associated with the idea of corpuscle. It is possible that this notion allows, while remaining on the standard model, to maintain the coherence of a formalism based on wave-corpuscle duality. This duality is imposed in large part because of the position of the observer, whose perceptive and instrumental capacities are determined by the material and biochemical nature of his own environment.

In this context, mass nevertheless remains a fundamental indicator. It allows establishing causal and temporal relationships in the analysis of energy transfers between physical systems. The particles of matter can then be interpreted as confined wave packets, that is to say as localized kinetic energy configurations in closed quantum systems. In this perspective, these entities would not inherently have more mass than free excitations of quantum fields. An alternative interpretation would be to consider that it is precisely the binding energies associated with these confined wave configurations that produce the effect macroscopically interpreted as mass. In fact, numerous experimental results show that the binding energy contributes directly to the effective mass of physical systems, which is particularly evident in the gravitational effects associated with the curvature of space-time.

Furthermore, new data from quantum mechanics such as the collapse of the wave function, derived from the Schrödinger equation, as well as the phenomenon of quantum entanglement — involving a form of non-locality — invite us to reexamine our way of perceiving time and space. Indeed, in the quantum description, particles are formalized as wave states. Matter could then be interpreted as a set of wave packets resulting from entanglement processes occurring in the early phases of cosmic evolution, rather than as a collection of point corpuscles from an initial singularity still poorly understood.

In this perspective, the wave-corpuscle duality could be considered as a tool for understanding allowing to link the quantum descriptions of the infinitesimal to the familiar macroscopic reality, characterized by the existence of observable material objects under different physical states (solid, liquid, gaseous, etc.), after the primordial phase of nucleosynthesis that structured matter in the primitive Universe.

In order to distinguish these packets of entangled waves that are elementary particles, modern physics assigns them a set of quantized properties: quantum numbers, angular momentum, electric charge and spin. These parameters define the quantum state of particles and determine the possibilities of interaction, assembly or annihilation with other particles or their antimatter equivalents.

Although this representation of matter may be only an approximation adapted to our description capabilities, it provides a coherent modeling allowing the interpretation of numerous experimental observations at different scales. Moreover, nothing indicates that our way of describing these phenomena would be fundamentally different if we operated in another space-time frame comparable to ours.

In our interpretation of physical phenomena, we tend to extend possible quantum interactions by equating them with events related by causal relationships. This interpretation implies temporal succession and localized energy transfers in space. However, the existence of quantum entanglement suggests that certain correlations between systems can be established without a classical transfer of information occurring from one point in space to another.

This situation raises the possibility that some aspects of quantum dynamics take place outside the usual space-time framework, or at least on the boundary of it. Our daily experience, however, remains deeply rooted in a macroscopic universe where phenomena are interpreted in terms of movements, relative positions and temporal successions.

We give potentially possible or prescribed quantum interactions an extension that leads us to interpret them as related events involving effects from cause to traceable effects. It is in total contradiction with the fact that we are now at a stage where we can foresee that, with quantum entanglement, certain exchanges would occur without transfer from place to place. The problem is that few things seem within our reach, at this level of scale that is quantum mechanics. It seems that quantum dynamics occur outside or at the space/time boundary. Our reality is macroscopic, imposes movements and the lighting we have of it, is closely related to our condition as observers.

In this evolution of a Universe that we perceive as totally factual, everything boils down for us to changes in positions and relative movements, inducing a temporality of events. Difficult under these conditions to make the transition by changing scale between what tends towards the infinitely small unobservable and a Universe at the far reaches out of reach. We cannot, we avoid time as well as space and yet everything encourages us to do it.

297

XXI <u>About the difficulty of speak metaphysics</u>
(Without falling into the trap of spirituality)

It is necessary to make a more than formal distinction between symmetry and load. Symmetry refers here to the existence of two opposing forms of matter, matter and antimatter. Charge, for its part, refers to the positive or negative electrical properties that, within each symmetry, contribute to the equilibrium of atomic structures.

The question then arises as to how to represent an extremely dense set of interactions, made up of force fields in opposite symmetries, interacting continuously, without resorting to the familiar notions of space and time as they are described by classical relativistic physics. To make such a situation intelligible, it would be necessary to consider a physical description of a different nature. One can, hypothetically speaking, describe this approach as discrete physics. This would aim to integrate a dimension of the Universe that does not manifest directly in experimental observations. It would stand out both from quantum mechanics, developed from extremely sophisticated experiments, and the laws of relativity that structure our current cosmological model.

Such a perspective naturally leads to consider the hypothesis of a Cosmos of the multiverse type. This proposition may appear speculative, or even difficult to reconcile with certain interpretative positions of contemporary physics. However, if we consider the role of mathematics as a formalization tool, this hypothesis is not necessarily more abstract than some already accepted mathematical constructions. For example, the square root of a negative number — such as $\sqrt{(-1)}$, defining imaginary unit — does not correspond to any directly observable quantity. Yet this entity possesses internal consistency and proves indispensable in many mathematical and physical developments.

Similarly, some mathematical relations seem to suggest structures that do not necessarily fit into a simple representation of space and time. The inequality $A+B \neq B+A$, often evoked in certain non-commutative algebraic contexts, can be interpreted as an indication that an operation depends on the order in which the elements occur. Such a property implicitly introduces a form of sequentiality that can be assimilated to a temporal structure. If two operations produce different results according to their order of application,

298

this implies that the elements involved - here noted a and b - are not defined in an identical spatial or temporal reference frame. In this perspective, one could consider that certain fundamental symmetries escape the constraints imposed by classical relativity.

Modern developments in geometry have also modified a certain traditional representation of space. Non-Euclidian geometries, for example, allow the description of spaces in which two initially parallel lines can meet, which amounts to assigning an intrinsic curvature to the space. At the same time, contemporary arithmetic and algebra have considerably abstracted themselves from their initial material representations and are exploring logical structures whose physical interpretation is not always immediate. If these mathematical advances open up important prospects, they do not necessarily lead, in particle physics, to the experimental or theoretical developments initially hoped for.

To formalize these abstract structures, mathematics uses tools such as matrices, vector spaces, or operators. These formalisms allow to translate complex concepts into systems of manipulable equations. However, for an unspecialized observer, these constructions may appear as extremely complex sets whose solutions do not always provide a clear physical interpretation. In some cases, the results obtained raise more new questions than they provide definitive answers. *Some pages of equations suggest a labyrinth whose outcome leads back at most to the access portal.*

The situation can be compared, in a similar way, to an attempt to design an entirely new technological device without having a complete understanding of its working principle or an exhaustive description of its components. This analogy illustrates the difficulties encountered when trying to develop theoretical models in a field where experimental data remain partial.

From the point of view of scientific methodology, a theoretical proposal can only be considered valid if several conditions are met. On the one hand, it must be based on a hypothesis deemed credible, generally developed from axioms, postulates or theorems already established, and likely to produce results consistent with observations. On the other hand, this hypothesis must be replicable in nature, that is to say that the predictions it implies must be repeatedly verifiable in the different experimental situations considered.

299

When these conditions are met, mathematical methods based on combinations or associations of quantities can be considered validated in their field of application. However, significant difficulties arise when attempting to apply these methods to some fundamental problems of astrophysics and cosmology.

Among these difficulties are notably:

- the consideration of the notion of infinity or unbounded space in reasoning based on necessarily limited sets.
- the introduction of a space-time comprising more than three spatial dimensions, which profoundly modifies the relationships between physical quantities.
- the possibility of parameters still unknown but potentially necessary for a complete description of the phenomena.
- the hypothesis of quantum symmetries that may imply the existence of additional or parallel dimensions.
- the phenomenon of superposition of states, which introduces a fundamental indetermination in the description of quantum systems.
- the wave-corpuscle duality, associated with decoherence processes, which makes measurements all the more uncertain as the scale of the system studied is reduced.

In order to describe the relationships between physical quantities of different nature — energy, mass, impulse or displacement — it was necessary to have a formal language capable of expressing these relationships in a coherent and generalizable manner. Mathematics fulfils precisely this function: it constitutes a symbolic system making it possible to model complex interactions, to integrate the effects of relativity and, in quantum mechanics, to reconcile the undulatory and corpuscular descriptions of matter.

It should be remembered, however, that mathematical formalism was developed by the human being according to his own capacities of representation and reasoning. It is therefore an extremely powerful tool, but which remains conditioned by the limits of our understanding and by the available data. If mathematics allows for the exploration of both the infinitely small and the infinitely large, their effectiveness strongly depends on the quality of empirical information and the means of calculation available.

300

Despite these limitations, mathematics plays a central role in many fields of science and technology chemistry, physics, medicine, computer science or climatology. They allow not only to interpret observed phenomena, but also to guide research, analyze random situations and formulate projections concerning the future evolution of complex systems.

However, the question remains whether it is realistic to hope to formulate a system of equations capable of comprehensively describing all the parameters governing the state and evolution of the Universe. Even if such a formulation were theoretically possible, it would probably be of such complexity that its interpretation would exceed the analytical capabilities currently available.

If the scientific questions are numerous, the answers are generally based on two main sources: the interpretation of observed phenomena and the formal analysis made possible by mathematical tools. Observation constitutes the first level of access to physical phenomena, but it remains constrained by technical and instrumental limits that are becoming increasingly difficult to push back. Mathematics, on the other hand, represents the privileged tool for analyzing and synthesizing available data. However, despite the considerable increase in computing power and the rise of computer processing, some mathematical questions remain particularly difficult to solve. Problems such as the equations of Navier–Stokes, the conjecture of Hodge, the problem P versus NP, the hypothesis of Riemann or the conjecture of Birch and Swinnerton-Dyer illustrate the existence of persistent limits in our ability to establish complete demonstrations or provide general solutions.

In mathematics, the notion of measure is fundamental. In geometry, and particularly in non-Euclidean geometries, assigning a dimension—whether it is a length, a surface or a volume — assumes that the object under consideration can be inscribed in a defined space, within which an operation of measurement is possible. In other words, the precise determination of a quantity implies the existence of a sufficiently stable and delimited frame of reference to allow the application of coherent units of measurement.

This requirement poses a difficulty when applied to the Universe as a whole. If it is indeed unbounded and subject to permanent gravitational fluctuations, it becomes difficult to assimilate it to a globally measurable space in the strict sense. The situation becomes even more complex if we
301

consider the hypothesis of a Cosmos of multiverse type, consisting of a plurality of universes potentially independent from each other. The notion of an ensemble quickly becomes problematic: depending on the point of view adopted, this global system can be considered as devoid of definable dimension or, on the contrary, as possessing a potentially infinite dimension.

These considerations lead to questions about the very stability of physical measurements. Space and time, which usually serve as reference sites, cannot be considered completely invariant in a Universe subject to the effects of relativity. Determining a position or assigning a dimension implies, in practice, neglecting certain variations of the gravitational context that modify the structure of space-time. In the same way, measuring a speed involves ignoring numerous gravitational effects that may locally influence the curvature of space.

In a Universe whose topology is continuously modified by gravitational interactions, the measured quantities cannot therefore be strictly invariant. Gravitational effects involve, in particular, a joint modification of space and time, resulting in a relative contraction of distances and an apparent slowing down of the flow of time in certain regions with high energy density. Therefore, any physical measure should be understood as relying on locally defined units and valid in a limited context. These units constitute operational approximations that remain relevant in a restricted spatial domain and over relatively short durations.

In this perspective, mathematics has gradually established itself as a language particularly suited to the description of physical phenomena. Compared to ordinary speech, which often remains ambiguous and loaded with subjective interpretations, mathematical formalism offers a more precise and rigorous language. However, the increasing complexity of models can lead to systems of equations with a very large number of parameters, whose resolution sometimes depends on methodological choices or simplifying assumptions. In some cases, these choices can influence the interpretation of the results obtained.

The mathematical formalisms used in physics — numerical algorithms, differential and integral calculus, exponential series, infinitesimal methods or statistical treatments — have demonstrated their effectiveness in many fields. Nevertheless, their application to particularly complex systems also reveal certain limitations, in particular when the data necessary for a complete description are not available.

Any scientific theory is first and foremost an exercise aimed at proposing a coherent representation of a set of phenomena. These theoretical constructions must then be confronted with observation or experimentation in order to be validated or corrected. It is possible, however, that certain technical, experimental or even cognitive limits progressively restrict our ability to test certain hypotheses under satisfactory conditions.

These limitations are notably related to our ability to handle very complex abstract structures and the modes of formalization that we have. Any mathematical formulation indeed requires a sufficiently complete set of data to produce significant results. When an equation relies on incomplete or poorly known parameters, its resolution remains partial and does not allow for reliable conclusions. Many theoretical works encounter precisely this difficulty: the development of coherent hypotheses runs up against the absence of certain elements essential for their formalization.

Moreover, although mathematical tools allow us to correct or overcome the limits of our intuitive perception, they do not completely eliminate the risks of misinterpretation. At extremely large or extremely small scales, descriptive precision decreases and phenomena often have to be expressed in probabilistic terms. In quantum mechanics, the uncertainty principle formulated by Heisenberg precisely illustrates this fundamental limit in the simultaneous determination of certain physical quantities.

Quantum mechanics today is based on the idea of a continuum. But is it compatible with the use of infinitely precise numbers that give the feeling of splitting what cannot be separated or distinguished? The standard model is considered a particularly successful working theory. But when it comes to quantum mechanics, some leading physicists remain uncomfortable with its conceptual foundations. Some even think that the world is fundamentally deterministic even when it seems random and that reality depends on causes and effects. Our probabilistic laws at the microscopic scale would reveal the limits to our ability to understand.

A classic example of analytical difficulty is provided by the precise calculation of the evolution of a gravitational orbit. Rigorously determining the trajectory of a celestial body would, in principle, require considering all fundamental forces as well as all gravitational interactions likely to influence the structure of space-time. This complexity appears clearly in the three-body problem, for which no simple general solution exists.

In practice, any object — from the atomic scale to galaxy clusters — undergoes the cumulative influence of numerous gravitational fields generated by surrounding masses. These interactions can lead to continuous changes in trajectories, orbital planes and movement speeds. The disturbances become all the more difficult to predict when the system under consideration has a high energy density or when the time projection covers very long periods. The explicit introduction of time in calculations intended to predict the future evolution of a system generally increases the uncertainty associated with the results.

Thus, physical models can only be considered as fully reliable in relatively restricted space and time domains. The study of terrestrial climate systems provides an illustration of this situation. The attempt to predict the long-term effects of increasing greenhouse gases involves taking into account a very large number of interrelated factors, the precise influence of which sometimes remains poorly known. Systems of this type exhibit a high sensitivity to initial conditions, a phenomenon often illustrated by what is called the butterfly effect.

Despite these limitations, mathematical models and the resulting complex equations have allowed considerable progress in understanding natural phenomena. Even when their results remain partial or difficult to interpret, these tools constitute an essential foundation of the scientific approach. Without the advances they have made possible, the theoretical reflections developed here could not go beyond pure speculation.

Like the first deep-sea vessels leaving to discover the open ocean and unknown lands, the development of contemporary technologies, particularly in the fields of computer science, numerical simulation and algorithmic analysis, opens new perspectives for the study of the Universe. The continuous increase in computing power, associated with software capable of optimizing their own procedures through algorithmic learning — often grouped under the term "artificial intelligence" — hints at significant progress in the modeling of physical and cosmological phenomena.

These advances are however based on a field of reflection consisting of already established physical laws. Most of these laws rely on descriptions based on the spatial location of phenomena and their temporal succession. It is nevertheless legitimate to question the real capacity of current disciplines — physics, chemistry, mathematics, or astronomy — to fully elucidate the origin and ultimate foundations of the Universe. The study of fields such as quantum physics, the structure of black holes or the conditions associated with the Big Bang hypothesis could require theoretical models that are not exclusively dependent on the relativistic character of space and time.

Such an approach would suppose the existence of means of investigation capable of simultaneously exploring extreme scales, both in the domain of the infinitely small and that of the infinitely large. However, the experiments allowing access to these physical regimes are becoming increasingly complex and costly. Large particle accelerators and space observatories are essential instruments for basic research, but their development requires considerable investment and particularly sophisticated technological infrastructure.

In this context, the notions of "infinitely small" and "infinitely large" can sometimes be ambiguous. In quantum mechanics, certain particles are described as field excitations and can appear in virtual forms in some interaction processes. Similarly, certain cosmological hypotheses, such as those associated with the existence of a multiverse, envisage structures whose physical reality remains difficult to establish experimentally. These considerations sometimes lead to relativizing the operational scope of these concepts when they are used as general descriptive categories.

Estimating the age of the Universe also illustrates the difficulties associated with using temporal units of measurement in a cosmological framework. The claim that the Universe would be about 13.8 billion years old is based on cosmological models in which the year is used as a unit of duration. Now this unit is derived from the orbital movement of the Earth around the Sun and does not constitute, in itself, a universal fundamental quantity. More generally, certain physical constants — such as the speed of light c — are generally considered invariant in current models, but their role in cosmological evolution might depend on the global properties of the Universe at different times. In this perspective, the attribution of a precise age to the Universe, or the

305

determination of the observable horizon expressed in light-years, necessarily relies on a set of hypotheses relating to the stability of these constants and the properties of space-time.

Astronomical observations are themselves susceptible to be influenced by various physical phenomena. Gravitational lenses, for example, can amplify and focus radiation from distant objects. This effect can lead to an apparent change in the brightness or angular dimension of the observed objects, which complicates the interpretation of astrophysical data.

In inflationary cosmological models, the primordial Universe would have experienced an extremely rapid expansion phase. In such a scenario, regions that were initially close would quickly become separated over considerable distances. Therefore, the extremely distant objects we observe today could correspond to past states of structures that have since evolved or even disappeared.

Cosmic radiation reaching the Earth's atmosphere consists of both electromagnetic radiation — covering a spectrum from radio waves to gamma rays — and energetic particles. These particles can be elementary or composite and have very variable lifetimes. Some, such as protons, electrons or neutrinos, are relatively stable, while others, such as muons, mesons or free neutrons, are unstable and decay rapidly. Observations have also shown that certain nucleosynthesis processes can occur in these high-energy particle streams.

The diffuse radiation observed at the cosmological scale is however affected by the interactions it undergoes during its propagation. Emissions from distant regions can be modified by the presence of stellar objects, interstellar or intergalactic media, as well as various energetic astrophysical phenomena. The cosmic microwave background, often interpreted as fossil radiation from the early phases of the Universe, thus reaches us after passing through a complex environment that may partially alter the information it carries.

The analysis of the weak anisotropies of this radiation aims to reconstruct certain properties of the primordial Universe. However, this signal has traveled cosmological distances for billions of years, which implies that the initial information may have been modified by the successive interactions encountered during this journey. The question therefore arises to know to

306

what extent this radiation constitutes a reliable indicator of the physical conditions that prevailed in the early phases of the Universe.

In current cosmological models, the cosmic microwave background corresponds to thermal radiation characterized by a temperature of about 2,7 kelvins. This radiation can be interpreted as the vestige of an ancient phase of the Universe, but it continuously interacts with the intergalactic medium in which it propagates. It therefore represents composite information, resulting from both ancient conditions and more recent transformations.

More generally, astronomical observation is limited by the fact that light from the most remote regions of the Universe has not necessarily had time to reach us. Some stars observed today may have already ceased to exist, while other objects remain invisible due to their low brightness or the absorption of radiation by interstellar matter. This situation is at the origin of Olbers' paradox, who wonders why the night sky is not uniformly bright despite the presence of a large number of stars in the Universe.

Cosmological interpretations therefore rely largely on extrapolations from the observable Universe. If one considers a model in which cosmic expansion would not play a dominant role, it is possible that the structure of the nearby Universe provides a relatively faithful representation of its current distribution of matter and energy.

The propagation of radiation across the Universe is also affected by numerous interactions: gravitational fields, magnetic fields, collisions with particles or diffusion in interstellar media. These processes can modify the observed frequencies. Some models interpret the elongation of wavelengths as a consequence of the expansion of the Universe. Other hypotheses consider that this effect could be linked to a gradual decrease in the energy transported by photons during their propagation through intergalactic space.

In this context, the use of cosmic microwave background radiation to determine precisely the age of the Universe remains dependent on the theoretical assumptions adopted. The history of cosmological estimates also shows that the age attributed to the Universe was revised several times during the 20th century, as observations and theoretical models improved. These estimates are nevertheless based on the assumption that the temporal units of measurement used are universally applicable, which in itself constitutes a choice for the observer.

The advanced techniques of nucleo-chronology are not lacking in interest but they can hardly allow to give an age to our Universe. Too many nuclear cycles and an imperfectly understood nucleosynthesis can at most give some indications in the comparative evolution of our solar system and the galaxy it occupies.

A distant present which would be concomitant with our actuality, would appear to the distant traveler who would find themselves there, rather in the form of staged waves from infrared to ultraviolet, including visible light (between 380 and 780 nanometers wavelength) that is so familiar to us. But what does simultaneity mean in a Universe where everything is relativity?

XXII <u>The subject is catching cold</u>
(But is there anything to worry about!)

There is a direct relationship between the spectrum of radiation emitted by a body, the nature of the energetic interactions to which it is subjected, the temperature of its outer layers and the intensity of the microscopic agitation of its constituents. In a material medium, this agitation can be described, at the microscopic scale, by random scattering phenomena such as Brownian motion, which constitutes an observable manifestation of the thermodynamic behavior of particles subjected to multiple collisions.

These different indicators — electromagnetic spectrum, temperature and particle dynamics — actually refer to the same physical reality: that of energy in interaction, apprehended under different modes of observation. This relationship is notably described by curves linking the spectral intensity of radiation to the temperature of a body, such as those derived from the theory of thermal radiation. Such distributions indirectly inform about the statistical state of the system and the degree of entropy associated with the energy flowing through it.

The visible light range corresponds to relatively high temperatures and significant thermal agitation of the particles. When the emission spectrum of a body is mainly in the infrared, this generally indicates less energetic interactions and more moderate thermal agitation. Conversely, when the spectrum moves towards shorter wavelengths, close to the ultraviolet, it translates higher energy levels likely to lead to more intense interactions with the material environment.

If it is admitted that the total energy of the Universe is conserved in different forms, it is nevertheless possible to envisage that certain fundamental wave structures — designated here as EMW — have gradually lost amplitude and frequency during cosmic evolution. In this hypothesis, the radiative entanglement processes assumed to be at the origin of the formation of matter particles (fermions) would gradually become more difficult to achieve, the necessary physical conditions being less and less frequently met.

Moreover, when the intensity of an electromagnetic field increases in a thermal system, the spectral distribution of radiation evolves non-uniformly according to wavelengths. In many cases, the increase in intensity is accompanied by an increased contribution of long wavelengths compared to higher frequencies. This results in a shift of the maximum emission towards the red domain, suggesting that the progression towards ever shorter wavelengths encounters physical limitations.

In this perspective, an inverse transition leading spontaneously to a regime dominated by gamma rays of increasing intensity seems incompatible with the constraints of a finite Universe. Indeed, in a circumscribed cosmic system, no physical quantity can reasonably tend towards infinity. Historically, the lack of an adequate model to describe short-wavelength thermal radiation led to what was called the «ultraviolet catastrophe», a theoretical difficulty that preceded the emergence of quantum physics.

Before the establishment of the physical conditions described by the Planck scale, the Universe might have been plunged into a state difficult to characterize by the usual notions of temperature. In this speculative hypothesis, the initial temperature would not be defined in the usual thermodynamic sense and could be assimilated to a neutral reference form, comparable to a «cosmic zero». The moment associated with the Big Bang would then be neither hot nor cold in the conventional sense.

However, as soon as the first radiative interactions appeared, the phenomena of entanglement and interference between energy modes could have led to locally more energetic regions. These concentrations of energy would have developed in a general process of gradual dispersion. In this scenario, visible light would only appear at a later phase of cosmic evolution, before the overall spectrum of radiation gradually expanded to the infrared and then radio domains.

In this logic linking temperature, wavelength and entropy, it becomes possible to consider an alternative interpretation of the Big Bang. This could correspond, not to an extreme thermal event, but to a particular cosmological transition just like the collapse of a binary system of universes in quantum symmetry. In this hypothesis, Big Bang and symmetrical collapse would constitute two descriptions of the same fundamental phenomenon, assimilable to a «cold non-event» in a potentially recurring cosmic cycle.

310

Temperature, as defined in the centigrade scale or any other thermodynamic scale, is essentially a parameter for quantifying the average energy of microscopic interactions within a system. It is used to characterize the physical transformations of matter, notably transitions between solid, liquid, gaseous or plasma states. However, the perception of hot and cold remains a macroscopic interpretation related to human sensory experience. At the scale of particles, these notions have no intrinsic meaning: only the average kinetic energy of the constituents can be defined objectively.

Despite this limit, temperature remains a relevant tool for assessing the intensity of physical processes and for analyzing the global dynamics of the Universe.

- With the cosmological event associated with the Big Bang, space-time would open. The thermal energy would then become definable and quickly reach an extremely high level, strongly disturbing a pre-existing cosmological equilibrium state in a region whose location and spatial volume cannot be defined conventionally.
- In a very brief initial phase, this energy, still relatively homogeneous and without marked spectral structures, would manifest extremely high frequencies, possibly higher than those of currently observable gamma radiation. This limit could correspond to the so-called Planck scale, beyond which our physical descriptions become incomplete.

From this stage, the first radiative entanglements could appear. The notions of propagation and energy dispersion would then begin to acquire a physical meaning. The initially homogeneous energy could gradually concentrate in two symmetrical states, constituting the precursors of future particles and antiparticles. The opening of time would then correspond to the beginning of a process of energy dispersion which can give the macroscopic impression of a cosmic expansion.

In classical expansionist cosmological models, this phase is interpreted as the initial explosion of the Universe. However, these models do not always provide a fully satisfactory explanation regarding the ultimate origin or final destiny of the Universe.

The first assemblages of matter, at the origin of primitive stellar structures, are accompanied by a gradual decrease in the average temperature of the cosmos, although it remains extremely high for long periods.

In a Universe whose temporal evolution would tend to slow down gradually, elementary particles — identified today as quarks and leptons — structure matter and participate in the definition of three-dimensional properties of space as well as the temporal succession of events.

Some particles are mainly distinguished by their electric charge (like electrons), others by their extremely low mass (like neutrinos). Quarks are characterized by their fractional charge and their participation in the structure of hadrons. Gluons intervene as mediators of the strong interaction responsible for the cohesion of quarks within nucleons.

Moreover, certain particles — bosons — play the role of interaction vectors. Photons are the most familiar example for electromagnetic interaction. These particles have an integer spin, unlike fermions which have a half-integer spin.

Leptons generally behave like isolated particles, while quarks tend to cluster into composite structures under the effect of strong interaction.

The current inventory of particle physics identifies a large number of elementary and composite particles, each characterized by specific properties and a variable lifetime. From an undulatory perspective, these entities can be interpreted as configurations of entangled quantum waves which, by organizing themselves into atoms and then into molecules, participate in the formation of structures of increasing complexity.
In this speculative approach, the existence and persistence of these structures could be related to their ability to fit into stable and reproducible physical processes. In other words, only configurations compatible with the laws of evolution of the global cosmic system would be likely to be maintained durably.
This amounts to assuming that the Universe can only produce and maintain structures compatible with its internal dynamics. Any configuration not satisfying these conditions would have an extremely low probability of persisting in the long term.

The different elementary constituents of matter are likely to transform into each other under certain physical conditions. At the microscopic scale, current knowledge remains largely provisional, and many aspects of the basic structure of matter remain uncertain. In order to organize these observations, particle physics relies on a set of fundamental properties—such as electric charge, mass, spin, helicity or even "color" in quantum chromodynamics — which serve as parameters for classifying particles and describing their interactions.

These characteristics have made it possible to develop a description model of the Universe, in particular the standard model of particle physics, which has significant internal consistency. However, this theoretical framework still has limitations and paradoxes. Certain experimental observations or theoretical considerations could lead, in the future, to modify or abandon certain currently accepted hypotheses, or to reformulate certain fundamental theories.

Furthermore, there is no absolute guarantee that the Planck units — time, length, mass, energy, temperature and electrical charge—indeed represent invariant physical limits. Although they are defined from fundamental constants and considered natural scales for fundamental physics, it remains possible that they do not correspond to the smallest units actually accessible to physical measurement or description.

- Beyond the so-called "Planck wall" scale, it is generally accepted that current physical models lose their descriptive capacity. In the hypothesis considered here, the initially released energy could manifest itself as an energetic impulse whose dynamics would not be described by the tensor formalisms used in general relativity. This energy would then manifest itself in the form of very high frequency modes whose interactions and interferences could lead to the formation of the first primordial particles.

  In the initial phases of the Universe, extremely high temperatures - which can reach several billion degrees Kelvin - would promote interactions between these primitive particles, which could alternately combine or annihilate. Gradually, the first stable structures would appear. The lightest atoms would form, then associate in simple molecules. These sets would constitute immense gas clouds that would later form the first astrophysical structures, such as stars and planets.

313

In this context, primordial nucleosynthesis corresponds to the formation of the first light atomic nuclei. Subsequently, chemical processes would allow the appearance of molecules composed of different atoms, stabilized by covalent bonds. Parallel to these transformations, the average temperature of the Universe would continue its decrease.

- As the Universe evolves, particles and emerging structures locally tend to cluster under the effect of their interactions, particularly gravitational. *This process can be compared, analogically, to the formation of hailstones resulting from the progressive condensation of droplets in a cooled cloud.* Gravitation is gradually becoming one of the dominant mechanisms in the large-scale organization of matter.

The Universe can then appear as the seat of a dynamic resulting from two simultaneous trends: on the one hand, a global expansion associated with the Big Bang model, and on the other hand, local gravitational collapse processes that recall, on a small scale, the dynamics of a Big Crunch.

In this perspective, the Big Bang can be interpreted as the apparent manifestation of a generalized expansion of space, often pictured as a radial divergence from a very dense initial state. Conversely, Big Crunch-type phenomena correspond to localized processes of matter concentration, leading to the gradual formation of increasingly massive structures.

The particles and material flows coming from different regions of space-time participate in the rotation of these first material groupings. These processes take place in an environment where the average energy density is gradually decreasing. Gravitational, electromagnetic and nuclear interactions then begin to play differentiated roles in the organization of matter. Despite the existence of locally highly energetic regions, the average temperature of the Universe continues to decrease.

- The progressive condensation of gas clouds constitutes a preliminary step for the emergence of a great diversity of material structures. Heavy and light particles interact, transform and assemble according to local physical conditions. The phenomena of thermal radiation and visible light production result from these energy interactions.

At a more advanced stage of this evolution, gravitational attraction leads to the formation of galaxies that gather much of the matter produced since the initial phases of the Universe. These galaxies can themselves gather into galactic clusters. In many cases, massive black holes form at the center of galaxies and play an important role in their internal dynamics.

In the regions close to these compact objects, gravitational effects become particularly intense. However, despite these phenomena of local concentration of matter, the global dynamics of the Universe continue to manifest what appears as a large-scale expansion. At the same time, the average temperature of the cosmos continues to decrease.

At different structural scales, orbital movements are frequently observed. Electrons are generally bound to atomic nuclei in localized quantum states. Analogously, planets gravitate around stars, and stars cluster into galaxies. In many galaxies, the orbital movements are organized around a central supermassive black hole.
Galaxy clusters also exhibit collective dynamic movements, sometimes comparable to rotations around a common centre of mass, materialized or not by a dominant galaxy. These clusters may themselves belong to larger-scale structures.
However, when the Universe is observed on a very large scale, its organization appears less regular. The distribution of matter takes the form of a cosmic network characterized by galaxy filaments separated by vast regions poor in visible matter.
In these large-scale structures, galaxies and galactic clusters seem to follow convergent trajectories along these filaments *such as our rivers with their tributaries and adjacent streams.* The electromagnetic and gravitational fields that accompany these flows of matter contribute to structuring and maintaining these cosmic circulation routes.

315

The most extensive intergalactic regions essentially contain residual radiation and highly diffuse matter, including dispersed baryons. Some of this matter could gradually be captured by extremely massive compact objects, referred to here as MMBH (mega-massive black holes), in a Universe whose average temperature continues to decrease.

On a very large scale, the distribution of galaxies and galactic clusters outlines a three-dimensional structure comparable to a vast filamentous weft. This organization, observed today, corresponds however to the image of a Universe as it was in the past, due to the time necessary for the propagation of light.

The large regions apparently empty of this cosmic structure are not totally devoid of physical content. They are filled with background radiation and extremely dilute particles. This radiation does not necessarily constitute a repulsive force in the dynamic sense of the term, but it contributes to the impression that these low-density regions extend relatively to areas more concentrated in matter.

At a very late stage in cosmic evolution, large numbers of extremely massive black holes could constitute the main residual structures of the Universe. As the apparent expansion continues and the intermediate structures disappear, these compact objects would seem to move away from each other at ever greater distances. In such a context, the average temperature of the cosmos would continue its gradual decrease.

In this hypothesis, the global dynamics of the Universe would tend towards a state of increasingly slow dispersion. Cosmic evolution would then appear as slowed down, a consequence of an environment whose average energy density would become extremely low.

- In a very distant future, each very massive black hole — referred to here by the acronym MMBH — could evolve into a state in which its accretion disk would have disappeared for lack of available surrounding material. Under these conditions, the effective temperature of the system would reach extremely low values.

  The environment close to these compact objects would then be characterized by an almost absence of macroscopic energy flows. It is possible to consider the existence of an extremely diluted state of matter, which can be described as an exotic plasma. This hypothetical medium could exhibit certain properties similar to those of collective quantum systems, with very high conductivity or behaviors close to superfluidity.

  In such an environment, the black hole would continue to capture residual particles still present in its vicinity. Moreover, the gradual decrease in the rotation of the black hole would lead to a weakening and then the disappearance of the magnetic field associated with it.

- Absolute zero (0 kelvin, being 273.15 °C) is generally defined as the lower limit of temperature in thermodynamic systems. It corresponds theoretically to a state in which the thermal energy of a system reaches its possible minimum.

  However, in physical practice, any material system always conserves minimal residual energy related to quantum fluctuations and zero-point energy. Thus, no real body can be totally devoid of energy or completely free of microscopic interactions.

  In this perspective, the notion of absolute zero can only be understood as a theoretical limit associated with our thermodynamic measurement scales, which historically rely on macroscopic phenomena such as changes in the state of matter (for example, water).

  In the speculative hypothesis considered here, the physical state closest to a true energy minimum could be reached within these MMBH, at a very advanced stage of cosmic evolution, when the Universe would have practically exhausted all forms of organized energy interactions.

The multiverse Cosmos — understood as the global set of possible cosmological structures — could not be described in a relevant manner in terms of temperature, which only has meaning in locally defined thermodynamic systems.

From this angle of analysis, the history of the Universe could be described, at least partially, from its thermodynamic evolution.

This interpretation differs significantly from the one most often accepted in the standard cosmological model. According to the latter, the gradual cooling of the Universe would essentially result from the continuous expansion of space, which would gradually dilute energy and matter.

This expansionist vision can appear coherent from the point of view of the observer located inside the cosmic system he observes. However, the paradigm of an unlimited expanding Universe, without a clearly defined spatial or temporal boundary, while originating from an initial event described as singular, still raises many fundamental questions.

It is for this reason that the reference cosmological model remains, in many respects, a theoretical model capable of evolution. Although it provides a very effective description of many observable phenomena, it still contains areas of uncertainty and tensions that show that our understanding of the global dynamics of the Universe remains incomplete.

Universe and thermodynamics

1 Energy, although it continuously changes form during cosmic evolution, does not disappear globally. It transforms between different physical manifestations — radiation, kinetic energy, potential energy, material structures—without total loss of the energy content of the system under consideration. This global conservation is consistent with the first principle of thermodynamics, according to which the total energy of a closed system remains constant despite the transformations it undergoes.

2   The thermal death of the Universe by collapse, implies in the absence of significant time, the cessation of any form of remarkable entropy within the MMBH. This does not contradict the second principle of thermodynamics which predicts an increasing and irreversible disorder for an isolated system. Indeed, if retrograde dispersion results in a decreasing entropy in our Universe, the latter, arbitrarily detached from its quantum symmetry, cannot be assimilated to an isolated system. In search of a relative balance, it happens that the disorder is «functionally arranged». This is the case of the black hole that sidelines space/time. It is also the case to a lesser extent of a living organism. Their statuses seem to locally violate the second law of thermodynamics. But on the other hand, if the nearby environment reveals an uncontrollable increase in disorder (accretion disk for the black hole, return to the mineral state for the living), it is only a step in a global process that will nevertheless lead to resorbing any form of entropy, no form of energy being able to manifest itself in a void space of matter and EMW.

3   With reference to the third principle of thermodynamics, we can consider that only a system consisting of a Universe and an anti-universe in symmetrical interaction can represent a complete system from the thermodynamic point of view. In this speculative interpretation, the progressive combination of these two components would correspond to a limit state in which the classical notions of temperature and entropy lose their physical significance. Such a state would refer to the idea of a multiverse Cosmos, that is to say, a broader cosmological model in which the observable Universe would only constitute a particular manifestation of a more general symmetrical system.

In a broader theoretical perspective, quantum thermodynamics should then consider the discrete interactions between states of quantum symmetry. Such a description would imply that particles can interact with their antiparticles according to wave configurations of the same amplitude but opposite phases, leading to destructive interference phenomena.
In the current Universe, the conditions necessary for this type of exact correspondence between quantum states appear extremely improbable. However, one can envisage that these conditions become progressively

319

feasible in the very long term, when the energy of the Universe would have lost the diversity of states and configurations which characterize today its dynamic evolution.

The corresponding quantum symmetries could then be confined in extreme regions of spacetime associated with gravitational singularities. Black holes —and, by theoretical analogy, white holes — could then represent the complementary manifestations of these symmetrical states.

According to this interpretation, each black hole could correspond to a physical system whose symmetrical counterpart would be similar to a white hole. The physical representation of a white hole, however, remains largely conjectural. Insofar as these objects would also be located outside the classical descriptions of space-time, their interpretation remains essentially theoretical.

An important question concerns the apparent growth of black holes. When a black hole absorbs matter and radiation, its mass increases and its event horizon expand. However, the external observer perceives this phenomenon only indirectly, mainly through energy processes that occur in the accretion region surrounding the black hole.

Quantum decoherence, which corresponds to the loss of accessible information for the macroscopic observer, could contribute to limiting our direct understanding of physical processes taking place near the event horizon.

If we consider that a black hole represents a singularity or a break in the classical structure of space-time, it is possible to envisage that what we interpret as an object occupying an increasing volume corresponds in reality to a modification of the spatial framework-temporal in which it is inscribed. In this perspective, the very notion of volume could become ambiguous when applied to regions where space-time ceases to be defined in a classical way. What we call the black hole would locally be an absence of space/time.

This local absence of space/time that black holes would actually be, acts like a bottomless cavity. It manifests itself as a low-pressure center drawing to itself any form of energy within its reach. This localized "void" of space/time prompts us to draw a parallel with the gravitational effects of massive objects drawing the surrounding matter

320

Ordinary matter, at the end of its gravitational evolution, can be confined in black holes. It is possible to consider, more speculatively, that antimatter can follow an analogous process, possibly in a distinct physical or dimensional context.

In this hypothesis, two possibilities can be considered. The first would consist in assuming the existence of two distinct types of objects: black holes associated with matter and complementary structures — sometimes assimilated to white holes — associated with antimatter. The second hypothesis, simpler, would consist in considering that black holes constitute the common final receptacle of matter and antimatter. In this perspective, these objects would represent the ultimate points of convergence of the energy transported by the two forms of matter.

At the atomic and molecular scale, matter like antimatter can form globally electrically neutral systems. Under these conditions, electromagnetic interactions between macroscopic structures made of matter and antimatter could be highly limited. This property could help explain the experimental difficulty of directly observing certain forms of antimatter on a large scale. It remains nevertheless possible that matter and antimatter share certain fundamental physical properties, notably their gravitational effects. The numerous black holes present in the Universe — most of which still escape direct observation — could represent the final structures in which the energy transported by matter and antimatter would gradually be concentrated.
Such a hypothesis, although speculative, presents the interest of proposing a conceptual framework likely to provide interpretative elements to several open questions of contemporary cosmology.

XXIII <u>The universe in terminal phase</u>
(Before he rises from the ashes)

Perception of energy phenomena and radiative evolution of the Universe

The concepts of heat, modification of appearance or brightness correspond to human sensory perceptions associated with certain physical phenomena. In natural or technological contexts, these manifestations are often interpreted as indicators of risk or energy intensity. They are, however, above all observable manifestations of the energetic behavior of matter and radiation, independently of the emotional or subjective interpretation that can be attributed to them.

From a cosmological perspective, one can consider that the long-term evolution of the Universe results in a gradual decrease in the average frequency of electromagnetic waves capable of transporting energy. If cosmic expansion continues indefinitely, the red shift phenomenon will lead to a gradual lowering of photon energy. The initially highly energetic radiation could then be observed, on cosmological time scales, at increasingly longer wavelengths.

In this perspective, high-energy photons could gradually correspond, in the visible spectrum, to radiation associated with green, then yellow and finally red colors. It should be remembered, however, that these colors do not possess any absolute physical reality: they only correspond to the perceptive translation that the human visual system attributes to certain wavelengths of the electromagnetic spectrum, especially when these waves are broken down by a prism.

As their frequency decreases, photons carry less individual energy, which reduces their ability to interact with matter, especially electrons. Very low-energy particles, such as fossil neutrinos from the early phases of the Universe, already exhibit extremely low cross sections of interaction with matter. They thus contribute, in the same way as very low frequency photons, to different cosmic microwave backgrounds.

It should also be remembered that the energy transported by a photon is proportional to its frequency. Thus, several low-frequency photons do not necessarily reproduce the effects of a single high-frequency photon, particularly in quantum phenomena involving specific energy thresholds, such as certain electronic transitions.

In a very speculative perspective concerning the ultimate evolution of the Universe, interstellar space could be dominated by an extremely diluted residual radiation, located at the long wavelengths of the radio domain. This radiation would constitute a very weakened energy trace of past astrophysical processes.

In the theoretical hypothesis introduced here under the term of TNMM, these radiative residuals could possibly be absorbed or trapped by energy-saturated cosmological structures, whose physical nature would remain to be specified.

Speculative hypotheses related to MMBH and the state of matter in black holes (who can affirm that astrophysics is a completed science?)

The following proposals concerning MMBH are explicitly hypothetical. Contemporary astrophysics remains an evolving discipline, and many extreme phenomena—notably those associated with black holes — remain imperfectly understood.

In the daily observable environment, matter presents itself in three main macroscopic states: solid, liquid and gaseous. These states correspond to different configurations of organization and interaction of the constituent particles, which occupy a given volume with varying degrees of cohesion and mobility.

At higher energy levels, matter can adopt a fourth state, plasma, in which atoms are partially or fully ionized. This type of state appears in several astrophysical contexts: primordial plasma from the young Universe, extremely dense plasma at the heart of stars or even plasma present in environments close to black holes.

We can consider that thermal and mechanical energies jointly participate in the evolution of these systems, with entropy characterizing the degree of disorder or energy dispersion. In highly ionized states, particles can be found in complex dynamic regimes where their collective properties dominate individual behaviors.

In the particular case of black holes, some hypotheses envisage that beyond the event horizon, particles could lose the properties that distinguish them in classical particle physics. In this speculative perspective, the internal state of the system could correspond to a form of extremely compact energy configuration, the current description of which remains unknown.

To give a simplified, although very approximate, representation—one could imagine that this state is similar to a confined or stationary waveform,

323

characterized by the absence of clearly defined oscillatory parameters such as measurable frequency or wavelength.

It is necessary to distinguish this hypothetical state from that of the stellar plasma, which corresponds rather to a dense medium composed of ions and free electrons subjected to extremely high temperatures and pressures.

Possible interpretation of black holes in this perspective

A black hole could be envisaged — hypothetically—as a system akin to an extreme plasma with certain properties analogous to superfluidity. Such a state would imply a particular radiative dynamic and could correspond to a configuration in which ordinary interactions become negligible.

In such a regime, some speculative models suggest that the description in terms of classical space-time could lose its validity, which would make direct analysis impossible within the current framework of general relativity. Any object crossing the accretion horizon of a black hole would then lose observable physical characteristics in outer space. Fermionic-like particles (half-integer spin), such as quarks and leptons, may no longer retain their usual distinctive properties (charge, spin or identifiable internal structure). The physical description of the system could then be reduced to the fundamental mass-energy equivalence.

In this extreme context, and by analogy with the supposed conditions of the primordial Universe before Planck's time, nuclear and electromagnetic interactions might no longer manifest themselves in their usual forms. Gravitational phenomena themselves could then adopt a different nature from that described by classical relativity. The gravitational effects of the black hole would result from the fact that it creates in the space/time we occupy, as it were, a partially open breakthrough into the multiverse Cosmos.

A possible interpretation—although highly speculative — would be to consider the black hole as an interface or discontinuity between our spacetime and a broader cosmological structure, possibly described as a multiverse.

Possible structure and peripheral dynamics of black holes

It is conceivable that the matter or energy contained in a black hole is not perfectly homogeneous. A hypothesis is to consider the existence of a peripheral region with a slightly lower density than that of the system core. In this boundary area, some dynamic processes could remain. Collective movements could then return a fraction of matter or energy to the accretion disk surrounding the black hole. Under the effect of the rotation and inertia of the disk, some of this material could be ejected along the axis of rotation, a phenomenon observed in certain astrophysical systems in the form of relativistic jets.

These jets could be associated with the presence of magnetic fields generated by the movements of the plasma in the accretion disc, a mechanism analogous to a dynamo effect.

Hypothesis on the internal magnetic structure

Despite the speculative nature of these considerations, it may be considered that an active black hole does not necessarily have detectable internal activity in the form of observable movements or interactions.

In this hypothesis, the system could contain two opposing potential magnetic configurations, associated with the two hemispheres defined by the axis of rotation of the accretion disk. These pseudo-vector structures could share a virtual convergence point located at the center of the black hole.

The ends of the axis of rotation of the accretion disk could then correspond to a configuration in which the same magnetic pole appears on either side of this axis.

In current astrophysical models, it is generally accepted that the black hole itself does not directly emit a magnetic field. The fields observed mainly come from the ionized plasma of the accretion disc, whose movements can generate intense magnetic fields by dynamo effect.

Hypothesis of an absence of internal excitation in black holes

A possible hypothesis is to consider that, in the internal regions of a black hole, no phenomenon could correspond to a dynamic energy field in the usual sense of field physics. Any electromagnetic or magnetic excitation would be impossible there. Observable interactions would then be limited

325

to phenomena occurring in the immediate vicinity of the surface defined by the event horizon.

If this hypothesis were correct, observed astrophysical phenomena such as relativistic jets accompanied by X- and gamma-ray emissions, associated with accretion disks of active black holes or the collapse of massive stars— would raise some questions. It would be appropriate to consider whether the current interpretation of electromagnetic processes associated with active black holes correctly describes the physical reality of these systems.

Analogy with the phenomena of superconductivity

To explore this question, we can introduce an analogy with some known phenomena of condensed matter physics. At very low temperatures, and under certain conditions of pressure and density, the atoms of a solid see their thermal movements greatly reduced. In certain cooled metals, a state of superconductivity appears, characterized by the disappearance of electrical resistance.

In this regime, electrons can move collectively across the crystal lattice without significant diffusion. The electric current then circulates in the form of a coherent flow of electrons, without dissipation of energy by interaction with the atoms of the network.

In such a system, the atomic nuclei remain relatively fixed, while electrons flow through conduction channels. The energy of the system is conserved but does not manifest itself in the form of heat dissipation or friction.

Speculative application to the case of black holes

By transposing this analogy — in a very speculative way — in the case of black holes, one could imagine a state in which internal entropic processes become negligible. In such a hypothetical configuration, the degrees of freedom associated with elementary particles would gradually disappear.

The space available for the displacement of the elementary constituents of matter could also become non-existent. Structures that previously corresponded to electrons, nuclei or elementary particles would no longer retain their physical identity.

A black hole could not generate an intrinsic magnetic field, the conditions necessary for the existence of moving charges being absent. The electromagnetic phenomena observed in the environment close to an active

326

black hole would then be essentially associated with the accretion disk and the surrounding plasma, rather than with the black hole itself.

A complementary hypothesis would be to consider that the electromagnetic fields related to the material captured by the black hole would be gradually neutralized by the physical processes occurring in the accretion disk. The magnetic structures observed could then result from the interaction between the disk plasma and the rotation dynamics of the host galaxy.

In this case, the magnetic field lines could form loop structures around the black hole, giving the impression of a spiral configuration converging towards the equatorial region of the system. This geometry could produce, from the observer's point of view, an apparent behavior reminiscent of a magnetic monopole, although such an object has never been experimentally observed.

Hypothetical characteristics of the internal state of a black hole

In this speculative perspective, a black hole could be described as a system with several extreme properties:

1. A superconductor-like medium, devoid of an intrinsic magnetic field, by analogy with the Meissner effect observed in certain superconducting materials.
2. An exotic radiative plasma, in which the distinctions between the different quantum degrees of freedom would become indiscernible.
3. An extremely cold thermodynamic state, characterized by the absence of detectable dissipative interactions.
4. A highly compactified energy condensate, which may exhibit properties similar to those of a quantum fluid, but without the possibility of flowing into an identifiable internal space.

Limits imposed by general relativity

Despite these extreme properties, black holes do not seem to be completely immune to the laws of general relativity, since their gravitational structure remains defined by the space-time in which they were formed.

327

It is possible that the gravitational collapse of matter cannot continue indefinitely beyond a critical energy density, corresponding to a state where the available space becomes negligible. Near this limit, the gravitational dilation of time would become extreme, giving the impression that internal processes are practically frozen.
In such a state, photons and massive particles could be described as converging towards an undifferentiated energy configuration.

Cosmological hypothesis: final collapse and quantum symmetry

In an even more speculative cosmological perspective, this state could represent an intermediate phase preceding a global collapse of the cosmic system. Such a process would recall the scenario of a Big-Crunch, in which the energy of the Universe would be refocused into a symmetrical configuration.
The global dynamics of the cosmos could be interpreted as an evolution between two symmetrical states characterized by opposite spatiotemporal chirality, resulting from the initial Big-Bang.
If the interior of a black hole corresponds to a region devoid of interstitial space and observable interactions, descriptions based on general relativity and quantum mechanics could lose their validity in this boundary zone.

Point of view of the external observer

For an observer located outside the black hole, the gravitational dilation of time implies that processes taking place near the event horizon appear more and more slowed down, until they seem practically frozen. The black hole then manifests as a gravitational singularity located outside the directly observable domain. Our Universe could see itself as a vast fragmented black hole in formation, from which nothing can escape, not even its symmetry of quantum origin.

Hypothetical evolution of a terminally ill Universe

In a very long-term cosmological evolution scenario, it is conceivable that the Universe would reach a state in which the structures described here as MMBH would no longer be powered by free energy.

In this context, the absence of accretable matter would lead to the gradual disappearance of accretion disks, and consequently that of associated electromagnetic phenomena. Surface interactions would then cease, and magnetic fields would become negligible.

If radiation and interstellar diffuse matter also disappeared, the remaining gravitational structures could tend towards a progressive clustering, leading to a strongly compactified cosmic configuration.

In the hypothesis of a cosmological system consisting of two universes in quantum symmetry, this evolution could be interpreted as a final phase of global collapse analogous to a Big-Crunch dominated by black holes.

Gravitational and geometric interpretation of black holes

Due to the absence of an accessible internal reference frame, it could be considered that it is not the black hole that moves in space, but rather that the surrounding matter and energy follow the trajectories that converge towards its gravitational horizon.

In this geometric perspective, black holes could be interpreted as extreme regions of space-time deformation, comparable to passages or discontinuities connecting different regions of a larger cosmic space.

Their evolution could lead, in the very long term, to a convergence of these extreme structures, restoring the global energy of the cosmic system to a more fundamental symmetrical configuration.

In this speculative vision, the observable Universe itself could be envisaged as a global gravitational system evolving towards a state of increasing energy concentration, analogous to a black hole-like structure on a very large scale.

Wormhole hypothesis and relationship between cosmic expansion and collapse

The concept of a wormhole, often associated with particular solutions to the equations of general relativity, is an interesting theoretical hypothesis. In some models, these structures would connect two distinct regions of space-time. They are sometimes considered to be able to appear in association with black holes, and to be linked to hypothetical objects called white holes, which would represent the temporal inverse of black holes.

329

In this speculative perspective, a wormhole could be interpreted as a geometric connection linking different phases of cosmic evolution, allowing to establish a link between the initial phase of expansion — associated with the Big-Bang — and a final phase of global gravitational collapse.

Radiative state of a terminally ill Universe

In a very long-term cosmological evolution scenario, an Universe reaching a so-called "end of life" phase could be characterized by an extremely low energy density. The photons still present in interstellar space would then have undergone a very significant redshift, leading to wavelengths comparable to cosmological dimensions.
In such a context, their individual energy would become negligible and their interactions with matter practically non-existent. The Universe would then be dominated by extremely diluted radiation, whose intensity would be close to zero on a local scale.
In the hypothesis introduced previously under the name MMBH, these residual photons could possibly be absorbed by extreme gravitational structures. The virtual absence of intermediate matter and radiation in space could promote a rapid convergence of these energy balances towards these regions.
This ultimate phase could be interpreted as an event of global energy reconcentration, manifesting itself in the form of a cosmological transition comparable, in principle, to a new expansion phase.

Hypothesis of a cosmological renewal

A second-generation Big Bang could then appear as the consequence of a previous global collapse. Such a process would erase the extreme thermodynamic conditions—described here as a « cold depression » — which characterized the final state of the previous cosmic system.
This interpretation leads to consider cosmological evolution as a possible succession of transformation cycles, in which each expansion phase would be preceded by a phase of energetic collapse.
In such a vision, the past states of the Universe would not be preserved in the form of a direct historical continuity. Each expansion phase would

reconstruct a new cosmic state possessing its own initial conditions and determining the future evolution of the system.

Conceptual limits in the context of a multiverse

If one adopts the hypothesis of a multiverse Cosmos, the notion of global location or movement loses much of its physical significance. The individual universes constituting such a set would not necessarily share a common space nor a global reference frame allowing to define distances or neighborhood relations.
In this perspective, the systems described as binary universes in opposite symmetry could not maintain direct spatial relationships comparable to those observed between astrophysical objects in our Universe.

Speculative dimension and role of theoretical exploration

The hypotheses presented here naturally depart from the currently dominant interpretative models in cosmology. They must therefore be considered as proposals, aiming to explore new possibilities rather than describing an established physical model.
However, the history of science shows that advances in knowledge often result from the confrontation between established theories and new hypotheses that challenge existing themes of reflection. The exploration of alternative cosmological scenarios can thus contribute to enriching theoretical reflection and opening new avenues of research.

XXIV <u>Why subtitle this book: tales and legends?</u>
(Would they try to make a silk purse out of a sow's ear?)

It was appropriate to propose a more evocative title, in line with the presented development. It is also worth remembering that many tales and legends have their origin in poorly understood or hardly accepted phenomena. These narratives often constitute attempts at interpretation where partial perception, imagination and the appearance of reality mingle.

The human being can be considered as a potential explorer and researcher, often without being fully aware of it. In this dynamic of understanding the world, modern physics has developed a representation of the Universe based on the existence of elementary entities called particles. The latter have been identified, classified and described according to a set of properties allowing the characterization of their interactions. Each type of particle is thus distinguished by a specific way of interacting with others and participating in interference or energy exchange phenomena.

However, it is possible to consider a different hypothesis from the most commonly accepted representation. Particles might not constitute distinct fundamental entities but correspond to particular manifestations of the same physical substrate. In this perspective, they could be interpreted as wave packets described by wave functions and placed in a superposition of potential states. The state actually observed would then result from the measurement process, that is to say from the interaction between the quantum system and the observer's environment. Quantum decoherence would precisely represent the transition between this description in terms of superposition of states and the appearance of a specific state compatible with macroscopic observation.

The reality that is familiar to us thus corresponds to an effective description, accessible at our scale of observation and in accordance with our perceptive capacities. It does not necessarily reflect the fundamental nature of phenomena, but rather the way in which they manifest themselves in a context compatible with our measuring instruments and modes of representation.

In this context, the notion of particle essentially meets a descriptive need. It allows to formulate a coherent representation of the observed phenomena in

332

accordance with a logic gradually developed by experience and experimental analysis. The representation of localized bodies, possessing spatial coordinates and evolving in a given space, thus constitutes a construction of the mind allowing to apprehend phenomena occurring at the quantum scale. The difficulty remains, however, in linking abstract descriptions of quantum mechanics to macroscopic phenomena that make up the observable environment.

The spatiotemporal environment in which these phenomena are described can be considered as a necessary structure for the organization of the experiment by the observer. Space and time provide a frame of reference for assigning meaning and coherence to observed phenomena. However, the fundamental processes likely to be at the origin of these phenomena—for example discrete interactions between particles and antiparticles — largely escape the intuition and direct analysis of the observer.

This difficulty led physicists to develop a structured theoretical model, represented in particular by the table of elementary particles and by the standard model of particle physics. This model provides a coherent description of the fundamental constituents of matter and their interactions. The introduction of the notion of particle allows in particular to avoid the use of the wave function, a central concept of quantum mechanics but often perceived as abstract and difficult to manipulate in common intuition.

In this representation, a particle is characterized by a set of physical parameters such as mass, electric charge, spin, helicity or certain specific quantum numbers such as color or flavor. However, these properties only really manifest themselves in the case of observable interactions. For example, the wave function associated with a free electron, isolated from any interaction with an opposing charge, would not directly reveal the existence of its electrical charge. This only becomes effectively observable through interactions with other particles. Thus, the properties attributed to particles appear largely as operational characteristics revealed by interaction processes.

In this perspective, the state of superposition that characterizes a quantum system—interpreted here as a wave packet that can be associated with a particle — depends on the observation context and the reference considered. The relativity of time and the absence of simultaneity for distant observers imply that different observers can describe the same system according to different states, each being linked to its own observation frame.

Moreover, if an elementary particle cannot be strictly associated with a classical spatial occupation, the spatiotemporal context in which it is described could essentially depend on the presence of an observer and the organization he imposes on the studied phenomena. In this sense, space and time would appear more as description structures than as intrinsic properties of the quantum system itself.

Under these conditions, the quantities used to define and quantify the energy of a particle—such as its speed, internal movements, acceleration or mass — can be evaluated differently by observers located in distinct frames of reference. These differences in evaluation are not necessarily perceptible to the observers themselves, but they reflect the fundamentally relative nature of any physical description.

It follows that the representation we construct of the Universe necessarily remains local and partial. Even when our analyses are based on very rich data sets concerning specific phenomena, they only give us access to a limited description of physical reality. At our scale of observation, it remains difficult to grasp the deep nature of the Universe, which could simultaneously present a multiple character in its manifestations and unified in its fundamental structure.

Dematerializing what we have become accustomed to representing, for ontological and descriptive reasons, in the form of objects or corpuscles is a destabilizing exercise. It is indeed difficult to imagine an object — understood as a set of particles — when these particles themselves are described, in quantum physics, as wave packets that can exist in a superposition of states and do not necessarily have well-defined spatial occupancy.

The main difficulty lies in the fact that our perception and our mode of intellectual representation are based on an irrepressible tendency to locate phenomena in space and to inscribe them in a temporal succession. Our local spatiotemporal referential thus constitutes the framework within which we organize the observation and interpretation of phenomena. Yet this framework imposes limits on our understanding: it provides a coherent representation of the observable world, but it increasingly appears that it only allows access to a restricted fraction of the underlying physical reality.

In its initial formulation, the Standard Model of particle physics was based on a relatively small set of fundamental entities. It included notably twelve particles of matter — the fermions — organized into different families. To these particles was added the photon, a particle lacking mass associated with the electromagnetic field. The photon plays a central role in electromagnetic interactions, ensuring the transport of energy and information in the form of quanta of the electromagnetic field. It thus acts as a mediator of interactions between charged particles.

By analogy with this mediating role of the photon, the theory has introduced other interaction vector particles, gauge bosons, in order to account for nuclear interactions and different fundamental forces. These bosons are supposed to mediate interactions between massive particles in a formalized standard model. However, one may wonder if this extension is not partly based on an analogical generalization of the role of the photon in electromagnetism.

As theoretical models and experimental observations became more precise, some phenomena predicted by relativity or the standard model continued to pose difficulties of interpretation. In an attempt to answer this question, various theorists have proposed the existence of new particles endowed with specific properties likely to explain these anomalies or theoretical gaps.

In this context, different hypotheses have introduced new entities such as preons, squarks, selectrons, sterile neutrinos, gluinos, photinos, gravitons or even the Higgs boson. The gradual lengthening of this list of hypothetical particles can be interpreted in two ways: either as a sign of a gradual deepening of our understanding, either as an indication that our current representation of the fundamental constituents of matter remains incomplete.

It is then legitimate to ask whether it would not be appropriate to return to certain more fundamental assumptions. One possibility would be to consider that matter particles are only phenomenological manifestations associated with elementary points of energy, without spatial dimension and independent of time as we conceive it. In this perspective, the quantum interactions and the different physical phenomena observed could be

335

interpreted as local manifestations of a more general process tending towards a cosmological equilibrium state.

Once rid of certain ideological representations introduced to facilitate modeling — but which nevertheless remain essential for a methodical approach — the image of the Universe that would result could deviate significantly from the one we are used to considering.

On the strong interaction

In order to explain the cohesion of quarks within nucleons — protons and neutrons — theoretical physics has introduced the idea of an extremely intense interaction, but with a very short range, called strong interaction. Quarks being themselves unobservable in the isolated state, the existence of this interaction has been deduced indirectly from the properties of hadrons.

In the Standard Model, this interaction is mediated by particles called gluons, classified among gauge bosons. Gluons are assumed to be devoid of mass and electrical charge in order not to modify the fundamental properties of quarks. They, on the other hand, possess a specific property called 'color charge', which leads to distinguish several types of gluons, generally described as eight possible states.

However, if one adopts a more radical interpretation in which quarks would be considered as virtual quantum entities belonging to a superposition of potential states, certain alternative hypotheses become conceivable. In this perspective, the distinctions between quarks of the 'up' and 'down' types could correspond to different manifestations of the same ground state.

Such an interpretation could lead to questioning the necessity of introducing specific mediating particles such as gluons. Another hypothesis would be to consider that hadrons—and in particular nucleons, made up of three quarks — should not be assimilated to structures located in a space in the classical sense.

Moreover, in the field of quantum mechanics, the notion of time does not necessarily play the same role as at the macroscopic scale. Before the Planck scale, time as we conceive it largely loses its physical meaning. It is only at higher scales — notably at the supra-atomic scale—that the notion of time becomes relevant in the description of physical phenomena.

336

In this perspective, fundamental quantum symmetry could play a determining role in the internal organization of nucleons. The observed properties of these composite particles could result from dynamic equilibria between superposed states, these equilibria intervening in a way not directly observable in the internal distribution of charges and energies.

Not all hadrons are stable. Some more complex configurations may exist briefly, such as tetraquarks (composed of four quarks) or pentaquarks (composed of five quarks). The mesons, on the other hand, consist of a quark and an antiquark. These different configurations all belong to the hadron family, but their stability varies greatly.

Most of these composite structures have extremely short lifetimes and are generally produced under special experimental conditions, for example during high-energy collisions in particle accelerators. It is theoretically possible to imagine other even more complex configurations—for example hadrons composed of six quarks — but their existence would probably be just as ephemeral.

These composite particles often appear as exotic configurations resulting from extreme experimental conditions. They do not necessarily participate in the stable structure of ordinary matter in the current Universe. Experiments carried out in facilities such as those at CERN nevertheless make it possible to produce and observe these transient states, which are in a way temporary quantum structures.
The study of these ephemeral particles aims to test the validity of the standard model and explore its possible limits. However, the gradual introduction of new particles into theoretical models also tends to complicate the overall description of the fundamental constituents of matter. At the current cosmological scale, stable matter seems to be essentially composed of nucleons — protons and neutrons—themselves formed of three quarks belonging to the first generation of fundamental particles. These structures appear as the elementary bricks of baryonic matter.

A question then remains open: why do quarks seem to possess lasting stability only when confined in groups of three within nucleons? What is the deep nature of the mechanism that ensures this containment?

It can be speculatively argued that this tripartite structure presents a formal analogy with the three spatial dimensions that frame all our observations. In
337

this hypothesis, the confinement of quarks in groups of three could correspond to a particularly stable energy configuration in a three-dimensional space.

In this case, the confined quarks would form an extremely compact collective energy structure, in which internal interactions would not necessarily correspond to distinct forces, but rather to a state of equilibrium within a complex energy field not possessing classical spatial extension. The introduction of a specific interaction called 'strong force' would then essentially constitute an artifice intended to describe this phenomenon without necessarily solving the fundamental question.

From a given reference point, the Euclidean space can be represented geometrically by a volume comparable to that of a sphere, whether it is of finite or indeterminate dimension. The sphere indeed has the property of being a closed surface which, for a given perimeter, maximizes the volume it can contain. The volume of a sphere is determined by the knowledge of its radius r, according to the relation $\frac{4}{3}\pi r^3$, which expresses a characteristic cubic dependence.

This relationship is not unlike the one that allows determining the volume of a cube on the side a, given by the expression a^3. The cube is also a remarkable geometric figure for its regularity, symmetry and the equality of its square faces. In both cases, the volumetric description is based on a cubic dependence characteristic of three-dimensional space.

Any portion of space can then be described as a scalar quantity likely to be associated with a symmetrical or regular volume. From a geometric point of view, a region of space can also be described as the three-dimensional extension of a base surface, which leads to mathematical representations relying on the product of three orthogonal dimensions. In a simplified form, this description corresponds to the product $L \times l \times h$ (length, width, height), equivalent to the product of three vectors defining an elementary volume.

Could such a geometric representation contribute to shed light on how we conceive certain fundamental structures of matter, for example the nucleon envisaged as a relatively stable composite particle consisting of three fundamental components? The reference to elementary geometric shapes such as the sphere or the cube may seem surprising in a quantum context. It testifies, however, to the fact that the descriptions used in quantum

338

mechanics and astrophysics are frequently based on analogies or mathematical structures derived from well-established physical laws.

This way of interpreting quantum phenomena is largely influenced by the observer's perspective and his way of structuring his reasoning. The representation we make of physical reality thus remains closely dependent on our ability to access different scales of magnitude. Yet these scales impose limits on our perception and understanding. It appears in particular that our position as an observer gives us access to a very partial description of a field — that of quantum mechanics — in which the notions of space and time do not necessarily have the meaning they have in macroscopic physics.

In this perspective, it may be useful to recall that elementary particles are often described, in theoretical models, as point entities without spatial extension. They can be assimilated to mathematical points characterized by a set of physical properties but not possessing an own dimension. These points essentially represent possible centers of interactions between energy fields, interactions that could result from early phases of the evolution of the Universe marked by a strong radiative entanglement.

In this interpretation, the particles would not correspond to material objects in the classical sense. They would also not represent localized regions of space and would not be directly subject to the temporality defined by relativity. The phenomena of quantum entanglement illustrate this particularity: they show that certain correlations can exist independently of distance in the classical sense.

At the subnuclear scale, the usual notions of distance, separation or rapprochement become difficult to interpret. A comparable situation appears in the description of the electron cloud of an atom. The electrons do not follow well-defined trajectories but occupy quantum states distributed around the nucleus. This probabilistic distribution contributes to the overall charge neutrality of the atom and gives the impression that electrons share certain orbital configurations rather than following determined individual trajectories.

In the standard model, strong interaction is generally described as a binding energy responsible for the cohesion of quarks within nucleons, and more broadly for the cohesion of nucleons within atomic nuclei. This interaction would represent a significant fraction of the total energy contained in a hadron.

339

However, one can consider an alternative interpretation according to which this phenomenon corresponds less to a force in the classical sense than to a particular form of quantum correlation. In this hypothesis, the strong interaction could be interpreted as a state of quantum entanglement connecting particles very close to each other—for example quarks within a nucleon or electrons bound to an atomic nucleus. This entanglement would allow these particles to instantly share certain physical properties, regardless of any classical notion of distance or information transmission.

In this perspective, the stability of a hadron could depend on the degree of coherence and completeness of the quantum correlations existing between its constituents. The stability of a composite structure would then result from a dynamic equilibrium between different correlated states.

In any case, the current state of the Universe results from an extremely long cosmic evolution, starting with the first phases of radiative entanglement following the initial moments of cosmic expansion. During this phase, primordial particles and their corresponding antiparticles would have been produced in initially comparable quantities.

Speculatively, one can imagine that the first massive entities likely to have appeared could have been kinds of primitive neutrinos, unstable massive primo particles, particularly energetic. Originally, without differentiated properties these primo particles could have, in the extension of the radiative entanglement phase, revealed a symmetry break. This symmetry break would have had the effect of generating pairs of particles (quarks and electrons) and antiparticles (antiquarks and positrons). These new quantum entities in a context of chiral symmetry, would then have been led to develop properties of complementary charges (between particles of matter) and additional charges (between particles and massive antiparticles), preserving charge neutrality at the atomic scale.

Thus, antimatter could constitute a possible component of what is today referred to as dark matter (see chapter XIV). When physical conditions allow matter and antimatter to meet, some of the energy resulting from their interaction can be converted into electromagnetic radiation or new particles. These radiations do not directly contribute to macroscopic gravitational effects, but they participate in the local redistribution of energy.

In the current state of the Universe, antimatter does not seem to interact directly with ordinary matter on a large scale. This situation makes its observation particularly difficult and leads to the apparent absence of antimatter in the observable Universe. Only certain indirect effects— notably gravitational — could possibly reveal its presence.

Faced with certain gravitational anomalies observed on a large scale, the dominant hypothesis has been to introduce the existence of dark matter of unknown nature. This hypothesis allows preserving the overall consistency of the standard cosmological model while accounting for observations.

An alternative approach is proposed by certain theoretical models such as the theory of axionic quark nuggets. The properties of antimatter would be described mainly in terms of fields, which leads to a less materialized representation of this component of the Universe.

The axion — a hypothetical neutral particle with very low mass — is sometimes invoked in this context. In some theoretical scenarios, it could decay into gamma photons. This description can be interpreted as a way to model the annihilation process between particles and antiparticles when they coalesce.

Such a process would be difficult to observe directly due to its brevity and the energy it releases. It could however occur in some extreme astrophysical environments, for example during the final collapse of massive stars leading to supernova explosions. In these events, the implosion of the stellar core is accompanied by a considerable release of energy, observable in the form of exceptional brightness.

On the weak interaction

The atomic nucleus consists of nucleons, that is to say protons carrying a positive electric charge and electrically neutral neutrons. The cohesion of these composite particles within the nucleus is generally attributed to the strong nuclear interaction, whose nature appears close to that which ensures the internal cohesion of quarks within nucleons. In our standard model, this interaction is described as being mediated by gluons.

However, nucleons are not immutable entities. They can change nature or leave the nucleus during nuclear processes such as beta radioactivity. In beta minus decay, for example, a neutron can transform into a proton while emitting an electron and an electronic antineutrino. This process

341

corresponds, in the description in terms of quarks, to the transformation of a quark of type down into a quark of type up.

In some theoretical descriptions, this type of transition is accompanied by the emission or absorption of intermediate particles making it possible to preserve the different physical quantities involved, such as the electric charge or the quantum numbers associated with leptons. To account for these transformations, the theory introduced the existence of massive intermediate bosons, called Wet Z bosons, considered as vectors of the weak interaction. This interaction is described as «weak» in comparison with the intensity of the strong interaction.

These gauge bosons are supposed to play the role of mediators in transformation processes between particles and in certain nuclear transitions. Their introduction helps explain a number of phenomena observed in decay processes and interactions involving leptons.
However, an alternative interpretation can be considered if one considers that any quantum particle is fundamentally characterized by a superposition of possible states. In this perspective, elementary particles from a common origin could be seen as entities sharing an identical fundamental structure. Therefore, the observed transformations would not necessarily require the explicit exchange of a material mediator but could correspond to an internal reorganization of pre-existing quantum states.
According to this hypothesis, elementary particles would not be strictly separated from each other in the classical sense. They would rather constitute local manifestations of a set of correlated quantum configurations. Any change affecting one of these entities could result in a correlated reorganization of the whole, without requiring a conventional information transmission mechanism.
The current observable Universe obviously does not correspond to the initial state of the primordial Universe. In the very first moments, the Universe is generally described as extremely homogeneous and isotropic, with no distinct structure that can yet be differentiated. In such a context, it is conceivable that certain quantum correlations established at these very early phases could have survived and still manifest themselves today in the form of entanglement phenomena.

Certain quantum information — corresponding notably to the intrinsic properties of elementary particles — could be shared on a large scale without requiring space propagation in the classical sense. This information

342

would be independent of the usual notions of space and time. The spatiotemporal descriptions introduced by relativity would then essentially correspond to a type of interpretation imposed by the observer to make these phenomena intelligible.

Quantum correlations, sometimes interpreted as forms of "teleportation" of quantum information, illustrate this non-local nature of quantum properties. This property has inspired the development of quantum computing devices, which exploit entanglement phenomena to perform certain types of calculations potentially more efficiently than conventional computers. However, the practical realization of such devices faces a major difficulty: preserving the coherence of quantum states in the face of disturbances introduced by the material environment.

In this perspective, it becomes conceivable that some information exchanged between particles does not correspond to an effective transmission in space, but to the expression of pre-existing correlated states. Such an interpretation could reduce the need to systematically introduce mediating particles like Wet Z bosons. The observed transitions could then result from internal adjustments of correlated quantum states, notably under the influence of the energy states associated with the particles.

The quantum configurations associated with the energy states of particles could thus contribute to ensuring the cohesion of nucleons within the atomic nucleus, in particular by participating in the global balance of charges and quantum properties. Nuclear interactions involving electrons, such as those observed in beta radioactivity processes, could also be interpreted as resulting from these quantum reorganizations, locally altering the distribution of charges and electron states around the nucleus.

On electromagnetic interaction

The atomic nucleus can emit or capture electrons while generally maintaining the balance of charges within the atom. Electrons occupy specific quantum states around the nucleus, often described as orbitals. Their distribution and organization contribute to the overall stability of the atom.

One of the central questions in atomic physics is to understand the mechanisms responsible for electronic transitions between different energy levels. These transitions can lead to the emission or absorption of energy in

343

the form of electromagnetic radiation. They also play a decisive role in the formation of chemical bonds, when electrons can be shared or transferred between several atoms.

The first studies on electrical and light phenomena led to introduce the notion of photon, defined as the quantum of the electromagnetic field. The photon is involved in energy transfer processes between physical systems and constitutes the mediator of electromagnetic interactions.

Unlike the bosons associated with strong and weak interactions — whose existence manifests mainly through indirect effects — photons can be observed directly in many physical phenomena. They intervene in a wide range of processes, from atomic interactions to large-scale astrophysical phenomena.

In a broader interpretation, photons can be considered as the elementary manifestations of the electromagnetic field, which constitutes a fundamental component of the energy present in the Universe. In this perspective, energy that does not manifest as matter could be essentially associated with different fields, among which the electromagnetic field plays a central role. During the early phases of cosmic evolution, a fraction of primordial energy would have given rise to matter particles through energy-matter conversion processes. This transformation could be interpreted as resulting from entanglement and quantum structuring phenomena of the initially distributed energy.

Energy that has not been converted into particles of matter would remain today in the form of energy fields present in space considered as «empty». This quantum void would actually be crossed by various fundamental fields, whose interactions with matter particles play an essential role in the dynamics of the Universe.

The energy states associated with electromagnetic fields constantly interact with matter particles. They can also be considered as establishing a link between the particles and their corresponding symmetries, possibly associated with antimatter components participating in vacuum energy.

In this context, the introduction of specific mediators for strong and weak interactions could be interpreted as a way to account for subatomic phenomena difficult to observe directly. It remains however possible that some of these interactions can be described, at least partially, as complex manifestations of electromagnetic interactions or more general field dynamics.

Photons, devoid of mass and electrical charge, represent the corpuscular manifestation of the electromagnetic field. Their interaction with particles

— which can be described as quantum wave packets — could play a determining role in many fundamental processes of quantum mechanics.

In this speculative interpretation, photons could be considered as privileged agents in the dynamics of energy interactions, participating centrally in the processes that structure matter and govern observable quantum phenomena in the Universe.

Integrate gravitation into a coherent ensemble model

The integration of gravitation into a unified model of fundamental interactions remains one of the main challenges of contemporary theoretical physics. One interpretation would be to consider that electromagnetic interaction occupies a central place in the set of recognized interactions between elementary particles.

In this perspective, the photon — or more generally the excitations of the electromagnetic field — would play a particularly structuring role. Devoid of mass and electrical charge, it can be considered as a fundamental mediator involved in many processes of energy transfer between particles, particularly between fermions.

Electromagnetic interaction is indeed omnipresent in observable physical phenomena, from atomic and molecular interactions to large-scale astrophysical processes. It is not limited to the production of light or the transport of electric current: it also constitutes an essential vector in the redistribution of energy within physical systems.

In a more speculative interpretation, this interaction could also intervene in the management of symmetry relations between quantum states. It could contribute to the dynamics by which some initial symmetries are gradually broken or reorganized in the course of cosmic evolution. In this perspective, electromagnetic interaction would indirectly participate in the mechanisms likely to correct certain fundamental asymmetries observed in the Universe, notably those associated with chirality phenomena.

If such a hypothesis were to be further explored, it would suggest that the global evolution of the Universe could be influenced by these mechanisms of progressive reorganization of quantum symmetries. Electromagnetic interaction would then play a structuring role in the energy processes that mark out cosmic history, up to the ultimate phases of the evolution of the Universe.

These reflections reinforce the widely accepted idea that the standard cosmological model, although extremely efficient on many points, remains incomplete. The history of science shows that theoretical models are continuously evolving with new observations and questioning. As the theoretical physicist Lev Landau pointed out, cosmologists can sometimes be led to revise truths long considered established.

The difficulty, however, lies in the ability to modify a set of reflections gradually constructed and validated by numerous observations. Contemporary physics is based on a set of specialized theories which, although partially segmented, form a relatively coherent global edifice.

The reflections presented here can sometimes seem speculative. They are nevertheless based on a set of experimental results and theoretical hypotheses widely discussed in the scientific literature. They essentially aim to explore certain theories likely to shed different light on the relationships between the fundamental interactions.

On the nature of the particle

A fundamental question is to determine to what extent the observed physical reality corresponds to the deep structure of phenomena or to a representation developed from our modes of observation. This question leads to examining more closely the very notion of particle.

In quantum mechanics, a particle can be described as a wave packet characterized by a wave function. However, for practical reasons related to observation and modeling, this entity is often represented in the form of a localized elementary corpuscle. This dual description leads to the image of a universe populated by extremely small entities that appear and disappear according to quantum interactions.

The trajectories and transformations of these entities remain largely inaccessible to direct observation. The quantum properties associated with these particles generally become perceptible only through specific interactions or under particular experimental conditions.

A comparable difficulty appears in the perception of electromagnetic waves. In many representations, these are described as successive wave fronts. In reality, these fronts correspond to interference structures resulting from the

346

superposition of a large number of elementary waves. The interaction zones between these structures can produce complex dynamic configurations, comparable, by analogy, to the instability zones observed in certain atmospheric phenomena when different air masses interact.

In physical modeling, the notion of point remains nevertheless a particularly convenient support tool. Even when its exact location remains uncertain, it provides a simpler reference than representations based on extended energy fields or diffuse structures. This trend illustrates the constant need of scientific thought to materialize and locate the studied phenomena.

However, current knowledge suggests that the elementary particle corresponds neither to a strictly localized point nor to a simple energy bubble. When a particle seems to disappear from an interaction process and later reappear in another form, it remains difficult to describe precisely what occurs between these two states.

If we consider that particles are not necessarily subject to a continuous temporality comparable to that of macroscopic physics, it becomes conceivable that they can be described in a framework where the classical notions of space and time lose part of their relevance. This situation recalls, by analogy, certain phenomena associated with the event horizon of black holes, where usual spatiotemporal descriptions become inadequate.

An analogous reflection can be conducted at the cosmological scale. In certain hypotheses, the Universe could evolve towards a final state characterized by an extreme concentration of energy. This state could be interpreted as a form of cosmic singularity in which observable structures gradually disappear.

From a speculative perspective, such a state could correspond to a transition towards a new cosmic phase, analogous to that which would have preceded the appearance of the observable Universe. Between the final collapse of a universe and the possible emergence of a new cosmic cycle, there could exist a phase devoid of observable events, corresponding to a dynamic belonging to a broader cosmological context associated here with the idea of multiverse.

In this context, the particle could be interpreted not as an autonomous fundamental entity, but as the transient manifestation of a more global process involving different configurations of quantum symmetry between universes or cosmological states.

Another way to represent the Universe would be to describe it as a complex set of interlocking domains of influence. The space would then appear as structured by a dynamic network of borders and areas of interaction in perpetual evolution.

In such a representation, the idea of loop structure appears at many scales. It is notably found in certain contemporary theoretical approaches such as loop quantum gravity or certain formulations of string theory.
In string theory, elementary particles are no longer described as points but as tiny one-dimensional structures capable of vibrating according to different modes. These vibration modes would correspond to the observed properties of the particles such as their mass or charge.
By replacing the notion of a point particle with that of an extended object, string theory proposes a representation which, in some aspects, meets the idea of a fundamental undulatory structure of matter. The particles would then be the manifestations of oscillation modes of a deeper energy structure. In composite particles such as protons, neutrons or mesons, quarks could thus be interpreted as sharing entangled wave states forming a collective structure. An analogy can be drawn with the electrons involved in chemical bonds, which can be shared between several atomic nuclei and thus contribute to the formation of molecules.
Some more elaborate theories, such as superstring theory or loop quantum gravity, seek to describe fundamental interactions from even more elementary entities. These extremely complex approaches aim in particular to provide a unified description of fundamental interactions, including the strong interaction, by requalifying elementary particles as different manifestations of a common fundamental structure.

Regarding the multiverse Cosmos

The concept of Cosmos multiverse implicitly refers to the notion of infinity, that is to say, to a set whose extension could not be spatially limited. Such a hypothesis, however, faces a fundamental difficulty: the idea of an infinitely large whole simultaneously implies that of an infinitesimally small one, these two limits constituting the ends of the same spatiotemporal continuum. In this perspective, such a set would contain neither beginning nor end in the traditional sense.

If it is also assumed that there is no absolute cosmic time scale, the temporality of our Universe could only be defined from the point of view of the observer who finds himself in it. Time would then become an emergent property related to local conditions of observation rather than an intrinsic characteristic of the Cosmos taken as a whole.

More simply, one can imagine that the multiverse Cosmos consists of a potentially unlimited number of universe pairs, organized according to complementary quantum symmetries. Within each pair, the internal interactions between these two symmetrical states would constitute the substrate of the phenomena described by quantum mechanics. These hypothetical exchanges would not necessarily be constrained by the usual categories of space and time, which would amount to considering that quantum symmetries operate in a more fundamental framework than that described by relativity.

Such a representation would also allow to consider a multiverse Cosmos that does not directly relate to the observable physical reality of our Universe. The hypothesis of discrete interactions between two quantum symmetric states would then constitute an attempt to approach a dimension of reality that remains currently inaccessible to direct observation.

At this stage of the reflection, theoretical development necessarily relies on a degree of extrapolation. The current limits of knowledge indeed imply that some propositions remain speculative. The history of science shows, however, that ideas initially deemed contrary to common sense have sometimes constituted important stages in the evolution of scientific paradigms.

Observation, determinism and probabilistic interpretation

The observation instruments we have at our disposal only give access to partial representations of reality, each depending on the methods and scales of observation employed. The results obtained often open up new questions rather than providing definitive answers. This situation partly explains why determinism is frequently questioned in certain fields of astrophysics and fundamental physics.

The famous remark attributed to Niels Bohr — according to which "if one believes having understood quantum mechanics, it is because one has not really understood it" — illustrates the difficulty of interpreting this

349

discipline. The complexity of quantum phenomena and the existence of many aspects still poorly understood have led to favor a probabilistic interpretation of their results.

Determinism does not necessarily disappear, it takes the form of a statistical determinism, expressed in terms of probabilities. This approach introduces some uncertainty into predictions, but it does not imply that the evolution of physical systems is governed by absolute chance. The principle of indetermination, formulated notably in quantum physics, rather reflects the limits of our ability to simultaneously predict certain properties of a system. Although relatively recent and deeply puzzling, quantum physics should not be interpreted as inherently more random than classical or relativistic physics. Admitting this possibility is a prerequisite for considering a unification of physical laws, despite the apparent divergences between current theories.

Such a unified theory would, however, remain structured by considerable differences in scale, from the microscopic phenomena described by quantum physics to the cosmological structures pertaining to relativity.

Causality and succession of states

In this perspective, the probabilistic approach constitutes mainly a modality for describing determinism in quantum physics. Every physical phenomenon, however complex, produces an effect that can be interpreted as the consequence—direct or indirect — of a set of previous conditions identified as its causes.

However, this way of seeing things raises a fundamental question: can we conceive that an event is not preceded by an identifiable cause?

Our representation of time is usually based on a succession of states, each one being both the effect of previous situations and the cause of subsequent states. The present state results from a chain of past events, and the future appears as the determined prolongation — at least partially — of this evolution.

The notion of the present nevertheless poses an ideological difficulty. It can be considered as a theoretical boundary separating a past that no longer exists from a future that does not yet exist. In practice, what we call the present always corresponds to an event that has already happened at the moment when it is perceived or measured.

350

Thus, the present moment can be interpreted as an approximation, corresponding to an extremely brief period between two almost simultaneous observations. The theory of relativity, by questioning the universality of simultaneity, also leads to relativize the idea of an absolutely defined moment devoid of duration.

Consequently, any attempt to forecast beyond a very short time horizon is inevitably marked by uncertainty due to the lack of sufficient information on all the parameters involved in the evolution of physical systems.

Hypothesis of a future-oriented causality

Time can also be interpreted as an ordered sequence of successive states. However, if one wonders what connects these different states, an alternative hypothesis can be considered: that according to which the future state itself would participate in the determination of the present state.

Such a hypothesis supposes to admit that the succession of states of a system is globally constrained by a final state, which would determine retrospectively the whole evolution process. In this speculative perspective, the final state of the Universe, possibly associated with a form of cosmic collapse, could be interpreted as the determining condition of a chain of previous events.

In other words, the cosmic process could be described as a progressive deconstruction leading to a final state, the latter playing the role of structuring condition of the whole evolution.

However, our inability to know this final state introduces a fundamental uncertainty into any attempt at description. This cognitive limit naturally leads us to conceive time as oriented from the past towards the future, because it is difficult to imagine a causality devoid of temporal direction.

Yet the idea of a deeper symmetry of time, in which causes and effects would not be strictly oriented according to a single direction, constitutes one of the hypotheses explored in certain extensions of relativity.

Physical organization of the Universe and exotic particles

The Universe can be compared, in an illustrative way, to an extremely complex physical system whose different elements interact according to structured relationships. Like a mechanism composed of interdependent

351

gears, each component of the system must have properties compatible with those of the others so that the whole remains coherent. Known physical laws thus impose strict constraints on the particles and interactions that make up matter.

However, under extremely high energy conditions, especially during interactions dominated by the strong nuclear force, it is possible to produce numerous transient subatomic particles. These entities often play an interface role in the processes of transformation of matter. Their extremely short lifespan makes their experimental detection difficult, which explains why their existence is generally only indirectly inferred from the traces they leave in the detectors.

Some of these unstable particles, often referred to as exotic particles, have unusual internal structures. While ordinary baryons are made up of three quarks and mesons of two bound quarks (quark–antiquark), some observed or theoretically considered configurations could have a different number of quarks, for example four, five or more, possibly resulting from complex assemblies of mesons. Other composite structures can also be considered.

These configurations, however, remain generally unstable and quickly disintegrate into simpler particles belonging to the standard model. In many cases, evolution leads to annihilation processes involving the corresponding antiparticles. The potential diversity of these transient states and their extremely short lifetime explain why they remain difficult to integrate systematically into current theoretical models.

Hypothesis of a cosmic determinism

Another hypothesis is to consider that the evolution of the Universe could be determined by a set of initial conditions and physical laws implicitly setting its global trajectory. In this perspective, the Universe would evolve according to a largely deterministic dynamic, whose outcome would already be inscribed in the fundamental conditions of its existence.

This approach aims to answer a classic cosmological question: why does the Universe possess the properties we observe it, and why its evolution follows this particular trajectory rather than another one? In a strictly deterministic interpretation, the Universe could evolve only in accordance with the constraints imposed by its fundamental laws and by its initial state.

Such a vision presupposes that the whole set of physical phenomena is deeply interconnected, so that the global evolution of the cosmic system does not result from a choice among several possible trajectories, but from a single dynamic compatible with the given physical conditions.

However, a difficulty arises when introducing the notion of conscious agents capable of making decisions perceived as free. Living organisms endowed with consciousness seem to have a capacity for action that introduces, at least locally, an unpredictable dimension into the evolution of systems. Arbitrary choices — comparable to the act of throwing dice — could thus modify certain future configurations.

However, even if these actions produce real effects, their consequences generally remain negligible on a cosmological scale. The local disturbances induced by these choices can be assimilated to minimal variations, comparable to the butterfly effects evoked in certain theories of dynamical systems.

If we neglect these marginal disturbances, the future of a physical system could theoretically be precisely determinable, provided that we have an exhaustive knowledge of all the relevant variables. In practice, the extreme complexity of physical systems makes this knowledge inaccessible, which leads to resorting to probabilistic descriptions.

Probabilities, wave function and paradox EPR

In this context, the wave function used in quantum mechanics is essentially a mathematical tool for describing the probabilities associated with the different possible states of a quantum system. This approach reflects less an intrinsically random nature than a limitation of our ability to accurately determine all variables involved in microscopic interactions.

The EPR paradox (Einstein-Podolsky-Rosen) illustrates the difficulties of data analysis associated with the description of quantum properties such as the position or velocity of particles. The observed correlations between entangled particles seem to challenge certain intuitions derived from classical physics, particularly with regard to the notion of locality.

It is however possible that what we interpret as randomness is in reality the consequence of a level of complexity too high to be described deterministically with current theoretical tools.

Superposition of states and quantum interactions

353

In some speculative interpretations of quantum mechanics, particles in superposition of states could be envisaged as belonging simultaneously to several possible configurations. We can then formulate the hypothesis that there exists a form of permeability between quantum symmetric states, allowing interactions that would not be strictly constrained by our classical representation of time.

In this perspective, certain quantum correlations could appear as if the particles were "communicating" outside the usual temporal constraints. It is, however, an interpretation aimed at accounting for phenomena whose deep-seated mechanisms are still poorly understood.

Measurement of the position and velocity of particles

The difficulty of simultaneously determining certain properties of a particle can be illustrated by examining the measurement methods used.

- Determination of position

The position of a particle can only be defined in relation to other physical objects serving as reference. In a three-dimensional space, it is expressed by coordinates associated with a given reference system. These coordinates correspond to a state of the system at a particular time, regardless of subsequent movements.

- Determination of speed

Speed, on the other hand, requires the comparison of several successive positions of a particle over time. It is defined as the ratio between a variation in position and a time interval. This measure therefore involves a dynamic description of the movement.

These two types of measurements rely on distinct experimental procedures, which partly explains the difficulty in simultaneously determining these two quantities with arbitrary precision.

Uncertainty principle and probabilistic description

In quantum mechanics, elementary particles are often described by wave functions associated with wave properties. This description leads to the uncertainty principle, according to which certain pairs of physical quantities —such as position and momentum — cannot be determined simultaneously with unlimited precision.

The situation becomes even more complex when attempting to determine several characteristics of a quantum system simultaneously, such as its energy state, position and dynamics. The measurement methods used can indeed disrupt the observed system.

Particles can be described as following several possible trajectories, each associated with a probability. Two particles from the same interaction can also form an entangled system, retaining correlations as long as they do not interact with other systems.

These properties lead to abandon the idea of a precise localization of particles in favor of a probabilistic description. This approach, although imperfect, nevertheless allows for structuring the study of quantum phenomena.

Role of the observer and reduction of the wave function

When a quantum system interacts with a measuring device, only one particular configuration among the set of possible states becomes observable. The other potential states are no longer accessible experimentally.

This phenomenon is often interpreted as a reduction of the wave function, whereby a system initially described by an overlay of states takes on a given state at the time of measurement (see chapter XXIX).

In this perspective, the observer and the experimental instruments are an integral part of the global physical system. Their interaction with the quantum system inevitably influences the observed state.

This situation is particularly encountered in the experiments conducted in particle accelerators and detectors, such as those operated by the CERN organization, where the experimental devices are designed to minimize these disturbances while allowing indirect observation of the studied phenomena.

Superposition of states and sharing of quantum information

In quantum mechanics, any particle can be described by a superposition of possible states. This property, highlighted notably in the discussions around the EPR paradox, implies that a quantum system does not necessarily present itself in a unique state before an interaction with a measuring device. The observed state then corresponds to one of the configurations compatible with the experimental conditions.
In this perspective, particles can appear punctually in a privileged state, determined by the observation conditions and by the dynamics of the ongoing interaction. This observable state would not, however, exhaust all the possibilities described by the wave function.
Unlike electromagnetic radiation, which locally interacts with matter particles by propagating in space, quantum information associated with fundamental particle properties — such as spin, the charge or energy — does not necessarily seem to be transmitted in the form of classical spatial displacement. One can then consider that certain quantum properties are correlated between particles sharing a common origin, even when these particles are spatially separated.

Quantum entanglement and correlations of common origin

The ability of certain particles to share correlations called quantum entanglement can be interpreted as the consequence of a common origin in their physical history. When two particles are produced by the same process, their quantum properties can remain correlated as long as no subsequent interaction disrupts this system.
In this context, the correlation is generally all the stronger if the particles have recently separated or if they have undergone few interactions with their environment. We can then consider that a set of particles from the same event forms a correlated quantum system.
By speculative extension, one could envisage that the whole of the matter of the Universe shares a common cosmological origin, going back to the first phases of cosmic evolution. If this were the case, all particles would possess, to varying degrees, a common physical history likely to have left traces in the form of very weak correlations.
Quantum non-locality does not mean that these correlations should be interpreted as a transmission of information faster than light. They rather reflect the existence of correlated quantum states that cannot be described

356

independently of each other. The classical notions of determinism and causality, built from a familiar spatiotemporal representation, must then be reformulated to adapt to the particular properties of quantum systems.

Particular case of photons and space-time structure

Photons exhibit some remarkable properties in this context. They can be produced in large numbers during the early phases of cosmic evolution, especially during periods of high energy density of the primordial Universe. Photons from the same process can form entangled states, maintaining measurable correlations even after their separation.

From the point of view of special relativity, photons move at the speed of light in a vacuum, the maximum speed accessible in our Universe. In the relativistic description, the eigentime associated with a particle moving at this speed tends to zero. This property sometimes leads to the intuitive idea that the photon does not 'perceive' the passage of time, even if this formulation remains an easy simplification.

These characteristics may give the impression that certain quantum correlations escape ordinary constraints of distance or duration. However, compared to current theories, these phenomena remain compatible with the fundamental principles of relativity.

If it is assumed that all particles originate from a common initial cosmological event, it is conceivable that some residual correlations may remain, although strongly attenuated during cosmic evolution. The successive interactions undergone by particles indeed tend to gradually disintegrate entangled states.

From this speculative perspective, it is not excluded that certain complex systems — for example some molecules sharing a common origin or environment — may manifest weak quantum correlations still poorly identified. Such a hypothesis would constitute a meeting point between quantum physics and relativistic physics, particularly in the study of large-scale gravitational interactions.

Measurement of motion and experimental limits

357

Determining the position of an object becomes all the more difficult as its speed is high. To measure its position at a given moment, it would ideally need a sufficiently fine temporal resolution to freeze its movement during the measurement.

When the speed of an object becomes very high, the time required to carry out this measurement can become comparable with the characteristic time of its displacement, which introduces uncertainty about its precise location. In the extreme case of a particle moving at a speed close to that of light, any attempt at precise localization becomes particularly delicate.

This difficulty appears clearly in the study of particles such as free electrons or neutrinos, whose speeds can be very high. The simultaneous determination of their position and momentum then becomes limited by the fundamental principles of quantum mechanics.

Uncertainty principle and statistical description

These experimental limitations are formalized in the uncertainty principle, according to which certain pairs of physical quantities cannot be determined simultaneously with arbitrary precision. Position and momentum are the best-known example of these conjugate variables.

In practice, the available experimental methods do not allow to measure simultaneously all the characteristics of a particle—such as its trajectory, speed, position and energy — with absolute precision. The description of quantum phenomena is therefore based on probability distributions and statistical averages.

Possible interpretation of the superposition of states

When a phenomenon remains partially unexplained, the state of superposition attributed to particles can be interpreted as a reflection of our ignorance of certain underlying mechanisms. In the same way that the laws of celestial motion could not be properly understood before the elaboration of relativity, it is possible that certain aspects of quantum mechanics may later reveal the existence of a deeper theoretical framework still unknown.

In this speculative context, one could consider the existence of symmetry structures still not observed, likely to influence energy exchanges between quantum systems. Some hypotheses suggest for example that energy interactions could appear as discrete quanta due to more fundamental properties related to the structure of quantum fields.

According to this interpretation, particles could present a dynamic and multiple nature, resulting from configurations of interacting energy fields. The observed transitions would then correspond to energy exchanges between different configurations of these fields. A chiral symmetry not open to observations would explain that all energy exchanges are perceived as occurring in successive packets or steps and not in a continuous manner.

Limits to the uncertainty principle

In theory, it is not strictly inconceivable to imagine methods allowing to connect simultaneously several characteristics of a particle, for example by describing its movement with respect to other trajectories in a global dynamic system. However, the mathematical and experimental complexity of such an approach remains considerable.

Moreover, the description of particles as wave packets involves the analysis of energy field fluctuations whose mathematical modeling is particularly demanding.

In this context, the uncertainty principle can be interpreted as the expression of fundamental limits in the simultaneous description of certain physical properties, but also as a reflection of the practical difficulties encountered when trying to gather a large amount of information about a quantum system without disturbing its state.

Finally, the very notion of simultaneity becomes problematic in a Universe governed by the principles of relativity, where events cannot always be defined as strictly concomitant for all observers.

Every material body is subject to interactions that can modify its state of motion. In a space-time characterized by permanent energy exchanges between physical systems, the hypothesis of a strictly rectilinear and uniform movement appears essentially as a theoretical idealization. In practice, gravitational, electromagnetic or other interactions induce

359

acceleration variations, positive or negative, which continuously modify the dynamics of bodies.

According to general relativity, the mass and energy distribution determine the local structure of space-time. Mass variations and their distribution thus generate a variable gravitational geometry, which makes problematic the idea of a universal simultaneity independent of the observer. Space-time can then be considered as a complex set of local frames of reference, each influenced by the gravitational fields produced by the different masses present in the Universe.

The image of a "pixelated" space-time can serve as an analogy, but it should not be understood as a true fragmentation of space. In reality, each mass exerts a gravitational effect that adds up to that of all the other masses. The resulting gravitational field corresponds to the superposition of these contributions. The reciprocal interactions between astrophysical objects or stellar systems then mainly depend on two parameters: their mass and their relative distance.

Time itself acquires a relativistic dimension. Its flow depends on gravitational conditions and the relative motion of observers. This idea meets certain intuitions developed by Ernst Mach, according to which local dynamic properties could depend on the global distribution of matter in the Universe. In such a perspective, the notions of absolute position, direction or speed lose their fundamental meaning and become dependent on the chosen reference frame. The measured quantities are then based on the interpretation specific to each observer.

At the microscopic scale, quantum mechanics challenges the classical notion of a physical quantity defined simultaneously with precision. The determination of the position and momentum of a particle obeys the Heisenberg uncertainty principle: the increase in precision on one of these quantities necessarily leads to a decrease in precision on the other. This limitation is related to the representation of the state of a quantum system by a wave function, a mathematical entity that does not directly correspond to a physical wave localizable in space..

The wave function only allows to calculate probabilities of measurement results. In the absence of a precise experimental context – including a measurement device and a determined spatio-temporal framework – the values associated with observables can therefore only be described in a probabilistic manner. This situation suggests that the operational concepts of space and time could be inseparable from the observation conditions themselves.

The unification of space and time has been a foundation of modern physics since Einstein's relativity. However, at the macroscopic scale, when a quantum system interacts with its environment and a measurement is performed, its state seems to be reduced to a configuration compatible with the observing capabilities of the experimenter. This process, generally described as quantum decoherence, corresponds to the apparent loss of state superpositions in favor of a limited set of experimentally accessible properties.

In this perspective, the actually observed state would represent only a partial projection among all the possible states described by quantum theory. In other words, some potential properties of the system could remain inaccessible to direct observation, not necessarily because they do not exist, but because they escape the available measurement modalities.

Despite these profound differences, contemporary physics is still trying to think quantum mechanics in continuity with general relativity. This approach raises the question of the relevance of a unified cosmological model based on theoretical frameworks developed for very different scales. It is possible that the physical laws describing macroscopic phenomena – in particular those concerning structured matter – are not directly adapted to account for the transition between the undulatory description of quantum systems and the appearance of the world corpuscular macroscopic as it is perceived after interaction with an observer.
In this perspective, it might be useful to explore less conventional approaches and examine more speculative hypotheses. Such an approach would consist in particular in considering that the physical reality accessible to observation is partly based on a limited interpretation, conditioned by the cognitive and instrumental properties of the observer himself.
Thus, what we call "reality" could correspond to a partial representation of a richer set of phenomena, of which only a fraction becomes effectively accessible given our modes of observation and interpretation.

To illustrate this reflection, one can resort to a thought experiment consisting of imagining a supercomputer designed as a functional replica of the human

361

being. This machine would be equipped with a set of sensors reproducing, as much as possible, the functions of our sensory organs.

The following devices could thus be considered:

- an acoustic and chemical detector for analyzing sound signals and certain volatile molecules, analogous to the functions of hearing and smell.
- a thermal sensor intended to measure temperature variations, corresponding to certain dimensions of the sense of touch.
- a device for measuring mass and density, comparable to a balance, allowing the evaluation of material properties of objects and their spatial relationships.
- a reference clock ensuring the chronological recording of events and making it possible to establish causal and duration relationships.
- a photoelectric sensor capable of analyzing the intensity of electromagnetic radiation, its spectral distribution and certain distance-related information, functionally analogous to vision.
- finally, a microscopic observation instrument intended to explore the structure of matter at fine scales.

All the data collected by these sensors would be transmitted to a central program responsible for their processing and interpretation. One might then assume that this computer system would analyze this information in a different way from the human brain, potentially stripped of the subjectivity inherent in our perception.

However, this hypothesis quickly encounters a substantive limit. The software responsible for interpreting data would necessarily remain marked by the theoretical choices, models and conceptual categories introduced by human designers. In other words, even if the raw data are recorded objectively, their interpretation remains conditioned by the rules and assumptions previously integrated into the system.

The information collected would therefore indeed be processed and decoded, but the question remains: in what form could it be returned? The fundamental entities described by physics – waves and particles – do not inherently possess the sensitive qualities associated with our perceptual experience. They are not, in themselves, colored, sonorous or odorous. They feel neither heat nor cold. Their physical properties relate to parameters such

362

as energy, frequency, mass, charge or momentum. Some of these quantities can vary or transform depending on the theoretical frameworks used, as illustrated for example by the relativistic equivalence between mass and energy.

Under these conditions, a computer designed to process these physical signals could face a fundamental difficulty. It would capture and analyze energy and information flows in forms that do not directly correspond to human perceptual categories. Deprived of any sensory experience similar to ours, its interpretation system could encounter difficulties in translating this data into a language understandable by the human being.

The question then becomes epistemological: how could such a system elaborate an intelligible model from potentially incomplete data or expressed in complex mathematical formalisms, even as we ask him to render these results in a form compatible with our own modes of perception and understanding?
In this context, it is possible that the operational logic of such a system, although inspired by that of its designers, may not necessarily be adapted to the nature of the phenomena studied. Certain constraints could introduce limitations in the interpretation of fundamental physical data.

This difficulty refers more broadly to the way computers are designed. Computer architectures and algorithms are largely inspired by human cognitive processes and rely on logical structures derived from our own reasoning mode. Therefore, designing an artificial intelligence truly based on cognitive principles radically different from those of the human brain constitutes a major challenge. If today's artificial intelligence can contribute to the evolution of mathematics, it cannot be totally free from the human framework, nor can it guarantee a "conceptual revolution." An AI learns from human data, manipulates existing formalisms, optimizes criteria defined by us. But it cannot change the basic concepts (object, number, relationship...), or invent a formalism independent of our cognition. It must and can only remain understandable, communicable, and verifiable by humans. Our brain needs a stable spatial or temporal framework to identify objects, locate them in space, track them over time, and build causal histories.

Indeed, the procedures we program to process information necessarily involve operations of selection, sorting and prioritization of data. These
363

operations may lead to prioritizing certain information at the expense of others, according to criteria that remain partly arbitrary or dependent on our initial assumptions.

One can then consider a more advanced technological hypothesis: that of a quantum computer associated with an artificial intelligence sufficiently developed to model and translate, in an intelligible way, the complexity of physical phenomena. Such a system could theoretically handle simultaneously a large number of possible states and explore information structures that are difficult to access with conventional computers.

However, the practical implementation of such a device raises considerable difficulties. Quantum systems are extremely sensitive to environmental perturbations. Background radiation present in space, as well as those emitted by the materials themselves, can cause decoherence phenomena that disrupt the functioning of quantum devices.
For this reason, it is likely that future computing architectures combine classical and quantum elements. Quantum operations could be limited to specific steps in information processing, while classical computer systems would ensure error correction and reduction of noise effects inherent in quantum technologies.

The major difficulty related to the design of computers exploiting quantum properties lies in certain fundamental characteristics of quantum mechanics, notably the questioning of the classical principle of locality. In classical physics, and more particularly in the relativistic description of macroscopic phenomena, locality implies that physical interactions propagate continuously in space and that one system can be considered distinct from another as long as they are spatially separated. This principle leads to that of separability, according to which the state of a compound system can be described from the states of its subsystems.

Now, in quantum mechanics, certain phenomena—particularly those associated with entanglement — seem to contradict this intuition. Quantum systems can exhibit correlations that are not explained by classical local interactions. This apparent non-locality challenges certain aspects of classical physics that structure our understanding of the macroscopic world.

Reality as it is perceived on a human scale largely results from a process called wave packet reduction, that is to say the transition between an
364

overlapping of possible quantum states and obtaining a single result during a measurement. Contemporary approaches based on the theory of quantum decoherence seek to describe how interactions with the environment gradually lead a quantum system to adopt a behavior compatible with the laws of classical physics.

The articulation between decoherence and non-locality nevertheless raises important questions. She suggests that macroscopic reality could be only an emergent description, related to the scale of observation and the methods of measurement available. In this perspective, the Universe would only appear to us through the interpretative categories specific to our condition as observers. The familiar notions of space and time could thus represent only effective structures adapted to the macroscopic scale, without necessarily constituting the fundamental elements of a deeper description of physical reality.

In this context, the development of quantum computers and artificial intelligences capable of exploiting the properties of quantum systems represents a considerable scientific and technological challenge. If these technologies reach maturity, they could constitute a form of interface between quantum mechanics and classical relativistic physics, by allowing the exploitation of certain quantum properties in concrete applications.

Such an evolution will probably require the design of algorithms of a new type, capable of manipulating, selecting and processing quantum information without causing the destruction of the superposition states that characterize it. These capabilities would be particularly promising in certain areas, notably for solving complex combinatorial problems.
However, the exploitation of quantum information also involves being able to interpret the results obtained from qubits — fundamental units of quantum information—and translate them into a format intelligible for conventional computer systems or for the user human.

A major problem lies in the physical stability of the systems that host these qubits. Even at extremely low temperatures, the atoms or physical devices supporting the qubits are not perfectly isolated from their environment. External disturbances can introduce errors in quantum calculations. These errors become all the more significant as the operations carried out are numerous, complex and cumulative.

365

To limit these effects, some approaches involve repeating the same operations a large number of times in order to statistically identify the most likely result. However, this strategy remains fundamentally probabilistic and does not guarantee absolute reliability.

Under these conditions, it is likely that future computing architectures rely on a combination of quantum computers and classical computers. Classical systems, based on binary logic, could be responsible for filtering, correcting and interpreting the results of quantum calculations using suitable algorithmic procedures. Despite these constraints, the progress made in this field opens the prospect of a new generation of potentially much more efficient computer technologies.

The main technical difficulty remains the fight against quantum decoherence. Theoretically, the complete elimination of this phenomenon would require a perfect isolation of the quantum system from any environmental disturbance: absence of radiation, extremely low temperatures and conditions close to an ideal vacuum. In practice, achieving such a level of isolation is extremely difficult if not impossible. Decoherence therefore appears as a largely inevitable phenomenon.

However, this same phenomenon also constitutes the mechanism that connects the quantum behavior of microscopic systems to the appearance of the macroscopic world. By allowing the emergence of stable and observable properties on a large scale, it makes possible the exploitation of certain quantum effects in applied fields such as space technologies, medicine or computer science.

In a more speculative register, some authors have advanced the hypothesis that the Universe possesses properties particularly favorable to the emergence of observers capable of becoming aware of its existence. This idea is generally referred to as the anthropic principle. According to certain formulations of this principle, the physical constants and cosmological conditions observed would be compatible with the appearance of a form of intelligence capable of questioning the structure of the world.

However, such an interpretation remains widely debated. The human being itself can be considered as the product of a continuous evolutionary process, sharing many biological characteristics with other species, notably primates. The observable differences mainly concern certain cognitive abilities:

development of abstract memory, increased ability to use symbolic language and to manipulate complex information.

These faculties have allowed the human species to develop models and develop sophisticated technologies. They have also led to the emergence of philosophical or metaphysical representations sometimes placing the human being at the center of the Universe (anthropic principle). Sad privilege that makes man the first predator, with a well-asserted ego. This ideology, centered on the human being, amounts for some to imagine a "supreme will" which would be the instigator of this design far from being won in advance. It evokes an old fantasy that even some scientists rally around. Indeed, it claims to explain, reassure and value all life endowed with a central nervous system that makes it question its raison d'être. In humans, it is a constant that nourishes their unconscious.

XXV <u>Our Universe is discreet about its age!</u>
(No marital status: no birth certificate)

In all directions of observation, the Universe presents the appearance of an accelerated expansion. Interpreted literally, this observation would mean that, to reach a given region of space, the necessary time would continuously increase, because this region would correspond to a past state of the Universe which would move away from us at an increasing speed.

However, general relativity indicates that time dilation depends on the local gravitational and energetic context. These effects can alter the way we interpret cosmological distances and apparent velocities in distant regions of the Universe. In this perspective, the age attributed to the Universe could depend strongly on the assumptions used to interpret the observations. It is therefore not excluded that this age is significantly different from the one generally used in the standard cosmological model.

When one adopts the hypothesis of a cosmic expansion, it becomes natural to assume the existence of an initial moment and a time evolution defined from this original state. However, the alternative idea of a Universe that would not be fundamentally expansionist, but which would constitute a globally circumscribed system although not bounded, remains conceivable. Such a representation may seem counterintuitive, but the history of physics shows that some initially controversial hypotheses— like the general relativity proposed by Albert Einstein — were eventually accepted after extensive theoretical and experimental examination. It should also be remembered that the latter did not initially adhere to the idea of a cosmic expansion.

According to the current dominant interpretation, the event called Big Bang would have occurred about 13.8 to 15 billion years ago. This estimation is mainly based on the measurement of the apparent velocity of recession of galaxies and on the retrograde extrapolation of these dynamics until an initial assumed moment. Such an approach implicitly implies the hypothesis of a Universe resulting from an initial singularity of zero dimension.

However, if we consider that an initial singularity would have given rise to an expanding Universe, this representation intuitively suggests a volumetric growth comparable to that of a sphere. Such an image could lead to
368

imagining an expanding center and boundary, which seems difficult to reconcile with the cosmological principle that the Universe is globally homogeneous and isotropic on a large scale.

In the alternative hypothesis considered here, the Universe would be characterized not by an inflationary expansion of space, but by a dynamic associated with a form of energy depression of space. The standard interpretation of cosmic expansion may not be the most appropriate description, and the age of the Universe may appear largely underestimated.

De Sitter's model of the Universe, often invoked to illustrate certain solutions to the equations of general relativity, assumes a continuous expansion of space. However, if the origin of the Universe remains unknown and is only described by the abstract notion of initial singularity, it remains difficult to assign with certainty a well-defined age to a cosmic system whose global dynamics could be different from that postulated by this model.

Moreover, the observation of very distant galaxies — which corresponds to the observation of the Universe in an older state—suggests that intense gravitational interactions, such as galaxy mergers and galactic black hole approximations, would have been more frequent in the past. Such a trend could be interpreted as an index of a cosmic evolution characterized by a progressive concentration of matter in certain regions, compatible with the hypothesis of a Universe evolving in a context of energetic depression of space.

The radius of the observable Universe is today estimated at about 46 billion light-years. This value corresponds to the maximum distance from which light information has had time to reach us since the beginning of cosmic history accessible for observation. It is therefore directly related to the speed of propagation of light, generally considered as the upper limit of transmission of physical information.

In an expansionist model, this distance could suggest that the observable Universe has an age comparable to this propagation duration. However, this interpretation implicitly assumes that the time units and physical conditions associated with light propagation have retained an identical meaning throughout cosmic history.

369

Now, if one admits that the energetic and gravitational structure of the Universe has evolved significantly over time, then the conditions under which light propagated in the primordial Universe could differ significantly from those observed today. At a time when matter was less structured and gravitational concentrations were more diffuse, the distribution of energy in the so-called empty space could be more homogeneous and less localized than it is currently. Massive stellar structures and strongly individualized galactic systems were probably less marked.

Under these conditions, the light-year—defined as the distance traveled by light in a year — constitutes a unit implicitly dependent on the relativistic context in which it is used. It cannot be interpreted independently of the global evolution of the Universe.

It is notably this context that leads to the observation according to which some very distant galaxies seem to present apparent recessive speeds higher than that of light. This phenomenon is generally interpreted as a consequence of the dynamics of space itself, rather than as a local displacement exceeding the relativistic limit speed.

It becomes difficult to establish a simple correlation between three often associated quantities: the observable radius of the Universe (about 46 billion light years), the rate of recession of galaxies interpreted as a cosmic expansion, and the estimated age of the Universe (about 13.8 billion years). The alternative hypothesis of a non-expansionist Universe but characterized by a retrograde dispersion dynamic with relativistic dominance rather than strictly radial, would make it possible to avoid this apparent tension between age, size and cosmic dynamics. The idea of a non-expansionist Universe in retrograde dispersion, more relativistic than radial (see illustration), avoids having to ask in this form, the question of the relationship between the age and the size of the Universe.

A large number of uncertainties and the insufficiency of available observational data still make it difficult to establish a fully reliable description of the origin and initial evolution of our Universe. In these conditions, it is not surprising that certain inconsistencies appear in the interpretation of light spectra from very distant objects, notably in the analysis of the red shift (Redshift). These divergences can also be related to the calculation assumptions traditionally based on the fundamental Friedmann equation, which describe a globally homogeneous and isotropic Universe within the framework of general relativity.

Moreover, the existence of the cosmic microwave background radiation, characterized by an average temperature of about 2,73 kelvin—identified as the Cosmic Microwave Background — does not fully account for the speed with which the first galaxies seem to have formed, if we simultaneously maintain the hypothesis of a Universe approximately 13,8 billion years old and if we consider the year as a strictly invariant unit of time in cosmic history.

Another approach is to attempt to estimate the age of the Universe from the particularly long half-life of certain radioactive isotopes. However, this method also has its limitations, as the elements used as chronological references could themselves come from the recycling of radioactive materials from previous cosmic cycles. In this case, the measured age would correspond only to a recent phase of transformation of matter and not necessarily to the origin of the cosmic system as a whole.

We are still largely unaware of what the Universe in its totality signifies, supposedly not bounded, compared to the portion accessible for observation. The history of astronomy also shows that the representation that observers have made of the Universe at different times has proved, each time, to be strongly limited by the available means of observation. It is therefore reasonable to assume that our current understanding also remains partial.

According to the Standard Model of particle physics, the lifetime of the proton — made up of first-generation quarks (up and down) —as well as that of the electron, would be extremely long in the absence of yet unknown decay processes. Theoretical estimates suggest durations that can reach at least several thousand billion of billions of billions of years. If the Universe really had only about 13,8 billion years, its lifespan would therefore remain very short in relation to these time scales. However, it is also conceivable that the Universe is much older than current estimates suggest. In this case, its longevity would greatly exceed the projections established from contemporary cosmological models.

In common representations, the event called Big Bang is often described as an initial explosion from a single point, dispersing matter and energy in all directions of space. One readily represents the Big-bang as the explosion of a very large dispersing firecracker contained and containing while following the most direct trajectory, that is to say, the one represented by rays starting

371

from a supposed "firing" point towards all directions of space. This intuitive image is based on an analogy with a radial expansion from a central point. However, a more abstract description could consist in considering the Universe as a manifestation of energy resulting from a state of broken symmetry. The diversity of shapes observed — particles, radiation, cosmological structures — would essentially result from the observer's point of view and from the local interactions that make these shapes perceptible. Such an approach would make it possible to describe the Universe without necessarily assigning it a defined initial dimension, nor assuming a volumetric growth resulting from a singular point.

This perspective naturally leads to a fundamental question: what is the position of the observer within the Universe? At present, no specific answer can be given. One of the few relatively robust conclusions of modern cosmology is that there probably exists neither a privileged center of the Universe nor an accessible frontier that would mark its limits.

To try to interpret this situation, it is necessary to recall certain physical principles. Gravitational effects result from the presence of mass or, more generally, energy. The gravitational interaction has a theoretically unlimited range and its intensity decreases with distance. In the hypothesis considered here, where gravitation would be deeply linked to electromagnetic interaction, these effects could be interpreted as the expression of a field of influence extending over the entire space-time.

If one imagines crossing the unmaterialized limits of the observable Universe, it is conceivable that the density of matter there becomes extremely low, to the point that the quantum interactions associated with material particles are no longer detectable. In such a context, only certain electromagnetic waves could possibly propagate beyond the cosmic region dominated by the gravitational interactions of our Universe.

This hypothesis leads us to question the very possibility of conceiving a conceptual boundary between our Universe and a larger cosmic whole, sometimes described as a multiverse. Such a boundary, if it exists, could be essentially abstract in nature and difficult to represent. Moreover, if the Universe were indeed infinite while coming from an initial singularity, its representation would not be easier.

In the very long-term evolution of the Universe, the interactions of organized matter should gradually decrease. The trajectories of celestial

372

objects, initially complex and strongly influenced by local gravitational interactions, would then tend to become more regular. In this context, relativistic movements could gradually adopt increasingly tangential directions with respect to the multiple gravitational horizons present in the Universe.

As the Universe cools and structures stabilize, dominant astrophysical events may be primarily related to the rapprochement and merging of black holes, particularly at the galactic scale.

Black holes, although they are an integral part of the Universe, cannot be considered as occupying space in the ordinary sense, insofar as they correspond to regions where the geometry of space-time is extremely curved. The global evolution of the Universe could thus lead to a minimal dispersion of the energy transported by electromagnetic waves, in a space progressively depleted of other forms of energy.
An energy-poor space could be interpreted as a space in which the regions of matter concentration — especially those associated with future massive clusters of black holes — would be separated by increasingly large distances.

Such an evolution could also help to explain why galaxies located at the observable limits of the Universe seem to move away from each other at apparent speeds higher than that of light. This impression could result in part from the fact that cosmological observations do not always fully account for relativistic effects associated not only with instantaneous space geometry, but also to the global and progressive evolution of the distribution of matter in the Universe.

XXVI <u>An open secret</u>
(Engraved in a past within sight)

Accelerated expansion or retrograde dispersion: elements of clarification

Let us return to the question of the accelerated expansion of the Universe and to the alternative hypothesis of a retrograde dispersion of energy, by first taking up some definitions allowing for the clarification of the analytical framework.

a) The present observational

From the observer's point of view, the space accessible to observation can be assimilated to an instantaneous representation, comparable to a frozen image. This representation corresponds to an observational present, limited to a relatively close environment.

However, this image also includes information from very distant objects, whose light reaches us after a sometimes-considerable propagation time. The observed space is therefore made up of a mixture of phenomena contemporary to the observer and traces of past states of the Universe, all the more ancient as the sources are distant.

In this perspective, the cosmological present accessible to observation remains essentially local, while most of the information related to distant regions corresponds to previous states of the Universe.

Moreover, the present time, as it is lived in this local environment, unfolds in a dynamic and unstable context, whose properties make it difficult to directly infer the global cosmological history.

b) Reconstruction of an evolution from an observed past

When one seeks to reconstruct the evolution of the Universe from observations from the distant past, two structuring hypotheses can be formulated.

On the one hand, the gravitational field can be considered a structuring energy medium, revealing a general tendency towards the aggregation of matter and energy.

374

On the other hand, the flow of time connects two distinct cosmological states:

- an ancient state, characterized by a greater dispersion of energy and matter, as it appears in observations from the distant past.
- and a hypothetical future state in which gravitational concentration processes could become more pronounced.

In this interpretation, time constitutes the parameter allowing to connect the degraded image of a strongly dispersive past with a speculative projection of a potentially more concentrated future.

Possible variations in the rhythm of time and perception of distances

One can formulate the hypothesis according to which the rhythm of time measured in the present would gradually decrease, under the effect of the evolution of the global gravitational field.

Conversely, in an older Universe characterized by a lower average gravitational density, time may have elapsed at a faster rate. If such a hypothesis were valid, the distances associated with objects observed in the past would appear, when related to our current time frame, larger than they were in their original context.

Evolution of the gravitational structure of space

The gravitational depression of space — that is to say, the structuring of the metric under the effect of matter and energy — does not appear directly in the observation of distant regions.

However, if one admits that matter was once more diffuse, it is plausible that the distribution of gravitational potentials has been more homogeneous than it is today. The local deformations of space-time would then have been less pronounced, producing an overall structure more uniformly leveled than that which we currently observe.

Conceptual difficulties associated with the notion of expansion

375

The notion of expansion generally implies an increase in the volume occupied by a system. Such a definition implicitly implies the existence of an external reference framework for evaluating this variation in size.
However, applied to the Universe as a whole, this idea becomes problematic: a Universe devoid of observable boundaries cannot be easily related to an external container that would allow measuring its expansion.

Future evolution of the observable Universe

Today, the observable Universe is limited by the cosmological horizon defined by the maximum distance traveled by light from the origin of the observable structures.

Two interpretations can be considered:

In the standard expansion model, some currently visible regions could, in the long term, become unobservable if their distance speed exceeds the ability of light to reach us.

In the alternative hypothesis of a progressive energy depression of space, the observable horizon could on the contrary remain globally stable. However, this stability would depend on the evolution of the cosmological metric, that is to say, on the way in which the distribution of matter and energy modifies the geometric structure of space-time and, consequently, the observation conditions.

Redshift, cosmic cooling and possible interpretations

The gradual decrease in the average temperature of the Universe as well as the redshift of light spectra (redshift) are generally interpreted as manifestations of cosmic expansion.
In the hypothesis examined here, these phenomena could also be considered as the consequences of an energy depression of space accompanied by a time dilation.
The dominant cosmological model, based on the hypothesis of an initial inflation followed by a continuous expansion, is based on a set of theoretical

376

and observational results which nevertheless present certain tensions, notably with the detailed analysis of the cosmic microwave background.

These divergences between observations and physical models do not constitute a new phenomenon in the history of cosmology.

Global dimension of the Universe and limits of estimates

In some extrapolations based on space curvature models, it has been proposed that the total Universe could be several million times larger than the observable Universe.

However, these estimates are based on assumptions that remain difficult to verify. Indeed, any assessment of the size or expansion of the Universe would implicitly assume an external reference frame or a broader unit of comparison, which is not accessible.

In the absence of reference units universally applicable to the entire cosmos, these estimates therefore retain a necessarily speculative dimension.

Limits of the Hubble constant as a measurement tool

The Hubble constant constitutes a cosmological parameter intended to relate the distance of galaxies to their apparent speed of separation.

It is commonly used to estimate the age and scale of the observable Universe. However, this constant is based on two magnitudes—distance and escape velocity — the evaluation of which itself depends on interpretative models.

In line with the previously mentioned questions concerning the structure of space, the dynamics of time and the interpretation of the Redshift, these two parameters remain likely to introduce significant uncertainties into cosmological estimates.

Observational tensions around the Hubble constant

Recent observations made using the Hubble space telescope have led to a new estimate of the Hubble constant, obtained by a method independent of the two traditionally used approaches. This measure results in a lower value than some previous estimates.

Until now, two main methods were used:

- one based on the observation of cepheid stars and supernovae, used as distance indicators in the local Universe.
- the other based on the analysis of cosmic microwave background anisotropies.

These two approaches have led to results that remain in disagreement, a phenomenon known as the tension of the Hubble constant.

Furthermore, other observations have suggested that cosmic expansion could be faster than what some previous models predicted. These discrepancies were first interpreted as measurement uncertainties. However, they could also reveal deeper limitations of the standard cosmological model, as previously mentioned.

In this context, a frequently advanced explanation consists in invoking the existence of a dark energy, supposed to account for the acceleration of cosmic expansion. This hypothesis, although operational within the framework of the standard model, remains largely phenomenological, due to the lack of precise physical identification of this energy component.

The central role of cosmic microwave background radiation

The standard cosmological model is largely based on data from the study of the cosmic microwave background (CMB). The cosmological parameters derived from these observations generally lead to a value of the Hubble constant lower than that obtained from measurements made in the local Universe.

This situation can be interpreted as an indication that the current rate of expansion of the Universe would seem higher than that deduced from the model based on the distant past.

Such a divergence can be explained by the fact that the standard model is essentially based on the analysis of a very old state of the Universe, whose observed image is necessarily indirect and difficult to transpose into current physical conditions.

378

In the hypothesis considered here, the cosmic microwave background would not strictly correspond to an immutable fossil radiation from the first moments of the Universe but could rather be considered as the current state resulting from a long evolution of radiation fluxes present in space. Since the early phases of the Universe — sometimes associated with the so-called limit of Planck's wall — these flows have undergone multiple interactions and interferences with matter and, possibly, with antimatter.

Possible interpretation of the Hubble constant

If the hypothesis of a cosmic expansion is maintained, the observed rate of expansion appears today difficult to fully explain on the basis of current physical models.

In this alternative perspective, the Hubble constant — generally interpreted as a measure of the rate of expansion of the Universe — could be considered differently: it could above all constitute an indicator of the level of energy depression of space, that is to say, the degree of energy structuring of the cosmic environment.

Fluctuations of the cosmic microwave background and space structure

The analysis of emission peaks of fossil radiation shows the existence of extremely low-density fluctuations in a Universe that appears globally homogeneous and isotropic on a large scale.

These local variations can be interpreted as irregularities in the energy distribution of space, corresponding to slightly more or less dense regions. In some areas of lower energy density, the propagation of radiation could be associated with an elongation of wavelengths, a phenomenon compatible with spectral shift observations.

The quantum vacuum and the example of the Casimir effect

In what is commonly called vacuum, quantum physics predicts the permanent creation and annihilation of virtual particle-antiparticle pairs.

379

The Casimir effect experimentally illustrates certain properties of this quantum void. When two neutral conducting plates are placed very close to each other, only certain wavelengths of the electromagnetic field can remain in the space between them. This restriction changes the energy density of the vacuum between the plates compared to that of the external vacuum.
This energy difference produces an effective pressure that tends to bring the plates closer together. The phenomenon can be interpreted as the appearance of an apparent attractive force, resulting in reality from an energy imbalance of the quantum vacuum.
In this perspective, the space separating the plates can be considered to be in a state of relative energy depression.

Energetic bursts and cosmic homogeneity

Some major astrophysical events—such as merging of black holes, supernova explosions or galaxy collisions — locally produce areas of energy over density and strongly modify the structure of the surrounding space-time.
However, the scale of these phenomena remains limited compared to the global dimensions of the Universe. It is therefore unlikely that they are sufficient to question the statistical homogeneity of the Universe when considered on a very large scale.

Large-scale structure and distribution of matter

Cosmological observations show that matter distributes itself according to a complex filamentary structure, often described as a network or "cosmic web" structure, sometimes compared to a "honeycomb" organization. Galaxy clusters are concentrated along filaments, separated by vast regions of low density called cosmic voids.
These variations in density correspond to more or less depressionary areas from an energy point of view.
It could be the same for antimatter, whose presence would generally only be detectable indirectly, notably through certain gravitational effects or during punctual annihilation events.

Re-examination of the hypothesis of a one-time Big Bang

These considerations lead to relativize the simplified image of a Big Bang assimilated to a punctual singularity which would initially have concentrated all the energy of the Universe.

In a strictly inflationary scenario starting from an initial volume close to zero, one might expect to observe today radial density gradients or concentric structures resulting from expansion from an initial centre.

However, cosmological observations do not clearly highlight such a large-scale organization.

Limits of the explosive analogy

The idea of an expanding Universe is sometimes compared, by analogy, to a supernova-type explosion. However, this comparison has several limitations.

In the case of a supernova, the expelled matter disperses in a pre-existing surrounding medium and produces a highly heterogeneous structure, characterized by significant variations in density according to the distance from the point of origin.

The observable Universe, on the contrary, presents a remarkable statistical homogeneity on a very large scale.

Moreover, the very notion of cosmic expansion remains difficult to define if it cannot be related to any external reference in which this expansion could be measured.

All physical laws are formulated for a given scale domain: some describe quantum phenomena, others atomic, stellar or galactic systems. It is then legitimate to ask whether an analogous principle can be envisaged for what, at a more fundamental and hardly observable level, could intervene in the relations between quantum symmetries.

In the hypothesis considered here, we assume the existence of a discrete level underlying quantum physics, within which symmetries would not only constitute formal properties of equations, but dynamic structures defined by their relation to an opposite symmetry. In this perspective, each symmetry would only make sense relative to its counterpart, so that the physical states would be defined by a complementary or symmetrical opposition relation.

Modern quantum physics is notably based on three principles of symmetry generally considered fundamental:

> **1. Load conjugation symmetry (C)**
> To every particle corresponds an antiparticle characterized by opposite charges. In principle, the existence of a particle implies that of a corresponding antiparticle.
> **2. Parity symmetry (P)**
> The three spatial coordinates can be inverted simultaneously. This transformation corresponds to a mirror inversion of space and translates the idea of a space devoid of privileged center or edge.
> **3. Time reversal symmetry (T)**
> The fundamental equations of many interactions remain invariant if one reverses the flow of time. This property fuels the hypothesis of a possible temporal symmetry of physical laws, although its cosmological interpretation remains open.

The combination of these three transformations constitutes CPT symmetry, which plays a central role in quantum field theory. As proposed here, this global symmetry is not called into question, but some apparent local manifestations of asymmetry are interpreted differently.

Several hypotheses are then considered:

- The strong and weak interactions seem to exhibit certain symmetry violations. These phenomena could be interpreted as the expression of a chiral asymmetry, likely to be linked, directly or indirectly, to gravitational effects still poorly understood.
- A particle of matter could be considered correlated or entangled with a corresponding antiparticle, this relationship constituting a fundamental property of their existence.
- In the hypothesis of a cosmological couple formed by a Universe and an anti-Union evolving symmetrically, the directions of space could appear reversed between these two domains, as in a mirror transformation.
- If cosmological evolution were to lead to a final collapse of the Universe, this could be interpreted as a global dynamic linking the initial and final states, which would suggest a form of symmetry between the conditions of beginning and end.

The CPT symmetry — associating charge conjugation (C), spatial inversion (P) and temporal inversion (T) — provides a mirror representation of the Universe. It implies that antimatter constitutes, in principle, the symmetrical counterpart of matter, in equivalent quantity.

An effective violation of this symmetry would imply a fundamental asymmetry between matter and antimatter. Some experimental observations, notably related to the behavior of neutrinos in the weak interaction, have sometimes been interpreted as suggesting a possible violation of CP symmetry. However, these results remain difficult to interpret and are often associated with poorly understood properties of these particles.

From the perspective adopted here:

- The C- symmetry could be related to a general tendency of physical systems to seek charge equilibrium, especially during recombination processes in the primordial Universe and in electromagnetic interactions.
- The P - symmetry, corresponding to spatial inversion, could be associated with chirality phenomena, particularly in certain weak interactions. This chirality would not necessarily constitute a

383

fundamental violation of symmetry, but rather a particular manifestation thereof.

- T- symmetry could be interpreted, on a cosmological scale, as related to the global evolution of the Universe towards a final equilibrium state. In this hypothesis, a cosmological collapse of the Big Crunch type would represent a global dynamic linking the initial conditions of the Big Bang to a symmetrical final state.

Structured matter and structured antimatter would not be directly "miscible". They could however coexist in a symmetrical relationship comparable to a mirror effect shifted in space-time.

One possible manifestation of this symmetry is the ability of particles and corresponding antiparticles to annihilate each other, a process in which their energy is transformed mainly into electromagnetic radiation, often in the form of gamma-ray photons.

Some weak interactions also illustrate these properties. For example, electrons and neutrinos sometimes exhibit left helicity, revealing asymmetry in interactions. Some experiments suggest that the equivalence of interaction processes between certain quantum symmetries might not be strictly respected.

These observations raise the question of a possible violation of CPT symmetry, particularly in the processes of B meson decay. The mesons are composite particles made up of a quark–antiquark pair, generally of neutral overall charge and extremely short life. The observed anomalies could be explained by a subtle dissymmetry between matter and antimatter, interpreted here as a manifestation of chirality.

The interpretation of these phenomena remains however delicate, notably due to the experimental difficulty of directly observing antimatter in natural astrophysical conditions.

Finally, some astrophysical observations, such as the detection of intense gamma radiation at the center of certain galaxies, could be interpreted as the signature of processes involving particle-antiparticle annihilations. Interactions of this type are also observed in certain nuclear reactions, where annihilation results in the conversion of the mass of particles into high-energy photons.

The reflection proposed in this chapter is a speculative construction that explores the deepest levels of the microscopic scale. It does not modify previous developments but seeks to establish a connection between quantum mechanics, spacetime relativity and the cosmological hypothesis of a multiverse. The central hypothesis assumes the existence of discrete exchanges capable of ensuring a form of symmetrical relationship between our Universe made up of matter and a complementary domain.

In this perspective, there would be no absolutely watertight boundary between quantum mechanics and classical relativistic physics. Physical laws remain continuous in their principle, even if their effective forms vary according to the scale considered. The rules that describe phenomena evolve gradually as one move into the domains of the 'infinitely small' or the 'infinitely large'.

The most precise experimental observations, as well as the most advanced calculation methods, today provide important information but still insufficient to establish complete certainties concerning the ultimate structure of physical reality.

In the hypothesis of a cosmological system comprising two universes linked by quantum symmetry, the laws governing elementary particles, atomic structures, macroscopic bodies and stellar systems should be considered as fundamentally related. The current difficulties in reconciling certain theories could then reflect less a real incompatibility of physical laws than a fragmentation of their description. Theoretical unification does not necessarily imply the disappearance of scale distinctions: it rather supposes the existence of transitional relationships between different levels of description.

In this perspective, a possible unified theory should integrate rules specific to each scale domain while describing the continuities and transformations that connect them. These rules could be considered as part of a global cosmological process leading the Universe towards a state of equilibrium. Such a perspective would invite to consider physics not only as the study of structures and interactions, but also as that of cosmological evolution which connects the origin and destiny of the Universe.

As previously mentioned, our perception of the Universe can be biased by the interpretation we make of certain relativistic phenomena. For example, the curvature or depression of space-time can be interpreted as an increase in distances, which leads to read certain cosmological phenomena such as the trace of past events. Similarly, time dilation is not perceived subjectively as a slowing down of the course of physical processes.

Thus, although astronomical observation allows access to ancient states of the Universe, our understanding remains largely dependent on the local and present frame of reference in which we make these observations. This limitation constitutes an important epistemological obstacle in the global interpretation of the cosmological structure.

385

In the microscopic domain, quantum mechanics describes phenomena that cannot be easily interpreted within the intuitive framework of classical space and time. Interactions internal to composite particles, for example between quarks inside hadrons, are characterized by extremely fast and highly unclassical correlations. In some interpretations, these correlations can be described as exchanges of information that do not correspond to an ordinary movement in space-time.

Quantum properties can thus give the impression that certain information is simultaneously distributed among several constituents, a phenomenon reminiscent of quantum entanglement effects. Elementary particles then seem to escape the classic categories of location.

An often-mentioned example concerns the electrons in the atom. Their quantum description does not correspond to a defined trajectory but to a probability distribution forming a region around the atomic nucleus. This region can be interpreted as an effective limit of electronic influence, analogous to a horizon defining the structure of the atom.

Our representation of the Universe is essentially located between two observation limits: on one hand, the currently recognized elementary constituents of matter, and on the other hand, the observable cosmological horizon. This situation leads to a vision of the Universe circumscribed by these two boundaries, even though many aspects of physical reality could be located beyond these areas accessible to direct observation.

Just as we must imagine the Universe beyond the observable horizon, we are led to formulate hypotheses concerning the internal structure of elementary particles, which currently represent the limit of our capacity for experimental exploration. These entities are not observed directly; their existence is inferred from theoretical models and indirect experimental results, often formulated in an elaborate mathematical framework.

Thus, both below the scale of elementary particles and beyond the cosmological horizon, scientific knowledge largely relies on theoretical constructions supported by observational indices, but which retain a speculative dimension.

In this perspective, the photon can be interpreted as the quantum of an electromagnetic field resulting from very early phases of cosmic history. It could be considered as the remnant of an extremely intense energy regime that characterized the early phases of the Universe, close to the Planck era.

Today, the average energy intensity of electromagnetic radiation is much lower than that which prevailed during these first phases. Consequently, current photons no longer have the necessary energy to spontaneously generate, through mutual interaction, new massive particles.

In a simplified configuration, the photon can be considered as the quantized manifestation of an elementary electromagnetic field, associating electrical and magnetic components and not possessing a classical spatial dimension in the strict sense.

Such an interpretation suggests that the Universe might not present a fundamental limit for electromagnetic waves propagating independently of charged particles, even if their mathematical description remains formulated in the context of relativistic space-time.

However, for reasons of mathematical formalization, the energy of a photon must always be expressed using energy and time units, which shows how much our description remains dependent on the categories of classical space-time.

The fact that our representation of reality changes profoundly when we change the scale of observation illustrates the persistent difficulty in intuitively linking quantum mechanics and classical physics. In practice, our scientific descriptions often combine the two, sometimes implicitly.

If we consider the elementary particle of matter as the localized expression of a quantum state resulting from the initial conditions of the Universe, it becomes difficult to attribute to it a true spatial occupation or a defined temporality in the classical sense.

Some elementary particles, notably quarks, can associate in a stable way to form composite particles called hadrons. These structures result from particularly intense strong interactions. However, even these composite particles do not always have a simple spatial and temporal location in the quantum framework.

When these particles interact with leptons, especially electrons, a charge equilibrium can be achieved. Some composite particles then group together to form atomic nuclei, which, associated with electrons, give rise to atoms.

It is at this level of organization that a field of phenomena corresponding to our familiar experience begins to emerge, objects endowed with spatial extension, evolving in time and connected by cause-effect relationships.

By assembling atoms into molecules and then into macroscopic structures, matter gradually builds the physical context in which the phenomena described by relativity are inscribed. We can then consider that space and time as we perceive them constitute the macroscopic expression of a more fundamental quantum dynamics, in which the classical notions of localization and temporality do not play the same role.

The transition between these levels of organization, linked to the change of scale, leads the human observer to perceive as fundamental reality the space-time in which he evolves. We are necessarily immersed in it and build our environment and our understanding of the world from this frame of reference.
It is likely that human cognitive abilities are not fully adapted to the intuitive representation of processes governing the infinitesimally small and fundamental quantum symmetries. Quantum mechanics thus remains, in large part, a field where understanding relies on mathematical and conceptual models rather than on direct intuitive representation.
In applications related to the physical world, mathematics almost always relies on notions of space and time. In the most abstract applications (computer science, statistics, logic...), they are mainly based on structures and relationships, apparently detached from their physical meaning, but which ultimately leads to a spatial or temporal interpretation.
If time and space are derivatives and do not constitute the foundation of our universe, how can applied mathematics describe what represent constraints and forms of coherence for an observer who cannot exclude himself conceptually from space and time?
The problem is not reality, but the mathematics we use to describe it. We must stop confusing what equations allow us to imagine with what nature actually makes possible.

On the 'elasticity' of time

The question of a possible shift in time constitutes a recurrent theme in theoretical reflection. In contemporary physics, such a possibility can only be envisaged from speculative hypotheses inspired, notably, by general relativity. The following scenarios should therefore be understood as constructions of the mind rather than as physically established propositions.

Hypothesis of a shift towards the future

General relativity establishes that the intensity of the gravitational field influences the rate of time. For a distant observer, the stronger the gravitation, the more time passes slowly for the phenomena located in this gravitational field. This phenomenon is known as gravitational time dilation. In the extreme case of a black hole, gravitation becomes so intense that physical processes observed from the outside seem to slow down considerably near the event horizon. In some theoretical interpretations, an object or an observer caught in such an environment could undergo an extreme slowdown of its own time compared to that of the rest of the Universe.

If a hypothetical observer succeeds in later exiting such an extreme gravitational environment — highly speculative hypothesis — he might find that much more time has elapsed in the outer Universe than for himself. He would then discover a universe much older than the one he had left with the risk of disappearing if the universe, having reached its end, were to collapse. Such a situation would correspond, in theory, to a form of movement towards the future.

Hypothesis of a return to the past

A second hypothesis, even more speculative, is based on certain cosmological scenarios predicting a final collapse of the Universe. In such a model, the expanding Universe could eventually evolve into a contraction phase leading to an extremely dense state.

In this fictitious perspective, an observer capable of surviving up to this final phase would witness the gradual disappearance of existing astrophysical structures. If one prolongs this speculation, one can imagine that a new Universe could emerge from this extreme state, constituting a form of cosmic renaissance.

In a purely conjectural framework, such a process could be interpreted as the transition to a second-generation Universe, distinct from the previous one. For a hypothetical observer surviving this transition — highly improbable hypothesis—this situation could be assimilated to a form of return to an analogous past, although it is not the strict past of his original Universe.

These scenarios, however, assume extremely speculative physical conditions, notably the survival of an observer in extreme cosmological environments and the possible existence of a multiverse-type structure.

Regardless of these assumptions, the very notion of time remains deeply linked to observation conditions. In relativity, each observer has his

389

own time, measured by the clock that accompanies him in his frame of reference. There is therefore no single universal time valid for all observers.

The statement that time flows differently depending on velocity or gravitational field means that the measurement of time depends on the local physical context. Time thus appears as a local quantity, associated with the interactions and dynamic conditions specific to each region of space-time.
In this perspective, time can be interpreted as an intrinsic property of physical processes, whose value depends on the nature and intensity of the interactions considered. Relativity thus leads to abandoning the idea of absolute simultaneity between distant events.
To illustrate this relativity of time, one can compare systems subjected to different physical conditions. Processes characterized by fast dynamics and weakly constrained by gravitation will evolve differently from systems strongly linked or subject to intense interactions. In each case, the perception of time essentially relates to the internal development of the processes considered.
If we refocus reflection on human experience, it appears that our conception of time is largely linked to our biological and cognitive experience. *It's the story of the hare and the turtle. The first mounted "on springs" with its long legs, seems little affected by the terrestrial attraction and plays with distances. The second one seems heavy, stuck to the ground and is forced to move with a slowness that penalizes her. Both, however, are capable of completing, each in their own way, the same journey. If they ignore each other, they will however not be able to make any connection in terms of speed and their notion of time will be reduced to that of a distance traveled.* Time constitutes a comparative tool of measurement that allows the ordering of lived events and the establishment of causal relationships.
This construction is however limited by the scales of duration accessible to our perception. Extremely fast phenomena, well below the fraction of a second, like processes taking place over immense cosmological durations, largely escape our direct intuition.
That is why our representation of time relies on familiar categories — past, present and future — closely associated with our perception of a three-dimensional space.

A simple illustration consists of imagining the accelerated projection of a film lasting ninety minutes condensed in a few seconds. The unfolding of the

Similarly, if one imagines extremely fast cosmological processes, certain
hypothetical structures could appear and disappear on such short time scales
that they would remain unobservable in our usual temporal framework. In
this speculative perspective, cosmological systems — such as Universe/anti-
universe pairs envisaged in some models — could form and disappear at
rhythms that entirely escape our observation capacity.
Thus, our position in time inevitably limits our perception of cosmological
reality. We only have access to a restricted time window, corresponding to
the time and physical conditions in which we exist. Any attempt to broaden
this understanding is therefore based on theoretical models intended to
extrapolate beyond this immediate experience.

XXVIII <u>Hidden universe and semblance of reality</u>
(A chapter that accumulates the images)

Cartesian logic, which spontaneously leads us to search for causal relationships and coherent structures between phenomena, is not free from limits. It can even become a source of confusion when trying to grasp physical realities that exceed our ordinary intuitive abilities. Numerous examples illustrate our difficulty in conceiving situations that appear, a priori, counterintuitive:

- a finite Universe but devoid of identifiable material boundaries.
- the hypothesis of a system consisting of two universes in shifted quantum symmetry.
- the possibility, suggested by certain formulations of relativity, of an interchangeability between space and time.
- a physical vacuum not really empty, structured by energy fluctuations and phenomena comparable to pressure variations.
- particles described by states of superposition, characterized by probabilistic domains of influence.
- fundamental interactions interpreted as structures of the branes or membranes type.
- the hypothesis according to which cosmic expansion could result from an effect of observation or geometry.
- physical constants likely to vary at the cosmological scale.
- exchanges of information or influence without traditional material shifting.
- a macroscopically tangible material but fundamentally described by non-localized quantum entities.
- the hypothesis of a cosmos consisting of a set of potential multiverses.
- elementary particles whose description might not depend explicitly on time.
- or even black holes considered as quantum objects that can, in some cases, detach from the classical space-time framework.

The multiplication of complex models and apparent paradoxes can give the impression of a thorough understanding of physical reality. However, when models become too large or complex, they often lead to a compartmentalization of theoretical approaches, which can cause the overall view to be lost.

Conversely, an excessive simplification that ignores certain empirical or theoretical data would introduce a reductive bias. In a field where physical phenomena interact closely, no parameter can be fully isolated without risk of interpretative distortion.

These considerations lead to a fundamental question: what gives the Universe its strongly interconnected character? One way to approach this question is to consider gravitation — central phenomenon in cosmic dynamics — as a starting point for reflection.

Relationship between time, space and gravitation

- **Time**

According to the theory of relativity, the flow of time depends on the gravitational field and the state of motion of the observer.

An observer subjected to a strong gravitational field or significant acceleration sees the pace of his physical processes slow down relatively to that of an observer placed in a weaker gravitational field. This phenomenon, called gravitational time dilation, implies that clocks located in a strong gravitational field evolve more slowly from the point of view of an external observer.

For the local observer, however, this slowdown is not perceptible: all biological and physical processes proceed at the same internal pace.

Conversely, in a region where gravitation is weaker, the time passes more quickly relative to that measured in a stronger gravitational field. Comparisons between distinct frameworks then reveal measurable temporal differences.

- **Space**

General relativity also establishes that gravitation is associated with a curvature of space-time.

In an intense gravitational field or in an accelerated reference frame, the measured distances may appear contracted to an external observer. The local space can thus be described as more strongly curved or more densely structured.

393

From the point of view of the observer located in this gravitational field, these modifications are not directly perceptible, because his measuring instruments undergo the same transformations.

In regions where gravitation is weaker, the geometry of space appears relatively less curved, which also modifies the relationships between distances, durations, and trajectories.

The role of the observation framework

Any observer describes physical phenomena from a local reference frame. In this context, a body subject to constant acceleration may appear, from the point of view of an external observer, as possessing increasing energy, which can be interpreted as an effective increase in its relativistic mass.

The close relationship between space geometry and the flow of time leads to consider space-time as a unified four-dimensional structure, in which temporal and spatial variations are inseparable.

From a more speculative perspective, one can consider that, beyond certain scales — notably at the atomic or subatomic scale—space and time may emerge from each other, with each structure contributing to define the other.

Relationship between energy, matter and gravitation

Energy is omnipresent in the Universe. The different energy fields — electromagnetic, quantum or gravitational—constitute the dynamic structure of space-time. Thus, physical vacuum is not an absolute vacuum, but an environment characterized by permanent quantum fluctuations.

Matter can then be considered as a particular configuration of these energy fields. In this perspective, it would represent a state where energy is concentrated enough to produce a stable mass-bearing manifestation.

According to the mass-energy equivalence relation, this concentration of energy modifies the local geometry of space-time and thus contributes to the gravitational phenomenon.

In a more speculative interpretation, the formation of material structures could correspond to a transition phase of space-time itself, when certain conditions of density or energy are reached.

Such a process could, in some theoretical models, participate in a mechanism of rebalancing between different quantum symmetries of the Universe, a hypothesis that remains largely exploratory today.

394

Gravitation: engine of the Universe

In general relativity, gravitation can be described as the manifestation of deformations of the space-time structure produced by the presence of energy and mass. These deformations modify the trajectory of bodies and the radiation that moves through them.

In a more speculative perspective, gravitation can also be considered as the result of interactions between underlying energy fields, whose configurations would determine the local geometry of space-time. In this approach, it is possible to draw a parallel with certain electromagnetic phenomena. In the same way that the opposite poles of a magnetic dipole interact, certain quantum symmetries could interact with each other through fundamental energy structures.

In this hypothesis, electromagnetic waves could play a structuring role in gravitational dynamics. They would then constitute a mediating element participating in the organization of energy fields responsible for spatiotemporal deformations.

When the gravitational intensity increases—for example near a very massive object — the geometry of spacetime becomes strongly curved. This curvature can be interpreted as a progressive concentration of energy in a region of space, leading to the formation of a deep gravitational well. *This is the case with a trawl net whose peripheral mesh size is less affected by the mass of caught fish than the few central meshes that concentrate the entire product of the fishery..*

In some speculative theoretical models, an extreme accumulation of energy and curvature could lead to the appearance of particular topological structures of space-time, sometimes referred to as wormholes. These hypothetical structures would correspond to regions where the geometry of space-time allows connections between distinct domains.

In this perspective, gravitation would not be limited to a simple attractive interaction between masses but would constitute a fundamental structuring mechanism of the Universe, participating in the organization and evolution of its large structures.

Electromagnetic waves as possible mediators of gravitation

Charged particles possess a magnetic moment, generally attributed to the combination of their electric charge and spin, an intrinsic property associated with their quantum angular momentum.

At the microscopic scale, the magnetic moments observed result from interactions between particles of opposite charges and electrical currents associated with their movement.

However, the question arises whether the magnetic dipolar state we observe in macroscopic matter really constitutes a fundamental property of the individual particle, or if it is an emergent property resulting from the collective organization of many particles.

If one considers the elementary particle as a packet of entangled quantum waves, it is possible to consider — as an assumption—that its ground magnetic state may differ from the magnetic dipole observed at larger scales. Some theoretical models suggest the possibility of a magnetic monopole, i.e. the existence of an isolated magnetic pole.

Such a property would imply that the corresponding antiparticle has a monopole of opposite sign. The dipolar state of observable matter could then emerge only from the supra-atomic scale, when collective interactions between particles lead to the formation of bipolar magnetic structures.

In classical electromagnetism, the magnetic fields observed always result from moving electric charges, which naturally leads to the formation of dipoles and closed field lines. This feature seems to exclude the existence of magnetic monopoles under ordinary conditions.

However, some theories of great unification predict the existence of magnetic monopoles in extreme physical conditions, notably at very high energies, close to those that prevailed in the primordial Universe.

Particular physical contexts could also promote the appearance or stabilization of such phenomena. For example, black holes, characterized by extremely high energy densities and physical conditions radically different from those of the surrounding environment, constitute environments theoretically conducive to exploring these hypotheses.

In this type of object, the internal dynamics of collapsed matter could lead to electromagnetic field configurations that are difficult to observe directly. The description of a magnetic field potentially frozen in a region where eigentime is strongly slowed, however, poses significant fundamental problems.

Another possibility would be that magnetic monolinearity constitutes a fundamental but discrete property of certain elementary particles, observable only in their minimal energy state.

396

If the existence of magnetic monopoles were confirmed—whether in the framework of elementary particles or that of black holes — this would imply a partial revision of the standard cosmological model and could strengthen the idea that gravitation has a more quantum origin deep that currently described. Such a hypothesis would also have the advantage of reducing certain difficulties related to the differences in scale between quantum physics and gravitation.

Wave packets and particle structure

To illustrate the idea that a particle could be described as a coherent set of quantum waves, an analogy can be drawn with acoustic phenomena.

In a musical system, each note corresponds to a particular vibrational frequency. When several notes are emitted simultaneously and form a consonant chord, their respective frequencies combine to produce a stable and harmonious sound structure.

In such a chord, the different notes become difficult to distinguish individually, because their superposition creates a coherent overall vibratory structure.

By analogy, an elementary particle could be interpreted as a stable configuration resulting from the superposition of several fundamental wave modes. The entanglement of these modes would then form a coherent system, whose global properties — mass, spin or angular momentum — would result from the combination of these elementary vibrations.

In this perspective, the quantum angular moment associated with particles could be understood as the expression of an entangled primordial wave system, forming a stable dynamic structure whose inertia would be an emergent property.

XXIX <u>Decoherence and its metaphysical interpretations</u>
(A theory that revives debate and disturbs understanding)

The wave function consists of assigning to particles of matter, in particular fermions, a mathematical description analogous to that of a wave packet. This formalization allows to account for the wave-corpuscle duality highlighted in quantum mechanics.

The wave function is not a physical wave in the classical sense. It constitutes an abstract mathematical construction whose role is to represent the quantum state of a system. As such, it should not be confused with interactions between charged particles that fall under electromagnetism. In standard interpretation, the wave function describes a probability distribution associated with the different possible values of physical observables.

A quantum particle does not necessarily have a well-defined position or a determined trajectory in the classical sense. It is rather described by a set of presence probabilities in space and time. Mathematically, this state can be represented by a probability field whose amplitude varies according to physical conditions and interactions with other systems. When several wave functions interact, their interference modifies the probability distribution associated with possible states.

An important difficulty of quantum mechanics is that this probabilistic description leads, in principle, to abandon the classical notions of well-defined positions and trajectories. However, any experimental observation requires the use of measurable quantities such as position, velocity or energy, whose definitions are largely based on the existence of physical constants. There is therefore a tension between the probabilistic description of quantum states and the concrete modalities of their measurement.

In this perspective, the wave function can be interpreted as a mathematical representation of all possible interactions in which a quantum system can participate. It can also be understood as the description of a wave packet resulting from the superposition of elementary states capable of being entangled.
In principle, any physical entity — whether elementary particles, atoms, molecules or even macroscopic objects—can be described by a wave
398

function. However, it is to be feared that for massive objects, the equation thus formulated:

$$i\hbar\frac{\mathrm{d}}{\mathrm{d}t}|\Psi(t)\rangle = \frac{\hat{\vec{P}}^2}{2m}|\Psi(t)\rangle + V\left(\hat{\vec{R}}, t\right)|\Psi(t)\rangle$$

be devoid of

practical meaning.

The wave function associated with an antiparticle can be considered as the counterpart of that of the corresponding particle, in accordance with the fundamental symmetries introduced in quantum field theory.

A central mathematical formulation of this description is provided by the Schrödinger equation. This equation makes it possible to represent, in a non-relativistic framework, the temporal evolution of the wave function and therefore of the probabilities associated with the different quantum states accessible to a particle.

A major difficulty arises from the fact that particles can be in an overlapping of quantum states. This superposition is not directly observable: any measurement carried out on a system only reveals one result among the possible states, selected according to the experimental conditions and the type of observable considered.

The transition between the superposition of quantum states and the observation of a single state is generally described by the phenomenon of quantum decoherence. Decoherence corresponds to the process by which interactions with the environment gradually make the different components of quantum superposition incoherent. This phenomenon is often associated with what is called wave packet reduction.

In experimental practice, the properties of elementary particles are inferred from traces or signals detected in specialized devices, such as detection chambers or particle detectors. These devices require very controlled experimental conditions, but they cannot completely eliminate the interactions with the environment that contribute to the decoherence of the observed system.

The experimental environment therefore plays a decisive role in any quantum measure. Ideally, the analysis of the results should consider all the parameters likely to influence the dynamics of the observed system. However, it remains difficult to precisely assess the effect of the measuring devices themselves, which constitute complex physical systems that can modify the initial conditions of the experiment.

399

Under these conditions, the interpretation of experimental traces relies largely on statistical methods and probabilistic models. The results obtained then depend on analysis algorithms and theoretical hypotheses allowing to link the observed signals with the properties of the studied particles. This dependence on interpretation models constitutes one of the major difficulties of quantum physics and contributes to the complexity of building a unified theory.

Quantum entanglement and inheritance of cosmological processes

The particles currently observed can be considered as the result of a long cosmological history marked by numerous physical processes. Among these are notably the primordial radiative entanglement, the decoupling of electromagnetic radiation, the recombination of electrons and nuclei, as well as the different phases of nucleosynthesis. These processes led to the formation of particles with differentiated properties.
However, when particles have the same fundamental properties, they can be described as belonging to the same quantum class characterized by identical states. In some theories, this identity can be accompanied by persistent quantum correlations, called quantum entanglement.

Entanglement is manifested by the existence of statistical correlations between the results of measurements carried out on distinct systems. These correlations may persist regardless of the distance separating the systems concerned, as long as the conditions allowing them to be maintained are not destroyed by interactions with the environment.
In this perspective, a modification of the quantum state of an entangled particle can result in an observable correlation with the state of another particle of the same system. This phenomenon does not correspond to a transmission of instantaneous information in the classical sense, but to a structural property of the global quantum state of the system.
These characteristics appear particularly marked at microscopic scales, where the classical notions of space and time cease to provide a fully adequate description of physical phenomena.
It is however likely that the degree of entanglement between two particles evolves over time. Local interactions with the environment can gradually reduce initial quantum correlations. Astrophysical energy events—such as certain supernova explosions — could also contribute to modifying or destroying these correlations in certain contexts.

400

At the macroscopic scale, complex systems evolve all the more independently of each other as they are spatially separated and their direct interactions become negligible. Emergent properties of systems can be described in an approximately autonomous manner. The non-separability highlighted in quantum mechanics and the principles of relativity are therefore not necessarily contradictory: they fall under different descriptions corresponding to distinct physical contexts. Quantum mechanics applies mainly to microscopic phenomena, while relativity describes large-scale phenomena in a continuous spatiotemporal framework.

This distinction can be summarized as follows:

> - Quantum mechanics aims to describe the possible states of microscopic systems using wave functions and state superpositions. This mathematical formalization accounts for probabilistic phenomena and non-local correlations, but it does not provide, in its standard formulation, a complete relativistic description of these systems. She ignores relativity.
>
> - Classical relativistic physics seeks to describe the global evolution of the Universe from its observable contents — matter, radiation, and energy—in a dynamic space-time.

These two approaches thus correspond to different levels of description of physical reality.

Undulatory description of particles

In the quantum description, a particle is not assimilated to a material point with a well-defined trajectory. It is described by a wave function representing a probability distribution associated with the different possible values of its observables. This representation can be interpreted as a wave packet resulting from the superposition of elementary quantum states.

In the case of composite systems, the quantum state can be described as an overlapping or entanglement of several wave packets. The classical representation of the atom—in which electrons orbit a nucleus — is

401

essentially an instructional image intended to render intelligible a fundamentally non-classical phenomenon. The electron corresponds rather to a quantum state characterized by certain probability distributions associated with the negative charge that balances that of the nucleus. There is no defined orbit in the sense of a traceable trajectory.

This description based on wave functions makes the intuitive representation of phenomena more difficult than in classical physics. As the physicist Richard Feynman pointed out, quantum mechanics remains a theory whose interpretation remains deeply counterintuitive, even for specialists.

Complexity of macroscopic systems

In our macroscopic experience of the world, we are led to describe physical objects as possessing definite properties — position, velocity or shape— resulting from an effective reduction of possible quantum states.

In principle, any physical system can be described by a global wave function. This includes not only elementary particles, but also complex systems such as living organisms. However, for these systems with an extremely high number of degrees of freedom, the corresponding wave function becomes so complex that it loses all practical use.

A macroscopic object can thus be considered, from a theoretical point of view, as a very complex set of interacting entangled wave packets. However, permanent interactions with the environment lead to extremely rapid decoherence of possible quantum states. For this reason, massive objects always appear in well-defined macroscopic states and not as quantum superpositions.

Decoherence and the role of observation

Decoherence corresponds to the process by which the interactions of a quantum system with its environment gradually destroy the phase correlations between the different components of the wave function. This phenomenon leads to the appearance of apparently classical macroscopic states and is often associated with the notion of wave packet reduction.

In this context, the observed state of a system corresponds to one of the possible outcomes described by the initial wave function. The experimental conditions and the environment play a determining role in the selection of actually observable states.

402

Thus, the macroscopic reality we perceive can be interpreted as the result of decoherence processes that make quantum superpositions practically inaccessible to direct observation. The environment in which the observer is located acts as a physical filter that selects certain stable macroscopic states. This situation is also due to the fact that our measuring instruments and analysis methods are based on classical concepts — position, duration, energy, location—which are themselves defined in a macroscopic spatiotemporal framework.

It should be emphasized, however, that the act of observation does not modify the fundamental properties of the Universe in the sense that the observer would create physical reality. The phenomena observed possess an existence independent of their observation. What is at stake concerns rather the conditions under which certain properties become accessible to measure.

If we consider only directly observable and measurable phenomena, the description we obtain of reality may appear partial or restrictive. Physical theories necessarily rely on a set of hypotheses, models and ideological constructions allowing the interpretation of observations.
In this perspective, the space-time in which we describe physical phenomena constitutes the framework within which our observations make sense. These depend on the properties of the instruments, the experimental methods and, more generally, the cognitive and technical abilities of the observer.
Recent advances in physics suggest, however, that the structure of reality could surpass the classical descriptive model. Some theories envisage the existence of more fundamental levels of description in which familiar notions of space and time might not play the central role they occupy in macroscopic physics.
The status of the human observer, a product of biological evolution and endowed with limited cognitive capacities, inevitably imposes a particular context on our understanding of the Universe. Our representations of the world result from a combination of intuition, empirical experience, theoretical knowledge and technological progress.
When exploring the foundations of physics, many questions remain open. The paradoxes and limitations of current theories suggest that our description of reality may be incomplete.
The recognition of these limits nevertheless opens up the possibility of a gradual broadening of our understanding. The difficulty consists in developing new theories capable of overcoming the constraints of our

403

current representations, while remaining compatible with experimental observations.

The question then remains as to how far it is possible to extend our models and methods in order to explore more deeply the structure of physical reality, beyond the limits imposed by our current observation framework.

More than a consequence of the intrusive approach of the observer, the collapse of the wave function seems to have to be considered as a general and unrecognized phenomenon of permanent state transition. This process is realized all the more fully as the observed object is massive and therefore complex to decipher. Allowing for some speculation, how to describe the real world? We constantly refer to fragmented and counterfeit information. These are the ones provided by our cognitive functions and which have vocation above all, to manage our essential needs in an environment that we dress to our liking. The hardware configuration that we give to any massive body, is due to the fact that the collapse of the wave function is for us an inevitable process in our need of understanding what our cognitive functions tell us.

Perhaps, without us being aware of it, this presentiment of a Universe fundamentally dematerializable in wave functions, is rooted in the collective unconscious. This would explain a well-established belief associating a non-physical entity (soul or spirit) with a fleshly envelope devoid of sustainability. We can make simpler with this paradigm-fiction suggested here and in which the Universe has for humanity, nothing of a procreator endowed with discernment. The big flaw of cosmology, in its non-anthropogenic sense, is that it promises nothing and does not give hope. The least we can say is that it is rather distressing for the morale, the image and the future of man. We understand that it does not seduce more than that the common man.

What distinguishes man from the monkey and the monkey from the fish is an ability to memorize and a depth of reflection related to increased longevity. Moreover, another advantage lies in our ability to project ourselves into a more or less distant future. But obviously, the future shows its limits and the past leaves few vestiges. A real frustration!

With Plato, Galileo, Newton, Planck, Einstein, Hawking and many others, how many theories have succeeded each other, each one making its contribution and its particular logic. All have enriched and often challenged the ideas of their predecessors. Since nothing can be considered definitive, why should it be otherwise today?

We guess a sort of intellectual resistance imposed at all times, by a minority of specialists, on ideas which are confirmed for some but are called to be invalidated for others. One of the latest is the Hawking's interesting theory dealing with the evolution of black holes. Both wrongly as rightly, Einstein would have said, that «superb mathematics (often marked by simplifications) could lead to build an abominable physics. A major reason for this is that the space/time metric cannot have a certain value because of relativity. This relativity, in a way, has a double effect because of symmetry, making any measure immediately taken, immediately invalidated.

Who can say that some of the most recent theories will not match the errors and aberrations that have nourished our history and have since been denied? Of course, the reflection developed here does not in any way claim to erect itself in truth. These transgressions can no doubt run counter to a scientific mind convinced of the pre-eminence of mathematical models and rules which have been the subject of a broad consensus, on a logic on the margins and more or less rebellious. But how many theories, initially recused, have made possible to advance knowledge. The increasingly complex and expensive investigative tools that have allowed to validate many assumptions are beginning to reveal their limits. Furthermore, it is highly unlikely that our scientific processes and observational tools are adapted to the "expertise" of a virtual multiverse Cosmos as proposed here. So, the open question to conclude would be:
Which ideas developed here deserve to be supported and on which points should they be corrected or invalidated?

"Errare humanum est, perseverare diabolicum," the past generations used to say with derision.
Any remark, objection or controversy is welcome, in so far as it would help to nourish this reflection and to take up again ideas which, for some, may have perhaps shock the reader.

405

Everything starts from the hypothesis that there exists a potentially indefinite multiplicity of binary systems of universes, organized according to quantum symmetry. From this point of view, these systems would form pairs capable of recovering, transforming and restoring energy without depending on the classical notions of location, distance or movement in space.

In this perspective, each symmetrical energy binomial would appear and disappear independently of a global cosmic framework, without lasting interaction with the other structures of the Cosmos envisaged as a multiverse. Such a representation may seem out of step with the empirical experience we have of our observable Universe.

The notion of infinity, considered as a reference tool in mathematics, seems difficult to apply directly to the Universe we inhabit, generally described in cosmology as finite but not bounded. It also seems problematic when transposed to the hypothetical scale of a multiverse, because it implicitly implies a certain idea of space. Similarly, the notion of eternity — understood as an infinite duration—seems hardly compatible with the idea of a virtual multiverse Cosmos in which particular universes could appear and disappear, and with the widely accepted hypothesis that our own Universe has a cosmological beginning. In this hypothesis, observable space and time could be interpreted as emergent properties resulting from a discrete and chiral quantum symmetry of the matter constituting our physical world.

It should also be remembered that living beings, including human beings, constitute only a particular form of organization of matter. From the biological point of view, an organism can be described as a complex aggregate of molecules structured in such a way as to maintain its organization, ensure its self-preservation and reproduce. In this very vast set of possible configurations of matter, living structures likely represent a relatively marginal category.

The ability to be aware of one's own existence is generally considered a prominent feature of the human species. However, the study of animal
406

behavior suggests that some forms of consciousness—at least in elementary forms — also exist in other species. This property could result from a process of acquisition, processing and storage of information in the form of electrochemical activities within the central nervous system, and more particularly in the brain, consisting of a dense network of neurons interconnected by synapses.

A parallel can be drawn with the functional architecture of a computer: in both cases, information is recorded, processed and combined according to certain operating rules. Nevertheless, this analogy remains limited. The biological brain is characterized by an adaptive learning ability, continuous information enrichment and dynamic interconnection of neural networks, which results from individual experience and interactions with the environment.

The information collected comes from both direct experiments and multiple interactions with an environment likely to represent both constraints and opportunities for survival. Sensory stimuli and the responses they trigger gradually lead to the formation of automatisms. These allow for quick reactions, often reflexive, which can then become more complex and gradually become more elaborate and thoughtful.

This behavioral evolution strengthens the adaptation of the organism to its environment. A major difference between a living organism and a machine—including in the case of artificial intelligence systems — lies in the fact that they are not derived from a biological process and do not possess either a genetic heritage or intrinsic preservation mechanisms or reproduction. In the absence of such evolutionary constraints, they do not spontaneously pursue survival or species continuity objectives and only interact with their environment through the purposes assigned to them.

Such an approach leads to a materialistic and agnostic interpretation of consciousness, close to a positivist perspective: the consciousness of existing would be an emergent property resulting from the organization and activity of the nervous system. However, this interpretation may seem unsatisfactory to those who consider the existence of immaterial dimensions or the possibility of a form of individual permanence beyond biological life.

The concepts of spirit, soul or consciousness have close connotations but are often used ambiguously. To clarify this, it is necessary to examine the

407

functioning of the central nervous system. An important part of brain activity is devoted to the regulation of vital functions — cardiac, respiratory, digestive, circulatory or renal — as well as the management of reflexes. These processes take place largely automatically and independently of our immediate consciousness, whether we are awake or not.

This reality has sometimes led to the simplified idea that humans use only a very limited fraction of their brain capacities, often expressed by the popular estimate of 10%. Current knowledge in neuroscience indicates rather that the entire brain participates in different functions, but only part of neuronal activity is directly associated with conscious cognitive processes.

Thanks to advances in anatomy and physiology, it now seems plausible that the traditional notions of mind or soul correspond, at least in part, to cultural and philosophical interpretations of phenomena related to brain activity. These artifices of thought can be understood as historical attempts to explain previously poorly understood brain functions.

In this perspective, consciousness could correspond to the brain's ability to process, memorize, and relate information collected by sensory systems during wakefulness. The information thus elaborated circulates, transforms and enriches itself thanks to social interactions, learning and the use of increasingly elaborate cognitive and technical tools.

The reality that we collectively perceive and build results from these cumulative cognitive and cultural processes. It has evolved over the course of humanity's biological and cultural history and profoundly differs from the perception of the world that could have the first primates from whom we descend.

Consciousness of existence could, from a biological point of view, be interpreted as the result of a process of natural selection favoring organisms capable of processing and integrating information effectively. In this perspective, cognitive abilities would be linked not only to the size of the brain, but especially to its functional organization.

The simple brain mass is indeed not a determining indicator of cognitive performance. The brain of the whale can reach approximately 7 kg and that of the elephant of about 5 kg, while that of the human being rarely exceeds

1,6 kg. These differences are explained in particular by the overall size of the organism and by the constraints imposed by the terrestrial living environment, aquatic or aerial. To compare cognitive abilities between species, neuroscience rather uses the encephalization coefficient, which measures the ratio between brain mass and body mass expected for a given organism. In humans, this coefficient is significantly higher than in most other mammals, suggesting an increased ability to process, select and relate a large amount of information.

In this perspective, self-awareness could be considered as an emergent property resulting from electrochemical processes taking place in the neural networks of the waking brain. These processes rely on the transmission and integration of signals within highly interconnected neural structures. Compared to current computer architectures based on a binary language and determined algorithms, biological systems present more flexible forms of learning and adaptation, relying on dynamic networks capable of reconfiguring themselves according to experience.

Several factors can contribute to explaining the high cognitive performance observed in contemporary humans: a significant density of neural connections, an organization into complex networks, genetic mechanisms favoring brain development, a great neural plasticity and, more generally, an increased longevity allowing the accumulation and transmission of information within human societies.

Since ancient times, human beings have sought to respond to fundamental existential questions by developing various hypotheses of a philosophical or religious nature. The notions of immortality of the soul, of metempsychosis or of the existence of a creator god have often served to give meaning to the consciousness of existing and to the finiteness of biological life. These representations can also be interpreted as cultural constructs aimed at alleviating the anguish linked to the prospect of death and individual disappearance.

From a speculative perspective, some might be tempted to establish a connection with certain concepts derived from quantum physics. By analogy, the persistence of a form of existence could be imagined as a transition to a purely undulatory state, in which the physical properties of an individual would be described by a set of wave functions rather than by a localized material structure. Such a hypothesis remains largely conjectural and does not rest on established empirical foundations.

It should nevertheless be recalled that, in contemporary physics, energy is not necessarily associated with a mass. The mass-energy equivalence relation $E=mc^2$ indicates that a mass can be interpreted as a form of energy. Conversely, some physical entities — like the photon — carry energy while being devoid of mass at rest.

In his role as a critical observer, the human being often tends to perceive himself as a form of conscious expression of the Universe itself. However, it is equally plausible to consider humanity as the contingent result of a set of particular local conditions, potentially fragile and unstable on a cosmic scale. The conditions necessary for the emergence of life — although they may not be exceptional in the observable Universe — nevertheless seem difficult to meet and probably remain precarious.

Human sensory systems, as well as the instruments we develop to extend our perception, are largely oriented towards the satisfaction of specific biological and cognitive needs. They only give us access to a limited fraction of physical reality. It is therefore understandable that our vision of the world retains a largely anthropocentric character. Despite these limitations, scientific research aims precisely to go beyond immediate appearances in order to gain a deeper understanding of reality.

Many scientific advances initially rely on direct or instrumental observation before being confirmed by controlled experiments. In cosmology, for example, the observation of very distant phenomena implies that we perceive events belonging to a sometimes extremely ancient past. These observations can be affected by different astrophysical effects, such as gravitational lenses or interactions between radiation and intermediate structures, which complicates the interpretation of these data and can produce partially distorted images of cosmic reality.

In this context, various speculative theories have been proposed to describe the global structure of the Universe or a possible multiverse. Among these are the hypotheses of multiple worlds, parallel universes or alternative realities. Other models envisage the existence of "wormholes", that is to say structures linking different regions of space-time by extreme deformations of relativistic geometry. These theoretical objects could, in principle, allow

410

extremely short journeys between regions very far from space-time. Their physical existence remains however hypothetical.

In the speculative perspective adopted here, black holes could play a particular role by allowing energy transfer to other systems of universes organized in quantum symmetry. This hypothesis, which constitutes one of the postulates of the proposed reflection, nevertheless largely exceeds the currently verified theories.

At certain extreme scales, the usual notions of space and time seem to lose their descriptive relevance. At the microscopic scale of elementary particles as well as that of black holes, internal phenomena remain largely inaccessible to direct observation and often appear 'frozen' from the point of view of an external observer. On the other hand, in intermediate domains — those that correspond to the observable evolution of our Universe — physical interactions are generally described in terms of relation to space and time.

The idea of describing all forms of energy mainly from the angle of their wave properties refers to certain interpretations of quantum mechanics, in particular to the notion of collapse of the wave function during a measurement. This approach can also be compared, in a speculative manner, to string theory. According to the latter, the fundamental constituents of matter would not be point-like particles but one-dimensional objects - 'ropes' - whose different modes of vibration would correspond to the various observed particles.

Superstring theory, however, is based on extremely complex mathematical constructions and remains highly speculative. It has many variants — open or closed strings, different categories of branes, varying numbers of additional dimensions — which testify to the lack of definitive consensus. The most well-known models generally introduce ten or eleven space-time dimensions in order to make the theoretical equations coherent. Let us take 3 spatial dimensions of the Universe, multiplied by 3 spatial dimensions «of antimatter» (to describe the combined gravitational effects) + 1 dimension of shared time (that is 2 temporalities in one) and we obtain the 10 or 11 dimensions of string theory. Such a calculation that extrapolates more than widely, is obviously a mathematical artifice without any real relevance. The theory of strings, by granting itself some freedoms, does not lack appeal.

411

Some attempts at interpretation seek to relate these extra dimensions to different physical properties, but these correspondences often fall more into mathematical constructs than established experimental results. It is therefore possible to consider these formulations as levers of thought allowing the exploration of certain theoretical avenues, without necessarily constituting a verified description of physical reality.

Any scientific demonstration is in principle based on a set of limited, reproducible and confirmed observations associated with coherent mathematical models. When these conditions are met, the resulting conclusions acquire a high degree of credibility and become difficult to challenge. However, even in this context the question of completeness remains. A theory can be firmly established while remaining incomplete, as many physical phenomena remain poorly understood or only formulated in the form of hypotheses.

The approach adopted in the reflection presented here does not strictly correspond to this classic method. It is more akin to a structured set of exploratory ideas than to a formal demonstration. This orientation is partly due to the inherent limitations of empirical observation and the mathematical or technical tools currently available to describe certain extreme physical realities. Human cognitive abilities, themselves finite, can also constitute a constraint in the development of fully coherent models.

The considerations developed — some of which explicitly fall within the scope of reasoned speculation — do not therefore claim to constitute a complete theory. They are rather part of a general hypothesis according to which the "whole" could emerge from a ground state containing no defined physical structure, but only potentialities of a virtual nature. In this perspective, scales smaller than those of elementary particles and larger than those of the observable Universe could fall into an essentially virtual domain. The physical reality we perceive would then be in an intermediate domain between these two limits.

However, the very notion of "virtual" poses a major difficulty: it does not possess, in the current state of our knowledge, direct physical representation accessible to observation or experimentation. This concept can nevertheless be considered as an attempt at interpretation intended to shed light on certain physical phenomena whose origin remains difficult to link to a primary identifiable cause.

412

It is also possible to consider this notion of the virtual as an intellectual construction, a pure product of imagination, rather than stemming from an empirically established foundation. In this sense, it illustrates the limits faced by any attempt to explain the ultimate origin of observable physical structures.

Such reflection necessarily leads to an open conclusion. It does not claim to provide a definitive answer to the question of the foundations of the Universe. Rather, it proposes a framework for reflection within which different hypotheses can be examined.
Everyone remains free to confront these proposals with their own interpretations or convictions. In such a complex field, absolute certainties remain rare and overly assertive positions can sometimes lack objectivity.
Throughout the history of science, many hypotheses have been formulated in an attempt to explain a cosmic environment particularly difficult to access for direct observation. The reflection presented here is in line with this intellectual tradition: it exposes certain divergences of interpretation while remaining generally compatible with the essence of established scientific knowledge. It thus aims to propose a reading grid likely to be confronted with empirical reality, while recognizing that no theory can claim to hold a definitive truth on a subject of such complexity.

Science, through its methods and applications, provides essential benchmarks for structuring thought and accumulating knowledge. It is based in particular on precise terminologies, databases and rigorous validation methods. However, imagination also plays an important role in scientific progress. Many major advances were preceded by theoretical insights or innovative hypotheses that were then formalized and tested.

In certain contemporary fields, notably in cosmology or fundamental physics, knowledge reaches particularly high levels of abstraction. The models proposed sometimes become difficult to access and only concern a relatively small circle of specialists. Even for the latter, a global and definitive description of the nature of the Universe remains out of reach at the moment.
Scientific research in these fields can be compared to a gradual exploration of largely unknown territory. Theories are developed, revised and sometimes abandoned over new observations or analyses. This process inevitably leads to sometimes intense scientific debates. The history of

413

astrophysics and cosmology shows that controversies between researchers have often accompanied major theoretical advances.

It is also common for initially contested ideas to be recognized when new data comes to confirm them. Conversely, long-held theories can be challenged when new observations reveal their limitations. The history of science thus reminds us that what appears to be firmly established at one time can be revised in the light of subsequent knowledge.

The reflection proposed in these pages is part of this critical spirit. It aims to formulate hypotheses and submit them for examination, while recognizing their provisional nature. In this context, error is not necessarily a failure: it often constitutes a useful step in the development of more relevant models.

It is likely that some future discoveries will lead to a profound transformation in our understanding of the Universe. These advances would be likely to go beyond certain currently accepted assumptions. The history of cosmology already illustrates how scientific paradigms can evolve over time.

In this perspective, the scientific imagination plays an important role when it remains guided by a coherent logic and by respect for established knowledge. Many major physicists — such as Albert Einstein, Paul Dirac, Werner Heisenberg, Niels Bohr, or John Archibald Wheeler—have largely mobilized their imaginative capacity to elaborate theories that then found a mathematical formalization.

Their work was often based on existing theories, sometimes confirmed later, sometimes corrected or abandoned. Even the most influential researchers may, however, encounter limits in their anticipations. Thus, Albert Einstein, despite his remarkable inventiveness, remained for a long time reluctant to accept certain cosmological implications, notably the expansion of the Universe or the physical existence of black holes.

This situation illustrates a profoundly human aspect of scientific research: it is sometimes difficult to question the convictions resulting from considerable intellectual work. Acknowledging a mistake can be perceived as a loss of credibility, when in reality it is a normal mechanism of scientific progress.

In quantum physics, certain interpretations also reflect the limits of our current understanding. The Copenhagen Interpretation, for example, highlights the difficulty of describing certain quantum phenomena independently of the observation devices that reveal them. At these scales,

414

the act of measurement often influences the studied system, which complicates any strictly objective description.

The physicist Max Born already emphasized that theoretical physics possesses, in certain aspects, a proximity to philosophical reflection. When the phenomena under study go well beyond everyday experience, the development of conceptual hypotheses becomes inevitable.

In this context, philosophy sometimes plays a complementary role by formulating fundamental questions to which science then tries to provide answers. However, when philosophical reflection completely detaches itself from empirical constraints, it can distance itself from the physical reality that it seeks to illuminate.

415

XXXI <u>Where it is about Nothing</u>
(A virtual nothing that leads to everything)

Physics, which is the basis of our standard cosmological model, was developed in a "context" that is familiar to us although imperfectly explored: that of vestiges of a past partially open to observation and a present of proximity that remains to be understood. Nothing says that the rules we have built from this, are proven and immutable. Although we lack hindsight, the physical laws that govern the precarious balance of matter should logically evolve like the evolution of our Universe. Thus, predicting future changes assumes a certain margin of uncertainty or more precisely, unpredictability in predictions.

The commonly adopted logic consists in wanting to explain the Universe on the basis of equations and mathematical formulations detached from any personal involvement and subjectivity. This approach, which has proven its worth, is unquestionably well-founded, although limited by the conception that we have made of science as a tool for exploration and interpretation. Mathematics, physics, chemistry, biology are the measure of our way of thinking. Applied to astrophysics, these tools that man has patiently developed, give meaning to what we struggle to understand about our reality, but is it enough to get to the bottom of things?

We would like to make science a charter made up of rules that are repeatedly verified and considered irrefutable. Without thought exercises, without imagination, without questions of a metaphysical nature, what science would be. But, without modern mathematics, an essential tool of scientific development, it would have had a limited development. This agreed language of development and prediction based on numbers, signs and symbols, gives us the means to reason and interpret in terms of quantity, value, relativity or causally. Could we without this artifice of thought give a profound meaning to what our senses reveal to us? On the other hand, this codified and abstract formalism moves us further and further away from an empirical vision from which we cannot detach ourselves from body and instinct. Indeed, we are destined to interpret everything in relation to our condition as living organisms closely conditioned by feelings of satisfaction, of frustration and of preservation. Our thought exercises have great difficulty in approaching an unknown reality, totally counter-intuitive. The more we try to blow up this screed which encloses us in a mirage of forms of reality, the more we accumulate paradoxes and abstractions.

Why does the Universe that we introspect show so many facets that seem to be out of tune with our reality? We are tempted to wonder if our Universe would not be itself in superposition of states (cf.: idea of state symmetry). This idea of a Universe that can only then be decohered, sums up our difficulty in distinguishing the true from the false when we seek answers to what makes us what we are. Who knows if we should not go further and open ourselves more to the imaginary? Due to lack of inventiveness, we have a natural tendency to look above all for evidence.

The idea of multiverse Cosmos is similar in some respects to that of multiple worlds but deprives it of the duplication aspect of alternative worlds advocated by H. Everett. This theory of multiple worlds implies that unconsciously, we would make choices that would make us evolve from one world to another in a succession of logical states. Would it not push the idea of decoherence a little further?

Are we using the appropriate logic? The one that would allow us to think outside the box, would it not be outside the major lines of research developed by technicians trained in excellence? Spectacular as they are, our decisive achievements have been achieved for the most part over the last 10 decades. This perception of an unsuspected world is too recent to conceal the extent of our ignorance and our embarrassment to connect between them phenomena difficult to explain.
For some, and this often more out of convenience than conviction, the truth can be only spiritual or divine. This simplistic and infantilizing view of our world dates back to the earliest days of humanity. Since then, man has evolved and his critical sense, long muzzled, has developed. For those who do not reject the real substantive questions, the Universe would make sense only for the observer who is its product and paradoxically questions the root cause, the reason for being and the destiny of this Universe so little intelligible to him? This is the thesis here.

If we get out of our most successful thought patterns, could not the answer lie in what we call the virtual? This point of view necessarily disturbs because the vision we have of our Universe exists only by the gaze of its observer. The latter can only be convinced of the reality that is imposed on him and in which he fits in completely.

417

When we dream or watch a film, our mind projects us into a world of fiction and our body often reacts by changing its emotional behaviour. The virtual then replaces a normality built around our lived. However, in astrophysics we are in the opposite situation. In the state of awakening, our lived reality is perceived through our history and conditioned by our senses while the true reality is to be found in the foundation, the unrecognized (and therefore virtual for us) nature of our Universe.

Like the idea of relativity (imagined, formatted by A. Einstein and initially rejected by the scientists of the time), the notion of virtual then becomes a key idea. It invites us to detach ourselves, by thought, from the too obvious realities of our good old planet, with the help of images or concrete case like for example to illustrate the relativity:

That of a traveler placed in a plane flying at 1230 km/h and who would throw a ball forward. the impetus given to this ball would be only about 20 km/h and therefore for the passenger, far from exceeding the speed of sound which is 1235 km/h. The plane is its reference point. Moreover, the ball does not emit the bang that it would produce in an open atmosphere.
However, for a stationary observer on the ground, the speed of movement of this same ball would be 1230 +20 or 1250 km/h; higher than the speed of sound. Thus, for this latter whose repository is given by a fixed point of our planet, the ball has travelled more distance in the same time or formulated otherwise, took less time to travel on an equal distance.

418

Similarly, an observer without any movement (theoretical hypothesis) in our galaxy, would find, in the case of the same event, a bigger lengthening of distances and a bigger shortening of time. Because in this case, the displacement of the earth must be considered. The repository is on the scale of the galaxy as a whole.

That would be just as much, of an observer outside our galaxy which should consider the movements of displacement and rotation of it within the galactic cluster of which it is a part.

In summary, a repository could be defined as the sum of the effects of synergy and gravitation specific to an observation point.

Each observer therefore has its own repository which is unique but is nevertheless almost the same for all on earth, because of close proximity. However, it will be different for a space traveller who should -even if it is not the case- perceive differently distances and time.

What may seem paradoxical is that the speed of motion of the photons (299,792,458 m/s) will be perceived as identical by any observer regardless of its repository. Indeed, distance (m) and time (s) vary together and the displacement/time ratio at this speed considered impassable, therefore remains similar for all, at the same time imagined universal T. This shared T-moment remains difficult to conceive, however, because of the lack of common shared repository that excludes any idea of simultaneity but also of universal chronology in the order of events.

The evolution of the Universe makes that the speed of light can, in the future, only tend towards a non-significant value. Indeed, as envisaged here, the condition for the final collapse is the absence of motion and radiation after that the energy fields have been absorbed by the MMBH.

One would be tempted to think that what we call «the void» should follow the disappearance of our Universe. It is to forget that true emptiness exists neither in the Universe (where energy without mass is associated with emptiness) nor in the multiverse Cosmos (where energy is presumed unrevealed).

Antiparticles make stealthy appearances in our Universe. On this occasion, the particles, confronting their symmetries, disappear from the landscape but the energy they carried is preserved in another form. Feynman had put forward the idea that the antiparticle went back in time in the opposite direction of the sister particle. Indeed, by annihilating itself with its symmetry, the energy it represented returns to its state before the radiation

419

entanglement phase or before the interaction that created the particle/antiparticle pair.

To continue on this idea, let us leave (difficultly) the Cartesian or Newtonian mode and approach the problem in virtual mode, in other words from "Nothing" …, or rather, from nothing understood here as a virtual conjecture symptomatic of latent energy and representative of a potentiality of unpredetermined states. This virtual can still be defined as a multitude of unforeseeable realities.

Heir to our history, our logic is formatted by and for observation. It purports to explain any event by a causal context of circumstances. And it must be noted that everything understood is always based on "something else" that we have previously understood and accepted, except to want to go back as we do here, to the origin of everything.

How to get out of this logic that perfectly adapts to our reality?
Unthinkable as it may seem, why not want to explain our Universe from Nothing, or more precisely « nothing else than latent or virtual energy». Because the logic of relying on something pre-established ultimately leads us to want to understand everything from a primary cause which, in any case, will remain inexplicable.
How to be more explicit on this notion of virtual Nothing?
The Nothing, we are talking about here, represents energy in the absence of time and space and refers to the idea of a virtual multiverse cosmos.

On this idea of « Nothing but only virtual », let us make a brief reference to mathematics:
If we start from 0, in other words from nothing, and cumulate in positive and negative all possible or imaginable numbers (admissible hypothesis in arithmetic), the final theoretical result would theoretically be equal to 0.
On the other hand, this operation, which presupposes an interminable process, can have different meanings at all stages of calculation:

- *whether you start or end with a positive or negative data*

- *according to the alternation of positive and negative numbers*

- *according to the completely random choice of numbers and their arithmetic meaning*

The result, <u>at any stage of the current calculation</u>, will be only exceptionally equal to 0.

420

It is the process of the calculations and therefore the time of the "events" (this succession of operations) that creates this illusion of non-zero result. If we remove the time factor, the result of this endless addition can only be equal to zero.

The difficulty lies in the fact that our form of thought is not capable of conceiving and describing, in appropriate terms, a unified theory for a Universe that turns out to be so different from the observable reality that we have built ourselves.

Cosmological Equilibrium could be described as an immutably stable, virtual, "latent state," but also as a continuum of universe binary systems in quantum symmetry. Faced with our reality, such a concept defies logic. We are in the most disconcerting abstract.

In the end, for those who refuse to associate this uncomfortable idea of « **All in Nothing that is only virtual** » with the concept of Cosmos multiverse, the question of the foundations of our Universe, remains open.
Curiously, speaking as we have just done, of the absence of space and time, far from evading the questions, proposes built-in solutions. Of course, this does not provide a fully satisfactory answer to the observer who has some difficulty in conceiving that it would, in the end, come from a virtual «Nothing».

A living organism, endowed with the capacity to think, would not in some way be the culmination of all this? This is not lacking in pretensions and joins this deep conviction for Man to have his place at the center of «everything».

It would be forgetting that no more than our galaxy and no more than our planet (observation point, privileged by force of things), we cannot be considered as the center or starting point of anything.
If we want to remain a little pragmatic, it becomes better not to speculate too much on this fantasy of anthropogenic Universe and to approach things from a less closed angle. However, we will come back to this in the epilogue.

XXXII <u>How to complete this reflection?</u>

(With a physics of a 3rd type, which cannot be more discreet)

How can we unify and reconcile what we consider to be confirmed achievements but without clear links, in our understanding of the Universe? A certain unfinished but globally consensus modelling has been built through a two-pronged physics:

- Classical physics, which is that of general and restricted relativity, is deterministic and is based on the gravitation of bodies in relation to time and space.
- Quantum physics that does not reject special relativity and is particle physics is considered too 'random' to take gravity into account. Space and time have no certain value. By assigning a uniform rectilinear motion to any object, it does not consider gravitational effects as does general relativity.

Quantum physics is thus led to ignore any spatiotemporal context and to relate on the one hand to the idea of superposition of possible states of particles and on the other hand to that of a necessary duality wave/corpuscle. In short, matter in its most fundamental entrenchments, seems to ignore time and space. This means that gravitational effects are not discernible on this order of magnitude.

All this suggests that we have probably not sufficiently expanded our scope of thinking. So, nothing really surprising that we cannot explain among others, mass and energy insufficiencies.

The black hole is a quantum singularity outside space-time that excludes itself as such, from our classical physics. However, a black hole remains because of its gravitational effects, an observable phenomenon that can be integrated into general relativity. From this double point of view, gravitation therefore also becomes a property of quantum mechanics. Our embarrassment comes from the fact that the horizon of a black hole does not let anything pierce from what is happening in the heart of it. However, we know that this screen is not an impassable border. The black hole is thus totally in this 2-pronged physics and marks the border with the multiverse Cosmos.

The gravitation that will lead to deconstruct the Universe seems to be present on any scale, even if the equations that define it in general relativity may lose their meaning in particle physics. Related to the discrete interactions of

423

an undisclosed physics, imagined at the root of what makes energy in rupture of symmetry, these equations even become totally irrelevant. The gap between our reality and what our most recent advances suggest is widening further. We are reduced to doubt, to question, and ultimately to try to conceive of what seemed unthinkable until now.

Building a global model from the sub-models of classical relativistic physics and quantum mechanics, invites us to imagine a context that we would have neglected until then and that would make them connect. One comes to think that there is a missing link or a necessary step to arrive at an archetype of physics representative of an expanded standard model. It could be, in a way, an interactive and virtual «base» devoid of a spatiotemporal dimension, underlying what we know of physical laws but determining them.

The answer would be discreet symmetry; it would explain all the oddities and paradoxes that taint the «enlightened side» of our Universe.

The concept of a binary system of universes in quantum symmetry leads us to imagine processes of discrete exchanges and interactions between two symmetrical states sort of "superimposed" or interacting in "parallel dimensions."

> What makes this arbitration between these two symmetries and which we could call <u>fundamental</u> physics, would have nothing mechanical. This unrecognized physics would neglect the location and time of events.

- **In classical relativistic physics**, an object (vehicle in motion for example) allows a traceability, in accordance with the vision of the observer. Speed and position can be put into equations. This classical physics describes in observation mode whether direct or indirect.

- **In quantum mechanics**, the elementary particles, wave packages that constitute matter particles and are observable by their incident effects in classical physics, do not have a certain position or speed of movement. This physics analyses in deduction mode, in close coherence with relativist classical physics.

In fundamental physics, (presented here as a 3rd type physics), <u>everything becomes informal.</u> There is no longer an object as a representation of matter but only discrete, unrecognized interactions

424

between quantum symmetries. These phenomena are not perceptible to the observer that we are. All interactions attributed to particles that we have great difficulty to describe through quantum physics, free themselves in fundamental physics from the laws on which we base our reality, to open on the Cosmos multiverse. This so-called fundamental underlying physics, based on the concept of quantum symmetry and which would suggest the foundations of the Universe, nevertheless allows us to build in this reflection links that would explain and make classical physics and quantum physics join. It cannot claim to be the bearer of apodictic truth.

Of these hidden phenomena, we are unable to make a mathematical transcription and can only stick to a statistical interpretation not significant, arising from calculations of probabilities. Despite this, we would like to give these totally counterintuitive concepts a spatiotemporal framework. This disconcerting exercise in the abstract, testifies to the limits of our understanding of a Universe that seems to exceed our sense of understanding when we change scales to speculate on the unobservable.

These 3 levels of physics; classical, quantum and fundamental, tell the story of our Universe. Such a triptych would be a global response relatively coherent, representative of this much coveted unified theory, of an All in "Nothing…other than Virtual". The laws that govern each of these three scales of physics, remain fundamentally linked and interdependent, as the pieces of a puzzle, although the nature of the phenomena they describe, makes think that they are specific to each of them. Changes in contextuality and scale are not foreign.

Galileo's realistic physics, Newton's pragmatic physics, Einstein's space/time physics, Planck's quantum mechanical …, the path is laid out that invites us to go beyond our particle physics. So, there would be a missing component to astrophysics, which would explain our Universe from its most remote foundations. We are not talking about a spiritual entity, but about an «ignored stage», necessary for our understanding and which would represent the « boundary zone » between the multiverse Cosmos and what we attach to the quantum world.
This additional fundamental mechanism does not seem really open to intellectual investigative tools such as our most advanced. The advanced ideas of non-locality, quantum entanglement, quantum decoherence as of

425

superimposition of states well reflect our embarrassment to conceive a Universe deprived of temporality and displacement in its last entrenchments (those of the infinitely small) as to predict a multiverse Cosmos of virtual nature.

A break in the cosmological balance (Big-bang) would be the cause of a metastable symmetry between particles and antiparticles. In any case, the revealed side (that of matter) to which we are attached, does not give us access to such a fundamental symmetry dynamic involving particles/antiparticles. These discrete, underlying interactions between quantum symmetries and justified by opposite quantum numbers and charge would be decisive in all phenomena affecting matter. But this matter of which we are made and which manifests itself to us on the scale of the observable, makes in some way screen. This is the idea adopted here and which would give more coherence and a new readability to a cosmological model that has become problematic in many aspects.

Gravitation is the physical phenomenon that we feel in the first place. Moreover, it affects all our behaviours. Remarkable on the macroscopic scale, it would find its deployment in quantum mechanics under the effect of electromagnetic force and would hide its origin in this so-called fundamental physics that represents the trade-offs between quantum symmetries. A virtual context made of «Nothing that seems to belong to our reality» is what would allow the best to take into consideration an «additional physics of quantum symmetry» by nature beyond our reach.

A long time ago, an Indian mathematician had suggested that the result of a finite number divided by zero, makes sense only if one substitutes for the idea of nothing, that of infinitely small. So, for example: the number 5 divided by an infinitely small, imagined number, would give an infinitely large but real value. Otherwise, without this extrapolation of the digit 0, the result of this division by nothing can be defined and remains unverifiable. Note that even by substituting for 0 an infinitely small value, no result multiplied by this almost 0 is equal to 5. This means that the number 0 only makes sense associated with an infinite value, contravening the idea of nothing as a total absence of anything. We can deduct from this that an infinitely small number is without remarkable spatial dimension. As such,

426

could it not then represent the fundamental unit almost indivisible and non-convertible that would be valid in all repositories and that would render the transformations by change of coordinate system obsolete (Lorentz transformations). That is kind of where we are headed with Planck's formula, this unit, which tends towards 0 and makes it possible to quantify the energy considering the undulating aspect of it. By pixelating the energy into insignificant fragments of space would theoretically mean dressing it with mathematically elusive values.

Paradoxically, starting from «something» imagined infinitely large or infinitely small, how to extract or produce anything of definite value except to distort the notion of infinity?
The notion of infinitely small as well as infinitely large would therefore tend to be superimposed on that of absence of all things. These «things» that make our physical reality, however, seem to want to inflict strange prohibitions to us.
This parenthesis on zero and infinity, two antinomic values, leads to a rapprochement with the concept of multiverse Cosmos without limit, indivisible and timeless. Let infinity merge with zero ($\infty \approx 0$), would validate the virtual nature of a multiverse cosmos that has nothing physical.

This idea of «Nothing» is clearly distinguished from that of nothingness which could be understood as the absence of energy potentially breaking symmetry. But, if we define the multiverse Cosmos as a form of latent energy in the virtual state, we cannot therefore report nothing in the literal sense of the word, in this discourse on the Universe. The idea of nothingness is totally a collective imagination.

This reflection proposes a rather coherent overall theory that would explain paradoxes, would rally differences, smooth out scale difficulties and give a constructed meaning to our observations. Given the growing difficulties that hinder the development of astrophysics, we are far from having the means to validate a theory of the Whole, whatever it may be.
Rich in advances, the recent period we have known, suggests a new era of research thought more in terms of probabilities and less in terms of certainties.

XXXIII Challenge Analysis
(But not necessarily objective)

<u>**Constructive points**</u> :

- Abandonment of the postulate of a creation out of nothingness (the starting point of this reflection)

- Aesthetics of a paradigm based on an upset balance of quantum symmetries and a form of unification of mismatched theories.

- Justification of energy/matter interactions.

- Concepts of infinity and eternity redefined.

- Rather «coherent» model of the evolution and creation of our Universe.

- Place of man, as of the rest, detached from any divine intervention.

- References to a lot of accepted physical assumptions or data.

- The Big Bang is no longer the unexplained central point of a mysterious container.

- Relative concordance of the 3 fundamental forces related to the gauge interactions in a theory of the «All in nothing other than virtual».

- Gravitation is no longer a dissociated «force» from the others.

- Adequacy of quantum mechanics and general relativity.

- Retrograde dispersion and expansion of the Universe are no longer confusing.

- A Universe with a constant «energy load» in a space in depression falsely called vacuum, is easier to conceive than a Universe in expansion, coming out of a supposed «point» of energy without precedence.

- The notion of superposition of possible states makes any particle a fundamentally and originally identical cloned model and of which each "copy" shares potentially «instantaneous» all or part of the same information.

- Particles that are considered to be entangled waves packets, out time: this gives grain to grind and gives hope for new advances.

- Quantum black holes representing the final stage in the evolution of our Universe.

428

- Broken symmetry in a cosmological Equilibrium (the notion of symmetry of particles is however not new)

- A multiverse Cosmos may at first glance seem difficult to imagine, even inconceivable for those who refuse to broaden their field of reflection

- A discourse that may appear, in some places, somewhat hermetic or negationist (these points are certainly to be deepened or reformulated)

- Knowingly limited references to mathematical data (this discipline, which opens doors, also shows its limits and is difficult in a context of chiral symmetry)

- Number of unproven assertions (but is everything demonstrable?)

- Finite universe, non-expansionist, all curves, not bounded, with gravitational effects without a marked limit (the notion of infinity can be left out)

- Frequent use of the conditional (one cannot be too careful on a subject that remains simply touched on, despite the most recent advances and many attractive theories)

- Use of shortcuts and metaphors to address the particle physics and the astrophysics (knowledge of these tools to be developed, but also for the sake of simplicity!)

<u>**Considerations on form:**</u>
The overall plan of this discourse may seem lacking in construction. It is, around a central theme, mainly a compilation of questions and ideas enriched, corrected and added over an unfinished reflection of 5 years. Have you noticed how often what we say, think, do, leave us feeling unfinished?

This explains some lengths, plenty of redundancies and image returns. Many reformulations reveal a real difficulty to develop in simple terms, certain ideas or concepts that can baffle in the first approach. No doubt, it would have seemed more authentic to use expressions such as Riemann tensor, vector of Killing, Noetherian symmetry, Hamiltonian constraint, Fourier series, Bayesian regression, anti Sitter space, operator of Laplace-Beltrami,

429

Majorana equation, Weyl structure, hermitic space, Bott periodicity, Maslow index, Abelian variety…, a scientific jargon that has fully its purpose but does not facilitate communication.

It should be recalled that a theory can only be validated if it is proven experimentally or scientifically. On the other hand, a theory is really refutable only if the demonstration that it is erroneous, is brought. Between the two, one can only doubt, while hoping that the uncertainty is lifted and the reflection ratified or invalidated.

<u>**Epilogue**</u>
(Or statement of unfinished business)

Can we venture to make predictions about the destiny of the human species? Some hypotheses do not bode well. It is enough to extrapolate from his past. An experiment with live animals in an organized society, in this case rats, seems to confirm what many sense nowadays. The experiment consisted of placing eight mice in a box with an open ceiling, offering plenty of space and protected from any predator. Food and water were provided at will. A year and a half after the beginning of the experiment, the population is 2200 rats and newborns survive with difficulty. The experiment will end only 5 years later. A majority of aggressive males spent its time feeding and killing each other while groups of females less and less willing to breed, remained away, marking the end of this declining society.

This renewed experience with the same destructive effects, tends to show that for an average individual caught in a society in excess, immediate and essential needs must be met as a priority, to the detriment, if necessary, of the interest of the group. This choice will be made all the more priority as the individual cannot stand out from a large group that ends up oppressing him. Denial, indifference, withdrawal into oneself, exacerbated competitiveness resulting in more aggressiveness, seem to assert themselves more in a population with high density where particular interest will take precedence over the collective interest. Groups sharing affinities, are showing growing difficulties in standing out. It all comes down to a question of sharing ideology and allocating resources. An overpopulated society that is poorly managed seems to develop an individualism which is difficult to reconcile with any form of democracy. It can only compromise social peace.

The human species seems to register into this category. The consequences of an unacceptable overcrowding are known but we are probably not fully aware of them: plunder of natural resources, pollution, social discrimination and segregation, race for weapons of mass destruction, conflicts of occupation or power. A characteristic distinguishes however, the human species from these rodent mammals: an ability to consider the harmful consequences of such an uncontrolled evolution and the threat posed by certain misguided technologies such as weapons of mass destruction. We would like this awareness to make a difference!

431

No one needs to be a diviner to understand that deleterious overpopulation ends up generating irreversible effects. In an overall overheated atmosphere and a climate disturbed by the greenhouse effect, the sharing of insufficient wealth risks becoming a vital issue, to the point of multiplying conflicts, destructions and pollution. If there is a point of no return, we may well be shortly about to cross it. It seems that for every generation, the future ends with it. It is in human nature. "After me the flood" reflects fairly well a generalized behaviour in a world now governed by the finance and competition.

If our future is largely based on more or less collective choices, it is difficult to state perfect consensus. Choices are most often made in a hurry, when they are not imposed. Major decisions are rarely intended to prevent or anticipate. There are several reasons for this:

1. All indications are that we are far from mastering the problems associated with rampant overpopulation and the resulting scarcity of accessible natural resources. Fossil fuels are running out, but we will not be fully aware of them until really, they begin to fail, widening the gap between developed and developing countries.
2. An overconsumption of natural resources beyond renewable energy and the resulting pollution has the effect of profoundly changing climate phenomena. In some parts of the world, the conditions conducive to life are already altered.
3. We are looking for, and that is our primary concern, ease into security. This pragmatic quest for a well-being that is far from being shared is our idea of happiness. Why should we feel concerned about what will happen after us or for the most altruistic after the generation that we leave behind us?

How many say that nothing will stop the earth from spinning. How many thinks that exponential curves tending towards a point of non-return are only theoretical models. It is not in our nature to project ourselves beyond a future at the end of which nothing concerns us anymore. Under these conditions, can we appeal to the freedom of responsibility of everyone?
Let us assume, however, that we reach as broad a consensus as possible on considering this uncontrolled growth. It is to be feared that the measures capable of breaking down this destructive dynamic will be poorly accepted by a majority of us. Therefore, it is not unreasonable to think that coercive measures will be necessary. Our future will then be that of a controlled

freedom in other words, a regulated democracy that rallies the whole human race. Does humanity really have other choices? The conditions being far from being met, we unfortunately put our heads in the sand too often.

More than 8 billion people on earth today. Of all this population marked by a predatory growth for the planet, only one holds more particularly, not to say exclusively, our interest: our ego.

This conviction of being unique, leads us to place ourselves preferably at the center of everything. We are all, at a time, spectators and actors, but everyone would like to be the only player and referee in this role play. That is why each of us, in our innermost being, can say:
"The Universe is present only because I conceive of it and I conceive of it because I exist in the short span of time that meets the conditions necessary for life.
This leads me to think that probably nothing exists really without me that represents this consciousness of a world that I have forged for myself. On the other hand, I cannot help but consider that I am an integral part of this Universe that makes my reality». "So, I would be going to come from something I designed..." Look for the flaw!

Let us now develop this idea without restriction of scale. Let us consider the multiverse Cosmos as an infinity of Universes, in an infinity of «dimensions», without limiting ourselves to the present moment. How many universes have evolved like ours to produce an intelligent life form and how many have aborted? Since nothing connects the numberless Universes that give meaning to the concept of multiverse Cosmos, the question cannot be answered.
And in our universe, how many planets like ours, what we call exoplanets, have been able, can or will be able to develop life at a stage at least as advanced as the one we know?

A near-infinite number of possibilities becomes in terms of probabilities, a near certainty. La life should statistically be present, in number, on other exoplanets, as at other past and future moments!
In that case, can it be envisaged **that what I am...** may exist or have existed without memory of any past, in that same Universe or any other binary system of universe in quantum symmetry and at any other moment. The physical appearance of another life form does not matter. Formulated

otherwise, why should we not be duplicated infinitely, without possibility of sharing these same answers.
We could compare this script to a puzzle in 3 D, fragmented to the extreme. Each thinking piece would represent a player who would play alone in his corner, ignoring the position of others and the progress of the game.
Since all communication is forbidden due to remoteness and temporal lag, what would be the raison d'être of this «universally» fragmented consciousness? The answer is in the multiverse Cosmos.

Thus ends a very singular draft of what could be the history of our Universe. It gives the reply to all these beliefs that abuse us by pretending to satisfy our ignorance. Religion is a compilation of pseudo certainties about things that cannot be seen and considered inaccessible to any scientific approach. Science proposes a pragmatic, open-ended model, based mainly on justified, recurrent and related observations, but also on realistic hypotheses as well as on concepts taken as a working basis. Whether you are gullible or critical, curious or contemplative, realistic or dreamy, manipulative or pragmatic, you will adhere to one or the other. These few lines answer also to those who are led to think the Cosmology in a logic that would explain the reason for our Universe by our presence. Clearly, this anthropogenic principle poses the problem upside down and is a fantasy as old as the world.

Of course, it is not forbidden to think that everything that has just been developed is akin to a form of philosophical therapy.
But who knows!

Contact and feedback: dominique.chardri@hotmail.com
Thank you for your comments and suggestions

The final word
(For an endless story!)

Without end, because this reflection has not ceased to be taken up, amended, supplemented through readings, exchanges and information from multiple sources. Most scientific breakthroughs bring as many questions as they offer answers. This incursion into a Universe that remains to be discovered gives the impression to climb a path more and more steep and less and less practicable.

The employee tone would like to be pragmatic, open and devoid of scientific claims and stripped of judgements of value. Respecting a relative coherence, took precedence over the rest, with doubt for alone certainty. We may wonder whether we should not think in a less conventional way by seeking points of support that are not the ones we would choose a priori.

With great application, while avoiding the immoderate use of the conditional, it was a question of particles, forces, link-lines, symmetrical dimensions, hidden interactions, singularities, fluctuating time until it disappears, of space that distorts until it disappears and other concepts so far removed from our reality.

These terms give visibility to phenomena that escape us essentially by clarifying as much as possible, the subject.

We need answers that are validated by proven experiences and confirmed by crossed observations. Although providing constructed and coherent answers, this theory of the Whole segmented by unavoidable levels of scale, does not fully meet these requirements and can only leave on a feeling of unfinished.

But what is the true truth?

Who would not admire these particularly brilliant researchers and thinkers - they are not so numerous - who mark this reconstitution of a Universe so concealer to us? They have inspired this reflection, which is not free of gaps and personal interpretations that call for controversy. We can only see emerging phenomena which have unrecognized links with discrete interactions predicted between quantum symmetries. We realize that we are more and more difficult to have appropriate tools to demonstrate mathematically or to test experimentally.

We perceive light (EMW) in wavelengths between 400 and 700 nanometers: our visual field quickly shows its narrowness. We are sensitive to sound at frequencies between 20 Hz and 20,000 Hz: our listening capacity does not

435

go beyond that. We manage to support an environment between +60° and -40° with contact time at extremes, quickly reached. Normally, we only perceive that proves to us within the limits of our feelings.

However, more and more sophisticated tools of technology and reflection, allow us to note today the presence of wavelengths, sound frequencies, temperatures, speeds of movements very outside these limits. This progress may suggest that we should be able in a long-term to understand what makes us what we are. This would mean also that, at a very advanced stage, science supported by technology would no longer have any reason to advance having proven and demonstrated everything that could be done. Is it not showing an excess of optimism to think that one day, the Universe will have no more secrets for us?

The human species in its evolution has endowed itself with cognitive capacities, which go in the direction of a better awareness but which remain primarily likely to meet our most present expectations and worries. The memories left by personal trials and the enrichment of a collective memory, unconsciously influence how we perceive what contributes to our living environment. In other words, we only see what we can see physiologically and more particularly what our most basic preservation needs, in particular, require us to see first. This peculiarity is not only peculiar to man. Since the unicellular organism, it singles out, to varying degrees, all forms of life and quickly shows quickly its limits.

Now, let us take this further. We should ultimately ask ourselves whether we have the potential to understand a reality so different from the soft vision that our condition imposes on us. Our recent immersion in the quantum "dimension" shows us that the macro world that is offered to our gaze is only a dressing to our liking of more complex phenomena and that we are only beginning to sense.

The subsidiary question would then be to know if a form of artificial intelligence disconnected from all felt, could give access to a cosmology that we feel so different from our standard model? Such a project would aim to make us discover, without a filter, the infinitely small as the infinitely large. there is no basis to assume that AI could contravene the collapse of the wave function and consider the superimposed states prescribed by quantum mechanics. But in any case, a future supported by machines equipped with artificial intelligence seems rather promising even if a risk of dependence that we will have to consider is not to be ruled out.

Let us return to earth! We bathe in space and we pass through time. But what becomes of space/time without observers? This absence of observer then amounts to depriving space and time of their raison d'être and the reduction of the wave packet then loses all purpose. Our understanding of the Universe would therefore result from a kind of mirage in which we would inscribe ourselves as an operator of quantum decoherence and by way of consequence inventor of temporality in a relativistic space to our measure. It can also be a way, even if it is particularly frustrating, to answer the great question: what is the Universe and what is really our place in it?

Since your reading has led you so far, it is time to confess that the objections, hypotheses and refutations formulated in this book have no other pretensions than to «shake the coconut tree». Should we not reconsider some of the default theories, assumptions and interpretations? These few lines are there to recall in particular 2 points put aside too often for scientific convenience:

- the mathematical uncertainty in quantum physics. Gödel's incomplete theorem has shown that any mathematical system capable of proving the basic theorems of arithmetic will necessarily include elements whose truth value cannot be proved using the axioms of the system in question. Which means that mathematics has its own limits, fundamental limits that restrict the use we make of it and can alter the interpretation we would like to give them.
- the problem of a «patchwork» cosmological model.

The difficulty is that we perceive and try to understand our environment by logical reference to everything that, in one way or another, represents or refers to the matter. This form of energy from which we cannot detach ourselves, does not allow us to understand and define what energy in the ground state really is. Answering it is essential to advance in the understanding of our Universe.

These few lines would like to shed new light, especially on this kind of problem and broaden the debate.

Thank you for taking the time to read me, by allowing me the right to error.

ISBN 978-2-9571506